INDUSTRIAL CIRCUITS
AND
AUTOMATED MANUFACTURING

INDUSTRIAL CIRCUITS
AND
AUTOMATED
MANUFACTURING

Clyde O. Kale
Professor of Industrial Studies
Moorhead State University
Moorhead, Minnesota

SAUNDERS COLLEGE PUBLISHING
A Division of Holt, Rinehart and Winston, Inc.

New York   Chicago   San Francisco   Philadelphia
Montreal   Toronto   London   Sydney   Tokyo

Library of Congress Cataloging-in-Publication Data

Kale, Clyde O.
  Industrial circuits and automated manufacturing.

  Includes index.
  1. Factories—Electric equipment.   2. Industrial
electronics.   3. Automation.   I. Title.
TK4035.F3K35   1988   670.42'7   88-6563

ISBN 0-03-013609-1

Requests for permission to make copies of any part of the work
should be mailed to:
Permissions
Holt, Rinehart and Winston, Inc.
111 Fifth Avenue
New York, NY 10003
Printed in the United States of America

9   0   1   2   016   9   8   7   6   5   4   3   2   1

PREFACE

This book is designed to be used in a four-quarter or three-semester credit course in industrial circuits or microprocessors and is primarily directed toward baccalaureate level industrial technology and industrial education programs. It may also be used in industrial engineering, manufacturing engineering, and industrial training programs. Graduates of these programs, for the most part, implement the production process: they take the design and convert it into a finished product. The book should also prove useful in post-secondary technical programs where the major course of study is not electronics but where electronics principles and applications as applied to manufacturing form an essential part of the curriculum. The book may also be used in secondary school electronics programs that include studies in industrial circuits, microprocessors, and manufacturing.

For the past decade the manufacturing industries in the industrialized world have been automating to increase production efficiency to compete in a world market. Manufacturing automation is a developing technology and encompasses engineering, marketing, inventory control, and factory floor production machines.

The objective of this book is to introduce the concepts and present the principles that make manufacturing automation possible. The book integrates many of the topics included in a traditional industrial circuits book with the basic principles of the microprocessor and automation.

It has become increasingly more difficult to add new technology to existing programs. Because of other academic requirements, new courses cannot be continually added to the curriculum. A solution is to delete older and obsolete content. This has its limitations as considerable content never fully becomes outdated. Another solution is to combine and integrate content where possible. That is what this book attempts to do; it integrates the principles of industrial circuits with the principles of microprocessors and automation.

A chapter dealing with fluid power is included. This technology is used extensively in manufacturing. Individuals entering the automated manufacturing industry must have a basic knowledge of the principles, devices, and terminology unique to this field. Although fluid power is usually taught as a separate course in mechanical programs, it is usually not a requirement in electronic curricula. The final two chapters are devoted to the techniques which make manufacturing automation possible. While some components of automation such as Computer-Aided Drafting (CAD), computer controlled machine tools, robots, and automated materials handling equipment are fully developed, wide-scale total factory automation has suffered from the inability of different types of automated equipment to communicate with one another. Local Area Networks (LANs) and protocols are just now being developed to facilitate this very important communications function. It is expected that students will have had previous introductory course work in passive, linear, and digital circuits.

I wish to thank the many individuals who aided me while writing the manuscript. Ms. Barbara Gingery, Electronics Technology Editor at Saunders College Publishing, a division of Holt, Rinehart and Winston, Inc., coordinated the activities and kept me and the reviewers on schedule. She made the task an enjoyable experience. Dr. Jonathon Lin, at Eastern Michigan University, reviewed drafts of the fluid power chapter and made several suggestions which were incorporated into that chapter. Manuscript reviewers included Mr. Gerard P. Colgan, Bowling Green State University; Dr. Lawrence Fryda, Illinois State University; Dr. John J. Jellema, Eastern Michigan University; Dr. Ronald Kovak, Ball State University; Dr. Douglas L. Pickle, Eastern New Mexico University; Mr. Roy H. Redderson, Georgia Southern College; Dr. Joseph F. Thomas, University of Wisconsin; and Dr. Ronald Tuttle, Kearney State College. Officials at Allen-Bradley, Intel, and Motorola graciously allowed the use of data sheets for selected products. Ms. Margaret Dickie, at Motorola, was especially helpful in supplying data sheets to fit the specific needs of the book. My wife, Sharon, proofread the manuscript. To all of you, thank you very much.

C.O.K.

July 1988

CONTENTS

CONTENTS

Chapter 5   *Introduction to Microprocessors*   **176**

CONTENTS

INDUSTRIAL CIRCUITS
AND
AUTOMATED MANUFACTURING

CHAPTER
1

CONTROL DEVICES AND CIRCUITS

OBJECTIVES

1. Define the term *switch*.

2. Identify the components forming a switch and discuss the terms used to describe the different types of switch configurations.

3. List eight types of switches and identify the general applications for each.

4. Determine the proper voltage and current ratings for a switch used in a given application.

5. Select the correct switch enclosure for a given application.

6. Identify three types of overload devices, discuss their general applications, describe their construction, and discuss the principles of their operation.

7. Define the term *relay*.

8. Identify the components forming a relay.

9. Define the terms *coil voltage, pickup current, coil resistance, maximum contact current, operate time, release time*, and *service-life rating*, as they relate to a relay.

10. Describe how switches and relays can be used to form AND and OR logic gates.

11. Discuss the composition of a relay-ladder diagram.

12. List three types of timers and identify the general application for each.

13. Identify four types of delay timers and discuss the principles of operation for each.

14. Discuss the operation of the bipolar-junction transistor switch.

15. Calculate the value of base resistor required to saturate a bipolar-junction transistor.

16. Define the term *thyristor*.

17. List six types of thyristor devices and draw the schematic symbol for each.

18. Discuss the principles of operation of the Shockley diode, SCR, GCS, SCS, triac, and diac.

19. Describe the gate requirements to turn on the SCR, GCS, SCS, and triac.

20. Identify two techniques used to turn off the SCR.

21. Discuss two methods used to trigger an SCR from an AC source.

22. Define the following terms as they apply to an SCR
   — average forward current $I_{F(AV)}$
   — reverse breakdown voltage $V_{BD(R)}$
   — holding current I_H
   — gate trigger current I_{GT}
   — maximum forward voltage drop V_F
   — switch time T_{ON} and T_{OFF}

1-1 INTRODUCTION

Approximately two-thirds of the wealth created in the United States comes from manufacturing. The manufacturing industries in the United States have been the envy of the world. They are a product of the industrial revolution which began in Great Britain in the late eighteenth century and made its way to this country approximately a century later. The industrial revolution replaced the cottage system of producing goods with the factory system. The cottage system utilized highly skilled craftsmen to produce custom-made goods of high quality and limited quantity. The factory system substituted machines for skilled labor to produce moderate-quality goods in great quantity. Large numbers of unskilled laborers were needed to operate the machines and perform ancillary activities.

Until the beginning of the twentieth century most Americans made their living farming. After the industrial revolution, the machinery produced by factories made farming more efficient. As a result, farms became larger and fewer individuals were required to produce crops. This led to a large number of individuals leaving farms to earn a living elsewhere. Many of these persons became the supply of individuals required to staff the manufacturing industries.

By the early decades of the twentieth century, U.S. manufacturing led the world in production quantity, quality, efficiency, and technology. The nation has continued to hold this lead, but its competitive edge has been blunted since the mid-1970s. Competition from abroad has seriously affected many of the nation's manufacturing industries. Labor is much less expensive in most overseas countries and in many of those countries the government subsidizes the production of selected goods. To remain competitive in a world market, U.S. manufacturers are turning to automation, a change that is sometimes referred to as the second industrial revolution. Automation increases production, reduces manufacturing costs, and improves quality.

This book deals with the devices, circuits, and systems that provide the technology to implement manufacturing automation. The content of this chapter is concerned with the passive and active control devices used in industrial circuits.

1-2 SWITCHES

1-2.1 Introduction

A switch is a device used for making, breaking, or changing connections in a circuit. As illustrated in Figure 1-1, a switch is made up of contacts and a conductive element, called a pole, which is used to provide continuity between the contacts. A wide variety of switches are available for an almost infinite number of applications. Switches can be categorized according to configuration, type, method of actuation, and application. These are discussed in the following sections.

Figure 1-1

Basic switch

1-2.2 Configuration

The design and construction of switches vary according to the number of poles, the number of throws, and the pole-contact schemes used.

Number of poles: This identifies the number of external conductors controlled by the switch. One set of contacts and one pole is required for each conductor. Switches may have one, two, or any number of poles, with one to three being the most common. These are illustrated in Figure 1-2 and are referred to as single pole (SP), double pole (DP), and triple pole (TP), respectively.

Number of throws: This identifies the number of possible actuated positions and consists of either one, single throw (ST), or two, double throw (DT). Either type can be used with single-, double-, or triple-throw configurations. The switches appearing in Figure 1-2 are all single-throw devices, while those shown in Figure 1-3 are all double-throw switches.

Pole-contact schemes: Switches are designed to have either open or closed contacts when not actuated, as seen in Figure 1-4. The switch contacts shown in Figure 1-4(a) are normally open (NO) and are closed whenever the switch is actuated. The contacts appearing in Figure 1-4(b) are normally closed (NC) and open when the switch is actuated.

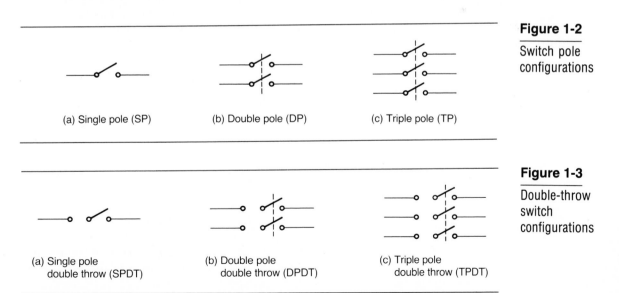

(a) Single pole (SP)

(b) Double pole (DP)

(c) Triple pole (TP)

Figure 1-2

Switch pole configurations

(a) Single pole double throw (SPDT)

(b) Double pole double throw (DPDT)

(c) Triple pole double throw (TPDT)

Figure 1-3

Double-throw switch configurations

Figure 1-4

Switch
pole-contact
schemes

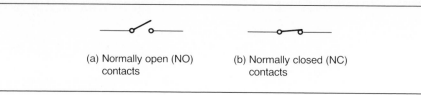

(a) Normally open (NO)
contacts

(b) Normally closed (NC)
contacts

1-2.3 Types of switches

This category groups switches according to their construction and includes the switches identified below.

Knife switches: The knife switch is made of a hinged copper blade that provides a connection between the switch contacts when forced between a spring-loaded jaw, as seen in Figure 1-5. These switches are often mounted in a nonconductive casing to prevent operating personnel from coming in contact with the switch mechanism. They may have one or more poles and be of the single- or double-throw type. The switch appearing in Figure 1-5 is an example of a double-pole, double-throw (DPDT) device.

Toggle switches: These switches are actuated by the operation of a lever attached to the pole through a toggle mechanism. They are designed with either single or multiple poles for single- or double-throw operation. A double-pole, double-throw toggle switch is shown in Figure 1-6.

Slide switches: Slide switches are actuated by a lever that is moved parallel to the plane in which the switch is mounted. As seen in Figure 1-7, they are available in either single- or multiple-pole form with either single- or double-throw capability. A typical slide switch appears in Figure 1-8.

Push-button switches: These switches are actuated by depressing a shaft extending from the switch housing, as seen in Figure 1-9. They are designed for either

Figure 1-5

Knife switch

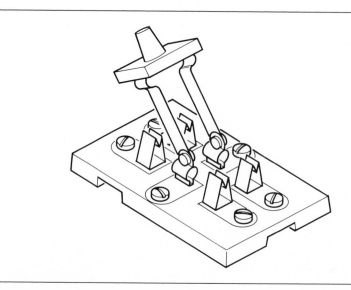

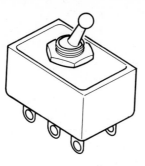

Figure 1-6
Toggle switch

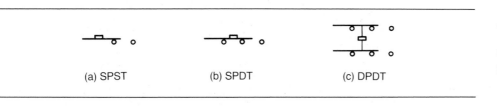

(a) SPST   (b) SPDT   (c) DPDT

Figure 1-7
Pole and throw
configurations for
a slide switch

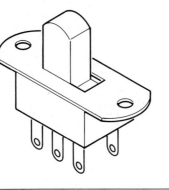

Figure 1-8
Slide switch

momentary or maintained contact. Actuation occurs when the shaft is depressed. The switch reverts back to its original unactuated state when the shaft is released in the momentary-contact switch. The switch remains actuated, until the shaft is again depressed and released, when maintained contacts are utilized. As seen in Figure 1-10, a variety of different pole and contact configurations are available.

Rotary switches: These devices are made up of one or more contacts, mounted on a shaft, that connect with multiple stationary contacts attached to a dielectric disk, called a wafer, as illustrated in Figure 1-11. Switching occurs as the shaft is turned and the rotating contacts make or break connection with the stationary contacts. Two different contact configurations are used—break-before-make and make-before-break. The rotating contact disconnects the present connection before making contact with the next connection when the break-before-make configuration is

Figure 1-9

Push-button switch

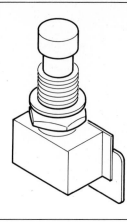

Figure 1-10

Pole and throw
configurations for
a push-button
switch

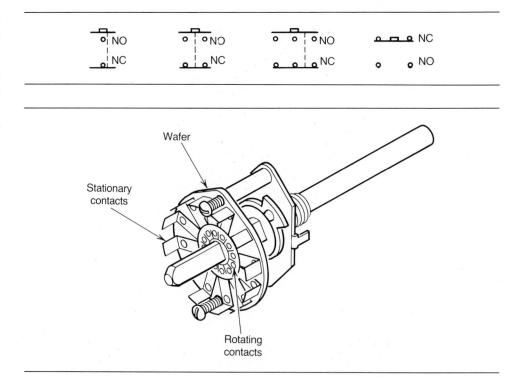

Figure 1-11

Rotary switch

utilized, as seen in Figure 1-12(a). The rotating switch contact makes connection with the next stationary contact before disconnecting the former stationary contact when the make-before-break configuration is used, as shown in Figure 1-12(b). Wafers can be stacked and additional rotary contacts attached to the shaft to increase the switching capacity of the switch, as shown in Figure 1-13.

Mercury-contact switches: Mercury-contact switches are composed of two stationary electrodes and a pool of mercury, all sealed in a tube. Mercury is a good

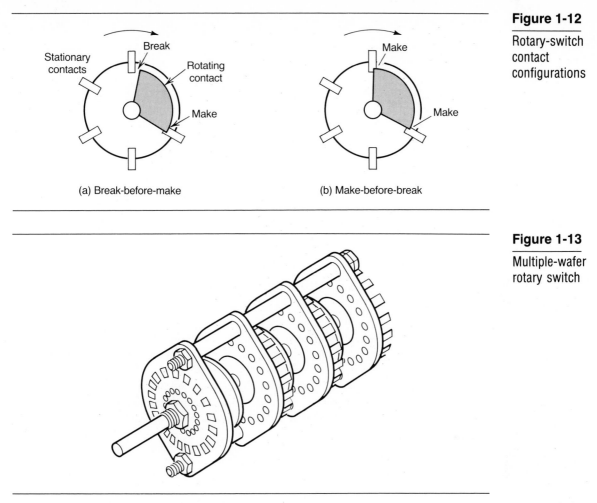

Figure 1-12

Rotary-switch
contact
configurations

(a) Break-before-make   (b) Make-before-break

Figure 1-13

Multiple-wafer
rotary switch

electrical conductor and acts as the switch pole. The tube is mounted on a hinged housing so that in the OFF position the mercury does not cover both electrodes, as illustrated in Figure 1-14(a). When the tube is raised to the ON position, the mercury covers both electrodes providing electrical continuity as shown in Figure 1-14(b). Depending upon the application, the electrodes may be made of iron, nickel-iron, chrome-iron, tungsten, molybdenum, or platinum.

Pressure switches: Pressure switches are actuated whenever a predetermined pressure is reached. Frequently used in fluid power systems, pressure switches are constructed by connecting switch contacts to the Bourdon-tube, bellows, or diaphragm pressure gages. (Pressure gages are discussed in Chapter 4.) The symbols used to represent pressure switches are shown in Figure 1-15.

Limit switches: These switches are used extensively in the manufacturing industries and may be used to turn on, turn off, or reverse the direction of machine-tool tables, conveyors, and robot manipulators. Several different types of limit switches

Figure 1-14

Switching action of
a mercury-contact
switch

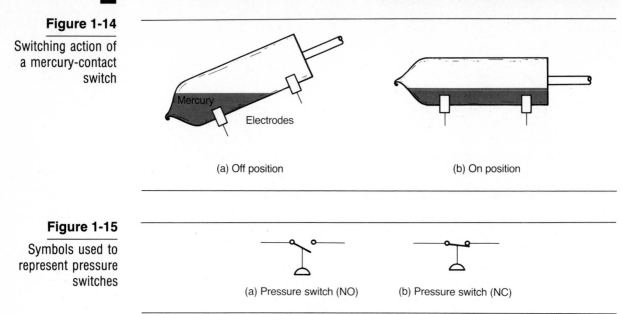

(a) Off position   (b) On position

Figure 1-15

Symbols used to
represent pressure
switches

(a) Pressure switch (NO)   (b) Pressure switch (NC)

are available, to meet a variety of switching applications. They can be classified according to speed of contact closure, type of actuation mechanism, and contact configurations.

Limit switches are designed to have either snap-action or slow-contact mechanisms. Snap-action switches snap over, or trip, when the actuation mechanism has traveled the required distance regardless of the speed at which it travels. Slow-con-

Figure 1-16

Push-actuated limit
switch

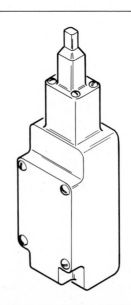

tact mechanisms are employed when the speed of the actuation mechanism determines the time required for the switch to trip.

Several different actuator mechanisms are available. These include the push, the lever, and the wobble-stick. The switch is actuated by a shaft located on the top or side of the switch housing, as shown in Figure 1-16, when a push mechanism is employed. The switch is actuated whenever the shaft is depressed. A spring causes the switch to become unactuated when the depressing force is removed. The limit switch is actuated by means of a lever clamped to a shaft when a lever-type mechanism is utilized, as seen in Figure 1-17. A flexible steel wire or nylon shaft, called a catwhisker or wobble stick, extending from the switch housing is used as the actuator when a wobble-stick mechanism is employed, as seen in Figure 1-18.

Limit switches are available in a number of different contact configurations. Some of the more common of those used with push-type switches are illustrated in Figure 1-19. The contact configuration for a lever-actuated switch appears in Figure 1-20. The switch is in the OFF position when centered and is actuated whenever the lever is moved to the left or right. The accompanying table shows the switch-contact states for the three lever positions.

1-2.4 Method of actuation

Switches may be actuated manually, mechanically, by motors, and by solenoids. Knife, toggle, slide, push-button, rotary, and mercury-contact switches are usually manually actuated. In some applications, a motor-driven cam or a solenoid may be used to depress and actuate a push-button switch for automatic operation. Motors are sometimes coupled to the shaft of a rotary switch, causing the shaft to turn

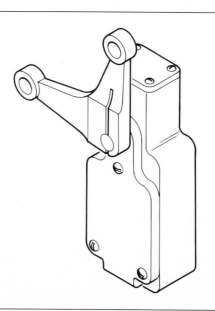

Figure 1-17

Lever-actuated
limit switch

Figure 1-18

Wobble-stick–
actuated limit
switch

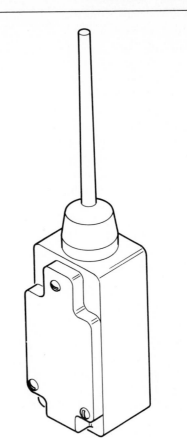

Figure 1-19

Contact
configurations for
a push-type limit
switch

whenever the motor is energized. Limit switches are almost always mechanically actuated. For example, the movement of a machine tool table to the point where the switch is mounted causes the switch to actuate.

1-2.5 Switch applications

Knife switches may be used for switching main power circuits and for starting and stopping some motors. Toggle switches are the most common type of switch and are used extensively for controlling low-power circuits such as lighting and small electronic equipment. Slide and push-button switches are mainly used to control

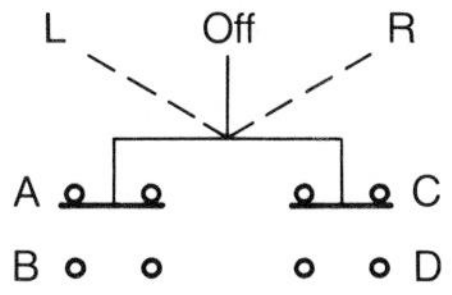

Figure 1-20

Contact
configuration for a
lever-actuated limit
switch

Lever Position	Contact Position			
	A	B	C	D
R	C	O	O	C
Off	C	O	C	O
L	O	C	C	O
C: closed				
O: open				

low-power electronic equipment. In these applications, slide, push-button, and toggle switches can be used interchangeably. Rotary switches are usually employed where several different circuits are affected when a change in the switch position is made. Because they have no moving internal mechanism and the contacts are protected from the environment, mercury-contact switches have an extremely long life span. General-purpose mercury-contact switches are designed to carry currents as great as 10 amperes and are used primarily in moderate-power switching applications. Heavy-duty mercury-contact switches are available for carrying currents in excess of 10 amperes and are often used in switching applications for large and heavy-duty industrial equipment. Mercury-contact switches can also be used as heat and centrifugal-force sensors. The movement of the mercury due to expansion or forces acting on it causes switch actuation.

Almost any type of machine-tool movement can be controlled with limit switches. These switches are also used extensively in the control of materials-handling equipment such as conveyors and overhead monorails. The applications for the switches just discussed are summarized in Table 1-1.

Table 1-1

SWITCH
APPLICATIONS

TYPE OF SWITCH	MAJOR APPLICATIONS
Knife	Switching main power circuits and starting/stopping some motors.
Toggle	Switching low-power utility and electronic circuits.
Slide	Switching low-power circuits in electronic equipment.
Push-button	Switching low-power circuits in electronic equipment.
Rotary	Switching several low-power electronic circuits simultaneously.
Mercury-contact	Switching moderate-to-high-power electrical circuits. May also be used as centrifugal and thermal switches.
Limit	Controlling the motion of mechanical devices.

1-2.6 Switch specifications

Switches have a variety of electrical and physical specifications—many of which have been standardized by the National Electrical Manufacturers Association (NEMA). Samples of each switch model are submitted to the Underwriters Laboratory (UL) by the manufacturer before being marketed to ensure that they meet these standards. The two most important specifications are the maximum voltage and current ratings. These must never be exceeded. When selecting a switch for a particular application, the designer must know the value of the maximum voltage that will be applied across, and the maximum current that will flow through, the switch contacts.

Switches are frequently mounted in areas where air cannot circulate freely around their mechanisms or housings. The maximum-switch-current rating value should be increased at least 25 percent to prevent excessive heat generation. This means that a switch that is to be used in a 220-volt circuit whose current is 45 amperes should have minimum voltage and current ratings of 220 volts and 56.25 amperes (45 A $\times$ 1.25). The switch selected may have higher voltage and current ratings, but never lower ones. A defective switch should be replaced with an identical type. If one is not available, a switch having a higher current (or voltage) rating may be substituted.

Switches used to start motors must be capable of handlng the maximum overload, or inrush, current that flows when the motor is initially started. This is usually six times the rated full-load current for AC motors and four times the full-load current for DC motors. For example, a switch used to control an AC motor that requires 12 amperes of current when running under full-load conditions should have a minimum current rating of 72 amperes (6 $\times$ 12 A). A switch used to control a DC motor that requires 8 amperes of current when fully loaded should have a minimum current rating of 32 amperes (4 $\times$ 8 A).

Table 1-2

NEMA SWITCH-ENCLOSURE CLASSIFICATION

NEMA TYPE	DESCRIPTION	APPLICATION
1	General-purpose	For general-purpose use.
2	Drip-type	For use in humid environments.
3	Weather-resistant	For outdoor use.
4	Watertight	For use in wet environments.
5	Dust-tight	For use in heavy dust environments.
6	Submersible	For underwater use.
7	Sealed-contact	For use where arcing may cause an explosion or fire.
12	Oil-resistive	For use where oil or grease are part of the environment.
13	Oil-type	For use in high-voltage applications.

The resistance of the closed contacts must be considered in high-current switch applications. Contact resistance must be low to prevent an unwanted voltage drop from appearing across the closed contacts and to prevent excessive heat from developing.

1-2.7 Switch enclosures

Switches are normally enclosed in a nonconductive material to protect operating personnel. In addition, switches are often mounted in locations where moisture, liquids, dust, and oil may be present. These materials cause contacts to short, stick, or wear excessively. Different types of enclosures have been developed to keep contaminants out of the switching mechanism. These have been standardized by NEMA and appear in Table 1-2.

1-3 OVERLOAD PROTECTIVE DEVICES

1-3.1 Introduction

Electrical energy, especially that coming directly from main power lines, can be extremely dangerous. A number of different devices have been developed for use in the automatic protection of life, property, and equipment. These devices automatically disconnect the circuit from the power lines if excessive load-current flows, a short-circuit develops, or a leakage path to ground occurs. These are discussed in the following sections.

1-3.2 Fuses

A variety of different types of fuses are available. These are grouped into two categories—plug and cartridge. The Underwriters Laboratory standards for fuses require that all blow promptly at overload and be capable of physically withstanding heavy short-circuit currents without emitting fire or exploding. Some fuses provide a short delay to prevent unwanted blowing whenever surge currents are present. The initial current in circuits containing loads such as lights, motors, and heaters is high when these circuits are first actuated. After a short period of time a lower steady-state current is achieved. Fuses used in these types of circuits should provide a short time delay before opening and are called slow-blow fuses.

Plug fuses have threads embossed on their base, as shown in Figure 1-21. Contact is made when the fuse is screwed into a receptical. This fuse is made of a strip of metal alloy that melts whenever excessive current flows through it. Plug fuses have little time delay and are primarily used in low-power household utility circuits where the surge current is minimum. They are available in sizes up to 30 amperes.

There are two types of cartridge fuses—the one-time and renewable. The one-time type is the more common and is shown in Figure 1-22. The fuse material is made of a wire or thin metal alloy mounted in a vulcanized paper or fiber cylinder. The cylinder is filled with a nonconductive powder that quenches the arc that often

Figure 1-21

Plug fuse

occurs when the fuse blows. Cartridge fuses are available in standard voltage ratings up to 600 volts with current ratings as high as 600 amperes. The internal fuse element is replaceable in the renewable cartridge fuse.

1-3.3 Circuit breakers

A circuit breaker is a switching device which automatically opens whenever excessive current flows. Unlike a fuse, whose protective element is destroyed, a circuit breaker can be reset after the fault that caused the excessive current has been discovered and corrected.

Circuit breakers are designed or are set to actuate at a predetermined current. Often, this is 125 percent of the maximum load current. A variety of circuit breakers are available for different load applications. Some are designed to respond instantaneously to overloads while others provide a delay before opening the circuit. Fast-acting circuit breakers are used to protect general-purpose utility and electronic circuits. Those having a time delay are used in circuits where high surge currents are present, such as motors. Some provide delays of up to 30 seconds. Two common types of circuit breakers are the thermal and the magnetic. The thermal circuit breaker reacts to heat created by current flowing through a control element.

Figure 1-22

Cartridge fuse

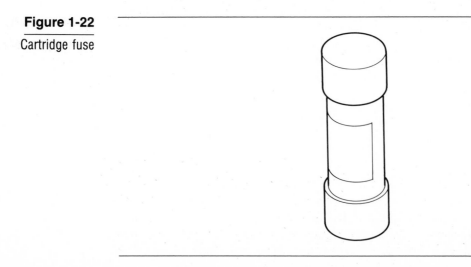

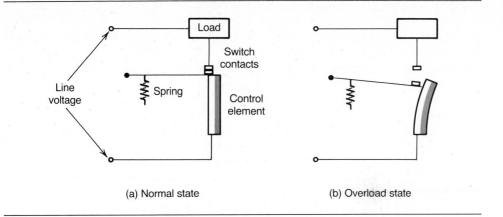

Figure 1-23

Thermal circuit
breaker

(a) Normal state

(b) Overload state

As seen in Figure 1-23, this element is composed of two dissimilar metals, which
have different temperature expansion coefficients, bonded together. As load current flows through the element, heat is produced, causing the element to warp but
not enough to open the contacts, as illustrated in Figure 1-23(a). If the load current
increases above the rated value of the circuit breaker, usually 125 percent, additional warping caused by an increase in temperature opens the breaker switch contacts, as shown in Figure 1-23(b).

A magnetic circuit breaker appears in Figure 1-24. This circuit breaker is actuated by the attractive force between an electromagnet and a tripping mechanism.
When the current flowing through the load is less than the rated value of the circuit
breaker, the magnetic force is not great enough to attract the tripping mechanism,
as seen in Figure 1-24(a). When an overload occurs the magnetic field increases,
actuating the tripping mechanism and opening the switch contacts, as shown in Figure 1-24(b).

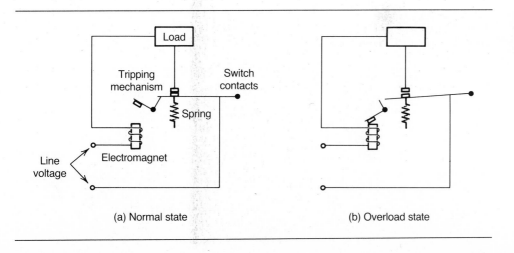

Figure 1-24

Magnetic circuit
breaker

(a) Normal state

(b) Overload state

1-3.4 Ground Fault Interrupters

Fuses and circuit breakers are not sufficiently sensitive to protect the human body if it comes in contact with an energized power line or circuit. Death occurs when approximately 75 milliamperes of current flows through the body. A Ground Fault Interrupter (GFI) is a very fast-acting switch whose contacts open whenever small leakage currents flow to ground. If the leakage current is caused by an individual coming in contact with an energized circuit, the switch contacts open before serious electrical shock occurs. Typically, GFIs are designed to actuate in approximately 25 milliseconds for currents as low as 30 milliamperes.

Several different types of GFIs are available for either single- or three-phase systems. A common single-phase type is the differentiator GFI appearing in Figure 1-25. The conductors running from the power source to the load pass through a toroidal coil whose sensing element is connected to a differential amplifier. When no leakage from the hot wire to ground occurs, the currents in both conductors are equal. The resulting circular magnetic fields about both conductors have the same amplitude and are 180 degrees out-of-phase and so cancel one another. This causes the output of the differentiator amplifier to be zero. Whenever a human body or other type of accidental load comes in contact with the hot wire, the current flowing through the hot wire increases. The additional strength in the magnetic field about that conductor causes a voltage to be induced across the toroidal coil, which appears at an input terminal of the differential amplifier. This voltage is amplified, thus actuating the relay that interrupts the current flowing through the hot conductor.

1-4 RELAYS

1-4.1 Introduction

A relay is a switch that can be actuated from a remote position. The device is composed of wire wrapped around an iron core, one or more stationary contacts, and one or more movable contacts mounted on a lever called an armature—all mounted on a steel frame, as illustrated in Figure 1-26(a). A spring, which holds the movable contact in its unactuated state, is attached between the armature and

Figure 1-25

Differential Ground
Fault Interrupter

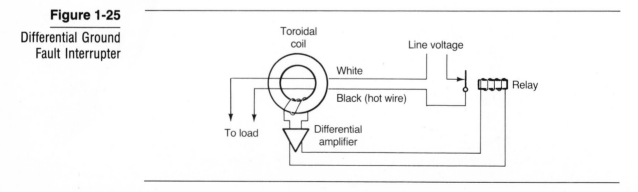

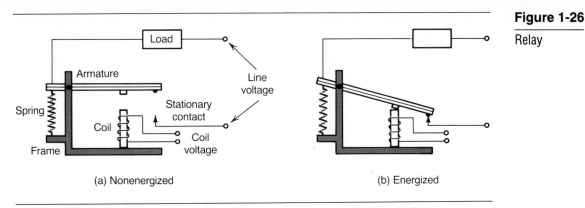

Figure 1-26

Relay

(a) Nonenergized   (b) Energized

frame. The relay is actuated, latched, or its contacts pick up whenever current of sufficient intensity flows through the coil and the attractive force between the core and armature overcomes the spring tension, as seen in Figure 1-26(b). The relay remains latched as long as sufficient current flows through the coil. When the current is reduced or removed, the core becomes unmagnetized and the armature is pulled up by spring action to its unactuated position.

1-4.2 Relay applications

Relays are usually used to actuate loads from a remote position for operator convenience or safety. The relay is mounted at or near the load and the switch is located in a place convenient for operating personnel. This is illustrated in Figure 1-27. A 12-volt source is connected to a relay coil via a switch. When the switch is closed the relay is energized, causing the movable contacts to connect with the stationary contacts. This completes the load circuit. Current flows from the main voltage source through the load, through the relay contacts, and back to the main supply. Relays designed to handle high amounts of power are called contactors. Contactors are constructed in the same fashion as conventional relays except that the insulation resistance of the coil is higher and the contacts will switch higher currents.

Relays may be used in conjunction with switches in locking circuits to turn machines on and off. Two push-button switches are utilized, one for starting the machine and the other for stopping it, as shown in Figure 1-28. Instead of requiring two voltage sources, one for the relay coil and the other for the load, the coil is connected in parallel with the load-supply voltage, as illustrated in Figure 1-28(a). The relay is energized when the momentary-contact start switch is depressed, al-

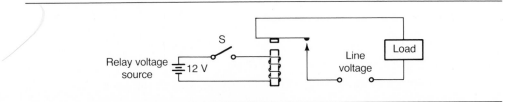

Figure 1-27

Actuating a load from a remote position

Figure 1-28

Relay locking
circuit

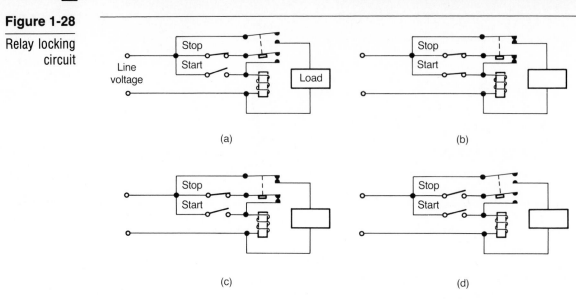

lowing current to flow through the coil. The relay contacts close and current flows through the load, as shown in Figure 1-28(b). When the start switch is released, the coil remains energized because of the current flowing through the stop-switch contacts, the closed relay contacts, and the coil, as seen in Figure 1-28(c). The load remains energized, or locked, until the momentary-contact stop switch is depressed, as shown in Figure 1-28(d).

Like conventional switches, relays are built with normally open or closed positions for either single- or double-throw operation and with multiple poles, as illustrated in Figure 1-29.

To represent relays such as those appearing in Figure 1-29, electrical symbols, as illustrated in Figure 1-30, are sometimes used instead of electronics symbols. These simplify the diagraming process considerably. The relay coil is indicated by a ramp-type symbol while parallel lines represent the contacts. Closed contacts are depicted by a slash line superimposed over the contact symbols.

1-4.3 Relay specifications

To ensure manufacturing standardization of relays, vendors, in conjunction with NEMA, have standardized the definitions of the variables associated with relays. These appear below.

Coil voltage: This specification identifies the voltage applied to the coil. Most relays are designed to be operated with DC voltage, although AC models are available. Typical DC-coil voltages include 5, 6, 12, 24, and 48 volts. AC-voltage models include 12, 24, 48, 110, and 220 volts.

Pickup current: Sometimes called pull-in current, this is the minimum current required to overcome the spring tension and actuate the relay.

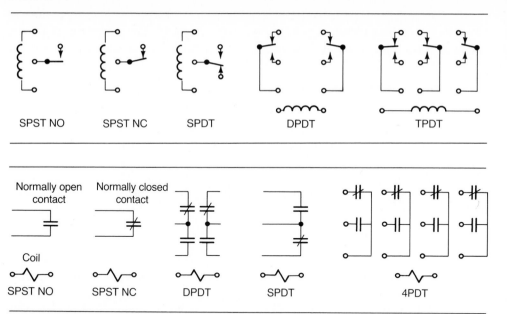

Figure 1-29

Typical relay pole and contact configurations

SPST NO   SPST NC   SPDT   DPDT   TPDT

Figure 1-30

Electrical-symbol representation of relay pole and contact configurations

Normally open contact   Normally closed contact

Coil

SPST NO   SPST NC   DPDT   SPDT   4PDT

Coil resistance: This is a measure of the resistance of the coil winding. Coil resistance ranges from approximately 40 to 10,000 ohms depending upon the coil-voltage requirement. The greater the coil-voltage requirement, the higher the coil resistance.

Maximum contact current: This specification quantifies the maximum current which can flow through the contacts. Relay contacts must be capable of carrying the current required by the circuits connected to them without producing undue heat and developing unwanted voltage drops. Maximum contact current may range from a few hundred milliamperes in small relays to several hundred amperes in contactors. Contactors used in electric-motor starters are usually rated in horsepower rather than current. A contactor capable of switching a three-fourths horsepower motor will have a three-fourths horsepower rating rather than a maximum current rating.

Operate and release time: Operate and release time identifies the maximum time required to actuate and deactuate the relay contacts. Operate time is typically 15 to 25 milliseconds, while release time for most relays is approximately 8 to 25 milliseconds.

Contact configuration: Relays may have either single or multiple poles and have either a single or double throw, as was illustrated in Figure 1-27.

Service-life rating: This rating identifies the minimum number of switching operations expected from a particular relay. The constant movement of the switching contacts and the arcing which sometimes occurs as the relay is operated causes wear and contact pitting. The relay will eventually wear out. Service-life rating may range from 50,000 to several billion with 5 to 50 million operations being typical.

Type of enclosure: Relays may be mounted with their contacts open, covered, or hermetically sealed. If the contacts are covered or sealed, the same NEMA classification as is used for switches is used to identify the type of enclosure.

1-4.4 Switch and relay logic circuits

The logic OR and AND gate functions can be accomplished using either transistor, switch, or relay technology. Numerous applications for switches connected to perform the OR and AND gate functions exist. If a motor must be capable of being turned on from two or more locations, switches can be connected in an OR gate configuration, as illustrated in Figure 1-31. An AND gate application appears in Figure 1-32. In this circuit, the motor will not operate until the ON–OFF switch, limit switch, and pressure switch are closed.

1-4.5 Relay-ladder diagrams

Instead of using conventional schematic symbols to represent logic circuits containing switches and relays, a relay-ladder diagram is often used. A relay-ladder diagram contains vertical rails and horizontal rungs much like a household ladder. The vertical rails represent the source-voltage lines while the horizontal rungs depict the devices and circuits connected across the lines. Relay coils and loads are represented by circles while switches and relay contacts are depicted by their normal electrical schematic symbols.

A circuit containing a switch, relay, and motor appears in Figure 1-33. The equivalent relay-ladder diagram contains two rungs, as seen in Figure 1-34. The top rung includes the switch and relay coil while the lower rung includes the relay contacts and motor.

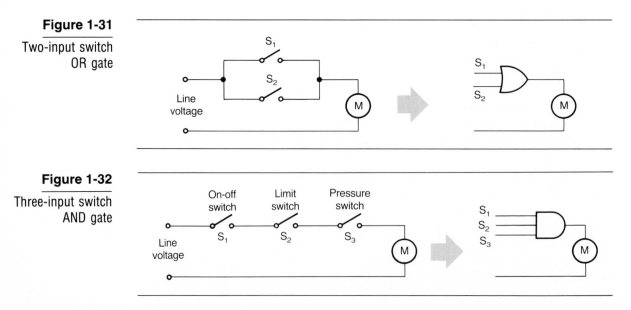

Figure 1-31

Two-input switch
OR gate

Figure 1-32

Three-input switch
AND gate

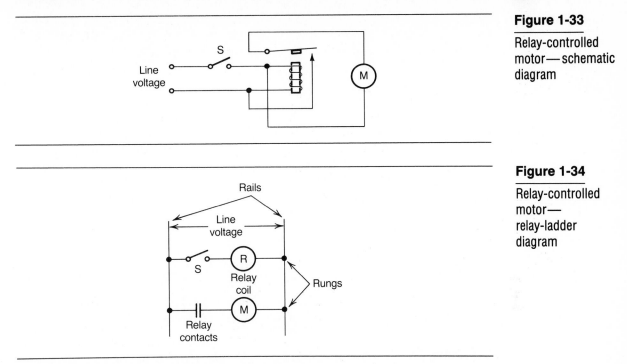

Figure 1-33

Relay-controlled
motor—schematic
diagram

Figure 1-34

Relay-controlled
motor—
relay-ladder
diagram

A three-input OR gate using relays appears in Figure 1-35. The motor is actuated whenever any one of the three relay contacts closes.

The diagram of a start–stop relay locking circuit appears in Figure 1-36(a). Three rungs are used in the equivalent ladder diagram shown in Figure 1-36(b). The first rung is composed of the start switch and relay coil. The stop switch, relay contacts 3 and 4, and the relay coil are all connected in series and form the second rung. Relay contacts 1 and 2 and the motor make up the third rung.

Machine tools that perform work activities on materials to produce a part and the conveyors and overhead monorails that move parts from one machine to another have until recently been controlled with logic circuits that incorporate switches, relays, and timers. In this type of controller, a relay picks up whenever the circuit in which the relay coil is located is completed. A typical circuit appears in Figure 1-37. In this circuit, a pressure switch controls relay 1 in the top rung. Contacts 1 and 2 of relay 1 and the control coil of relay 2 form the second rung. Relay 2 will not

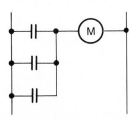

Figure 1-35

Three-input relay
OR gate

Figure 1-36

Start–stop relay
locking circuit

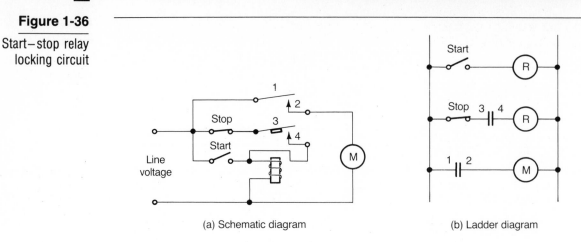

(a) Schematic diagram   (b) Ladder diagram

pick up until relay 1 has been energized. Relay 3, in the third rung, will not pick up until relays 1 and 2 have been energized. Relay 4, in the fourth rung, is initially energized and remains in this condition until relay 1 is actuated. Timers can be used with any or all of the relays to provide delays if required.

1-5 TIMERS

1-5.1 Introduction

Timing is extremely important in industrial electronics. Actuators, such as electric motors and fluid power devices, employed in automated manufacturing must be turned on when needed, operated for a specified length of time, and then turned off. In many applications the operation of production machines must be sequenced

Figure 1-37

Manufacturing
controller

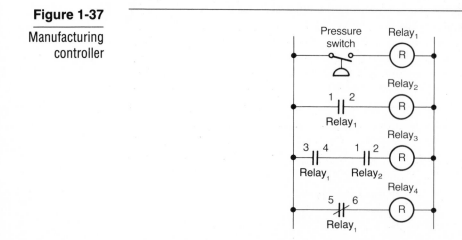

with the activities performed by other machines on the assembly line. Three categories of timers are available to perform timing functions—delay, interval, and cycle timers.

1-5.2 Delay timers

A delay timer is one that provides a delay between the time an event is initiated, such as the closing of a switch, and the time that the event is actually performed, that is, energizing a load. This type of timer may also be used to provide a delay at the terminating end of an activity. In this case a load or circuit remains energized after a switch is opened for the duration of the delay. The former application is called on-delay while the latter is referred to as off-delay. A number of different methods are used to achieve time delays. These include motor-driven delay timers, thermal delay timers, electronic time-delay circuits, and dashpot timers. Delay timers may be packaged as discrete devices or they may be included as part of a relay to provide a delay between the time a switch is actuated and the relay becomes either energized or deenergized.

Motor-driven timers: A motor-driven delay timer consists of a single-phase synchronous motor whose shaft is connected to reduction gears to reduce its speed. A lever or similar switch-actuator mechanism is connected to the reduction gear unit, as illustrated in Figure 1-38. When S_1 is closed, current flows through the motor by way of the NC contacts energizing the motor. As the motor slowly turns, the relay picks up when the actuator mechanism comes in contact with S_2. The NC relay contacts open, causing the motor to stop turning, and at the same time the NO contacts close, energizing the load. The load remains energized until S_1 is opened and the relay contacts drop out. At the same time, the motor shaft is returned to its original position.

Thermal delay timers: Several types of thermal delay timers are available. They can be grouped into two categories—bimetal and expansion. The bimetal timer operates on the principle that dissimilar metals expand at different rates when heated. As illustrated in Figure 1-39(a), the device is composed of two supporting dissimilar metal columns joined together by an armature. A heater wire is wrapped around one of the columns. When the switch is closed, heat created by current flowing

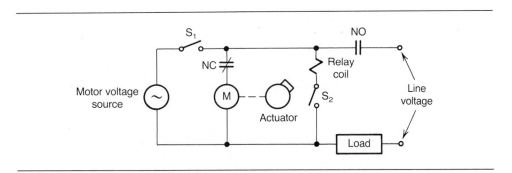

Figure 1-38

Motor-driven delay timer

Figure 1-39

Bimetallic delay
timer

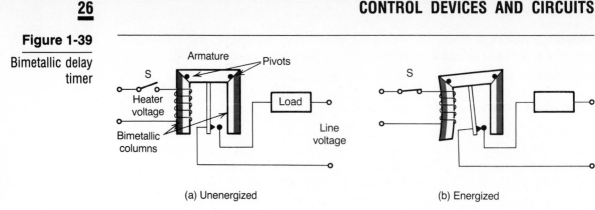

(a) Unenergized

(b) Energized

through the heating element allows that column to expand. The uneven expansion of the dissimilar metals causes the column to warp, forcing the contacts together as seen in Figure 1-39(b). The time delay is the interval occurring between the time the switch is closed and the time the load is actuated when the timer contacts make connection. Ambient temperature can affect the delay in this type of timer. This is compensated for by also using dissimilar metals in the unheated column. This allows the distortion created by ambient temperature to affect both columns equally and they cancel one another.

The diagram of an expansion delay timer appears in Figure 1-40(a). This timer is made up of a metallic column around which a heating element is wound, and an armature mounted on a pivot. When the switch is closed, current flows through the heating element, causing the metallic column to expand. This forces the armature downward, causing the contacts to make as shown in Figure 1-40(b). The delay oc-

Figure 1-40

Expansion delay
timer

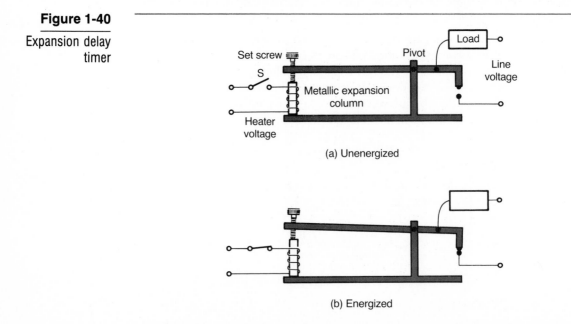

(a) Unenergized

(b) Energized

curling between the actuation of the switch and the closing of the timer contacts can be varied by adjusting the set screw.

Electronic timers: These types of delay timers utilize the charging time of a capacitor in an RC circuit to provide the delay. They operate on the principle that the time required to charge or discharge a capacitor in an RC circuit depends upon the size of the capacitor and resistor. The diagram of an electronic timer appears in Figure 1-41. The RC time-delay circuit is connected to the base of the transistor. As long as the switch is open, no voltage appears across the base terminal and the transistor is cut off. When the switch is closed, the capacitor charges through the resistor until the base voltage is high enough to turn the transistor on. The delay is the time that elapses from the closing of the switch until the transistor turns on. A rheostat can be substituted for the resistor to allow the delay time to be varied. Usually, only the time provided by the first RC time constant is utilized for the delay, as this is the most linear portion of the charging curve, and consequently provides the greatest timing accuracy. In some electronic time-delay stages, special circuits are provided, which expand the linear portion of the charging curve to allow longer delays.

Dashpot timers: Dashpot timers rely on the time that it takes a fluid to flow through an orifice to provide the delay. There are two types of such timers—those

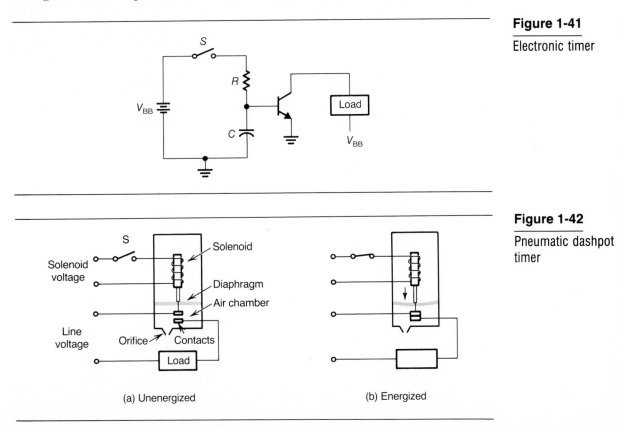

Figure 1-41

Electronic timer

Figure 1-42

Pneumatic dashpot timer

that utilize air as a fluid (pneumatic) and those that use a liquid (hydraulic). As illustrated in Figure 1-42(a), a pneumatic timer is made up of a solenoid, diaphragm, air chamber, and orifice. When the switch is closed the solenoid is energized, causing the movable core to push the diaphragm downward, forcing air through the orifice. The downward movement slowly continues until the contacts make, as seen in Figure 1-42(b).

The hydraulic dashpot timer is similar in appearance and operation to the pneumatic device except that a liquid, such as silicon, is used as the delay medium.

1-5.3 Interval and cycle timers

An interval timer is used whenever a load must be operated for a particular length of time. A cycle timer is one that recycles and performs its timing activities over and over again. Both interval and cycle timers are usually driven by a single-phase synchronous motor. A clutch is used to actuate the switching mechanism in an interval timer. When the timing event is completed, the clutch is disengaged and the motor returns to its original position. Disk cams are used in cycle timers, as illustrated in Figure 1-43. When the main switch (S_1) is closed, the motor begins turning and the cam lobe causes a spring-loaded switch (S_2) to close, energizing the load. The load remains energized during the time that the cam lobe is in contact with S_2. As the motor continues to turn the cam lobe moves past S_2, causing the switch contacts to open and the load to become deenergized. It remains in this state until the leading edge of the cam lobe again actuates S_2. In an interval timer, the timing interval is established by the size of the cam lobe.

Cycle timers may also be used to develop the sequential operation of several loads, as seen in Figure 1-44. This timer utilizes three cams that are mechanically linked. The angular position and size of the cam lobes determine the sequence and duration of load operation.

1-6 THE TRANSISTOR AS A CONTROL DEVICE

The bipolar-junction transistor is a control device. When connected in the common-emitter configuration, base (input) current controls collector (output) current. A small change in base current causes a larger change in collector current. This is how the

Figure 1-43

Cycle timer

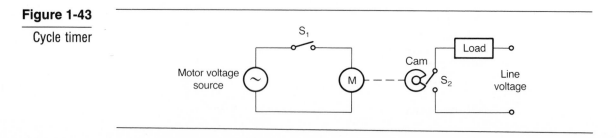

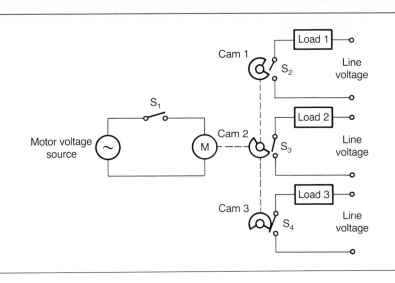

Figure 1-44

Multiple-cam cycle timer

transistor operates when connected as an amplifier. An input-signal current controls the output current and causes it to vary above and below the operating (Q) point.

The transistor can also be operated as an ON–OFF control device. In this application, it is operated as a solid-state switch or relay. Instead of having a single operating point, the transistor is operated either at cutoff or saturation. When cut off, the transistor is operated without bias, as shown in Figure 1-45(a). In this condition the transistor performs the function of an open switch or relay. The transistor behaves as a switch or relay in the closed position whenever a base bias potential V_{BB} large enough to cause the transistor to become saturated is applied, as seen in Figure 1-45(b).

Ohm's law is used to determine the value of collector current flowing in a BJT at saturation, as shown by Equation 1-1.

$$I_{C(\text{Sat})} = \frac{V_{CC}}{R_C}$$

Eq. 1-1

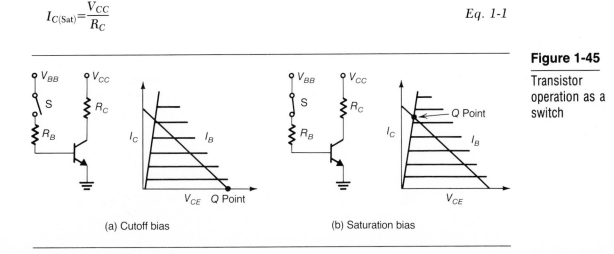

(a) Cutoff bias   (b) Saturation bias

Figure 1-45

Transistor operation as a switch

Figure 1-46

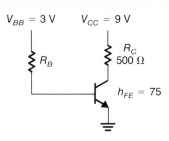

Since the DC transistor-current gain (h_{FE}) is the ratio of the collector to base current, the base-current value required to drive the transistor into saturation may be found by using Equation 1-2.

$$I_{B(Sat)} = \frac{I_{C(Sat)}}{h_{FE}} \qquad\qquad Eq.\ 1\text{-}2$$

For a given V_{BB} value, base current can be obtained by choosing the proper size base resistor, as seen by Equation 1-3.

$$R_B = \frac{V_{BB} - V_{BE}}{I_{B(Sat)}} \qquad\qquad Eq.\ 1\text{-}3$$

This process is illustrated in Example 1-1.

Whenever very fast switching times are required, a small capacitor (several hundred picofarads) may be connected across the base resistor, as shown in Figure 1-47. Called a speed-up capacitor, the reactance of the device is small enough to cause the base current to bypass the base resistor when the base voltage is initially applied. This high surge of current turns the transistor on and very rapidly drives it into saturation. After five time constants, the charge on the capacitor blocks any further flow of current, the base current all flows through the base resistor, and the capacitor no longer affects the circuit. When the base voltage is removed, the po-

Example 1-1

DETERMINING
THE BASE-
RESISTOR SIZE
TO SATURATE A
BJT

Problem: Determine the size of base resistor needed to cause the silicon NPN transistor appearing in Figure 1-46 to be driven into saturation.

Solution:

$$I_{C(Sat)} = \frac{V_{CC}}{R_C} = \frac{9V}{500\Omega} = 0.018\ A$$

$$I_{B(Sat)} = \frac{I_{C(Sat)}}{h_{FE}} = \frac{0.018A}{75} = 0.00024\ A$$

$$R_B = \frac{V_{BB} - V_{BE}}{I_{B(Sat)}} = \frac{3 - 0.7V}{0.00024A} = 9,583.33\ \Omega$$

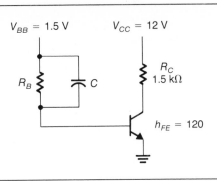

larity of the charge on the capacitor reverse biases the base–emitter junction, causing the transistor to be turned off very rapidly. The capacitor discharges through the base resistor and is free to bypass current again when the transistor is turned back on.

Transistor switches have a variety of applications in industrial circuits. They can be used to drive indicator lights that depict the status of a circuit or device, as illustrated in Figure 1-48(a). A relay coil can be connected in the collector to allow the relay to be operated electronically, as seen in Figure 1-48(b). A reverse-biased diode is usually connected across the coil to prevent the CEMF developed by the coil from destroying the transistor when the transistor is turned off.

The loads are connected in series with the collector for the circuits appearing in Figure 1-48. This type of connection reduces collector current for high-impedance loads, however. High-impedance loads can be connected in parallel with the collector as illustrated in Figure 1-49. In this application, the load is energized whenever the transistor is cut off and is deenergized when the transistor is saturated.

Switching transistors have traditionally been of the bipolar-junction type. Field-effect transistors have not been capable of handling the power required of some loads. This is beginning to change, however. There has been considerable recent development in power MOSFET devices. These transistors switch more rapidly than do the bipolar types because they are majority current devices and turn-on

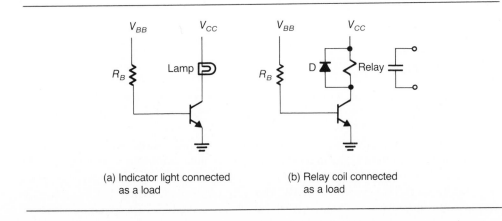

(a) Indicator light connected as a load

(b) Relay coil connected as a load

Figure 1-49

Transistor switch
with load
connected in
parallel

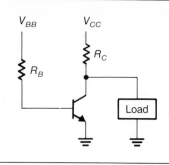

time is not affected by the recombination time of minority carriers. The switching speed of some of these power MOSFETs is extremely fast. These transistors have enhanced the ability of computers and microprocessors to control relatively high current loads.

Transistors used as switches have a number of advantages over conventional switches and relays. They are much faster. Actuation time is measured in milliseconds for switches and relays, whereas it is measured in micro- or picoseconds for transistor switches. Since transistors have no mechanical parts, they can be operated indefinitely. For the same reason they also have no contact bounce. The contacts in most switches and relays tend to vibrate back and forth for a few milliseconds when the switch or relay is first actuated. This phenomenon is called contact bounce. A transient voltage is developed each time the switch contacts make as they vibrate back and forth. Usually, this has no effect on the operation of the circuit or load being controlled except in digital circuits. However, digital circuits may mistake this transient pulse train as a binary word and react accordingly.

Although the transistor has numerous advantages over a switch or relay in switching applications, it does have some disadvantages. First, transistors are not capable of carrying the very large currents that flow through some switch and relay contacts. Second, there will always be a small collector–emitter voltage across the output terminals no matter how heavily the transistor is driven into saturation. Third, the amplitude of the voltage applied to the transistor is limited to the collector–emitter breakdown voltage.

1-7 THYRISTORS

1-7.1 Introduction

Thyristors are a group of semiconductor components that behave as open-circuit devices until a particular breakdown voltage is exceeded or until triggered, at which time they conduct, or latch, and become low-resistance conductors. These devices can be divided into two categories and include those which conduct in the forward direction only (unilateral) and those which conduct in both the forward and reverse directions (bilateral). Unilateral thyristors include the Shockley diode, the silicon-controlled rectifier (SCR), the gate-controlled switch (GCS), and the silicon-controlled switch (SCS). Bilateral devices include the triac and diac.

1-7.2 The Shockley diode

This is a two-terminal, four-layer, three-junction, PNPN semiconductor device. As shown in Figure 1-50, the device is equivalent to two transistors, a PNP and an NPN, with the two inner sections being shared. When the diode is forward biased, junctions J_1 and J_3 are forward biased while J_2 is reverse biased. Since J_2 is included in both Q_1 and Q_2, both transistors are at cutoff and the anode current I_A is zero. If the bias voltage V_S is increased, the avalanche breakdown potential of J_2 is eventually reached and the diode begins conducting. The voltage required for conduction to occur is called the forward breakover potential $V_{BR(F)}$. When this potential is reached, the voltage V_{AK} across the anode and cathode terminals decreases substantially, the anode current increases very rapidly, and the diode operates as a switch in the closed position. The bias voltage can be reduced and the diode will continue to conduct as long as the anode current remains above a minimum intensity called

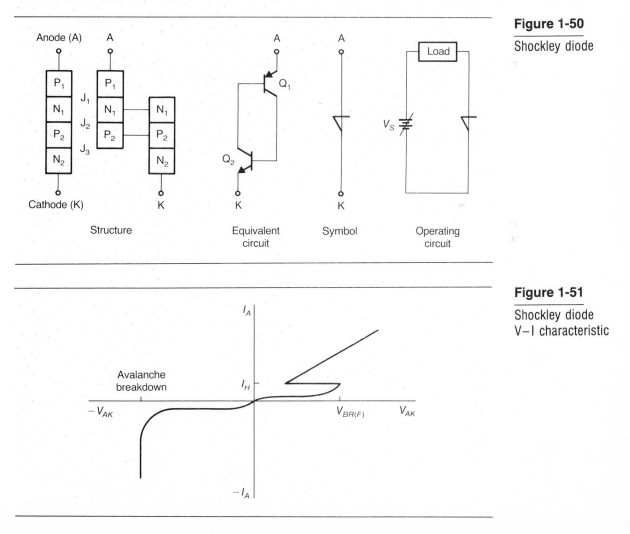

Figure 1-50

Shockley diode

Structure   Equivalent circuit   Symbol   Operating circuit

Figure 1-51

Shockley diode V–I characteristic

the holding-current value I_H. The diode drops out of conduction whenever the anode current decreases below this value. When reverse biased, the reverse volt–ampere characteristic curve is similar to that of a conventional PN junction diode. The volt–ampere characteristic for the device appears in Figure 1-51.

1-7.3 The silicon-controlled rectifier

The silicon-controlled rectifier (SCR) is a three-terminal, four-layer, three-junction PNPN diode whose forward breakover voltage can be varied. As seen in Figure 1-52, the equivalent circuit for the SCR is identical to that of the Shockley diode except that a gate terminal has been included to forward bias J_2. When the SCR is biased so that the anode is positive with respect to the cathode, J_1 and J_3 are forward biased and J_2 is reverse biased. Except for the minority carriers associated with J_2, conduction does not occur until the forward breakover potential $V_{BR(F)}$ is

Figure 1-52

SCR

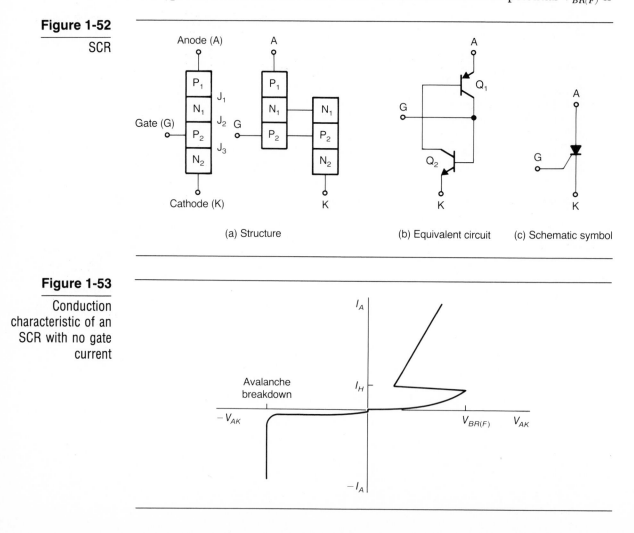

(a) Structure (b) Equivalent circuit (c) Schematic symbol

Figure 1-53

Conduction characteristic of an SCR with no gate current

Figure 1-54

Conduction
characteristic of an
SCR with gate
current

reached, as indicated in Figure 1-53. If a source is connected to the gate terminal to forward bias J_2, the forward breakover voltage required to cause conduction is decreased, as seen in Figure 1-54. There is an inverse relationship between the gate current and the anode–cathode voltage required to produce conduction. A family of volt–ampere curves can be obtained if the gate current I_G is increased in increments. The SCR exhibits the same characteristics as a conventional diode when reverse biased. Once the SCR has turned on, the gate current can be removed and the device will continue to conduct, due to the latching effect that is produced by the regenerative feedback caused by the cross connection of the base and collector terminals of Q_1 and Q_2. For this reason, gate current is often applied in the form of a short pulse. Since the gate does not affect the operation of the SCR after it has begun conducting, it cannot be used to turn the device off. Conduction is not affected by applying a negative voltage to the gate to reverse bias J_2. The only way to turn an SCR off is to reduce the anode current below the holding-current value. Two different methods are used to accomplish this—anode interruption and forced commutation. The anode-interruption method is illustrated in Figure 1-55. When opened, the switch in the anode circuit interrupts the anode current, causing the SCR to stop conducting.

Current is shunted around the SCR when the forced-commutation method is used. A transistor switch is connected in parallel with the SCR, as seen in Figure 1-56. The SCR is turned off by applying a positive pulse to the base of the transis-

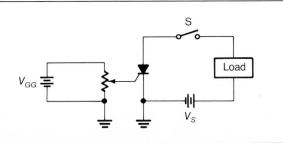

Figure 1-55

Anode-interruption
method of turning
an SCR off

Figure 1-56

Forced-commutation
method of turning
an SCR off

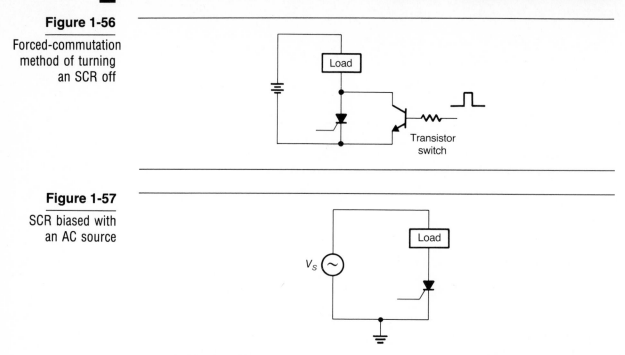

Figure 1-57

SCR biased with
an AC source

tor, causing it to become saturated. The resulting low-resistance path across the
SCR causes most of the current flowing from the bias source to bypass the SCR,
decreasing the anode current below the holding-current value.

The SCR is often connected to an AC source, as shown in Figure 1-57. In this
type of operation conduction is automatically extinguished every cycle, when the
sinusoidal anode voltage is sufficiently low to cause the anode current to drop below
the holding-current value. Some means must be provided to retrigger the SCR
each cycle. An effective method of accomplishing this is seen in Figure 1-58. The
voltage developed across the potentiometer during the positive alternation is applied to the gate, causing the SCR to conduct. The diode prevents the negative alternation from appearing across the SCR gate and cathode terminals. The SCR is

Figure 1-58

AC gate triggering
an SCR

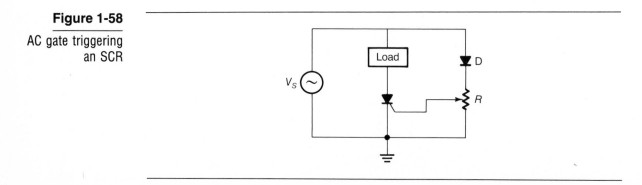

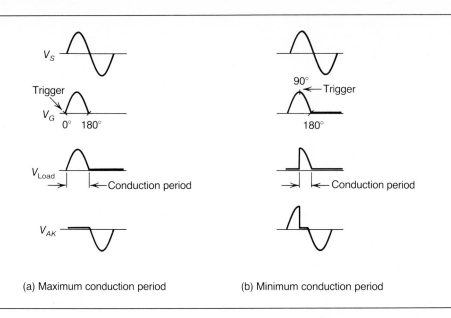

Figure 1-59

Maximum and
minimum
conduction periods
for an AC-gate
trigger circuit

(a) Maximum conduction period   (b) Minimum conduction period

essentially a half-wave rectifier when operated in this fashion. Unlike a conventional diode—which conducts for one-half cycle, or 180 degrees—the conduction period for an SCR can be controlled by varying the time at which the gate potential is applied to the SCR. For the circuit appearing in Figure 1-58, the conduction period can be varied from a maximum of 180 degrees (when triggered at 0 degrees) to a minimum of 90 degrees (when triggered at 90 degrees), as seen in Figure 1-59. Any conduction period between these limits may be obtained by adjusting the resistor.

For conduction periods of less than 90°, the trigger circuit shown in Figure 1-60 can be used. Called phase-control triggering, an RC phase-shift network is connected to the trigger circuit to reduce the conduction period. This trigger technique provides a conduction period ranging from a low of approximately 15 degrees to a high of about 170 degrees.

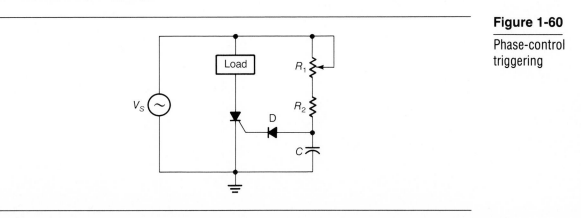

Figure 1-60

Phase-control
triggering

The advantage of using an SCR instead of a conventional rectifier diode in a rectifier circuit is that the load voltage and current can be varied efficiently. The load often consists of a motor whose speed must be varied or a lamp whose intensity must be changed. Either could be accomplished by connecting a rheostat in series with the load, but this is an inefficient method of achieving load control as the rectifier delivers full power to the rheostat–load combination for its entire conduction period of 180 degrees. Power not consumed by the load is dissipated by the rheostat and may be considerable for high current loads. When an SCR is employed as a rectifier diode, the load current can be controlled by varying the conduction period. This is why an SCR is called a controlled rectifier. Its conduction period can be controlled by delaying the time that the gate causes the device to conduct.

A half-wave–rectifier application for an SCR appears in Figure 1-61. In this circuit, the SCR is used to vary the speed of a DC motor by converting the AC line voltage into a DC potential whose amplitude can be varied. The output voltage and motor speed are adjusted by varying the resistance of R_2. This potentiometer, along with R_1 and R_3, forms a voltage divider network that develops the gate-triggering voltage. The diode connected between R_2 and the gate terminal prevents the negative alternation from appearing across the gate and cathode terminals.

In addition to rectifier applications, the SCR is often used as an electronic switch. As a switch, the SCR can change states faster and carry much higher currents than can its transistor counterpart. This technique is illustrated in Figure 1-62. The SCR is being used as a high-current switch. Two spring-loaded push-button switches are used to energize and deenergize the circuit. The contacts of the OFF switch are normally closed while those of the ON switch are normally open. Momentarily depressing the ON switch causes current to flow through the voltage divider network made up of R_1 and R_2. The voltage developed across R_2 provides the potential to turn on the SCR, and current begins flowing through the load. The ON switch is released and the SCR continues conducting. The circuit remains in this state until it is desired to deenergize the load. This is accomplished by momentarily depressing the OFF switch. This interrupts the SCR anode current, causing the

Figure 1-61

SCR half-wave
rectifier

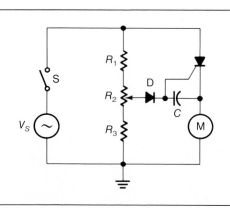

Figure 1-62

SCR switch

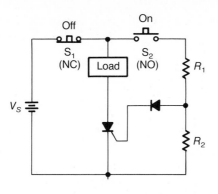

SCR to cease conducting. In this circuit, the SCR performs the same function as does the locking relay in the circuit appearing in Figure 1-28.

Like conventional PN-junction diodes, SCRs have operating variables that identify the maximum ratings associated with a particular device and its typical operating characteristics. These variables are all specified at a particular temperature and may be obtained from semiconductor data manuals. The more commonly used variables are defined below.

Average forward current $I_{F(AV)}$: This rating identifies the maximum DC anode current the SCR can carry when it is in the conducting state.

Reverse breakdown voltage $V_{BD(R)}$: This rating identifies the maximum reverse voltage that can be applied across the anode–cathode terminals without reaching avalanche breakdown.

Holding current I_H: This characteristic identifies the minimum anode current required to maintain conduction.

Gate trigger current I_{GT}: This characteristic identifies the value of gate current required to turn on the SCR.

Maximum forward voltage drop V_F: This characteristic identifies the maximum voltage drop across the anode–cathode terminals specified at a particular $I_{F(AV)}$.

Switch time, T_{ON} and T_{OFF}: These are turn-on and turn-off times, respectively, specified at a particular $I_{F(AV)}$.

1-7.4 The gate-controlled switch

The gate-controlled switch (GCS) is a unilateral device whose composition is similar to that of the SCR. It has the same turn-on characteristics as the SCR, but unlike the SCR can be turned off by applying a negative potential to its gate. It can also be turned off by interrupting or reducing the anode. The GCS is mainly used in low-power switching and rectifier circuits. It is not capable of dissipating large amounts of power as the SCR is. The symbol for the device appears in Figure 1-63.

Figure 1-63

Gate-controlled
switch (GCS)

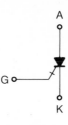

1-7.5 The silicon-controlled switch

The silicon-controlled switch (SCS) is a two-gate unilateral thyristor that has characteristics similar to those of the SCR. The device has both an anode and a cathode gate, as depicted in Figure 1-64. Either one of the gates may be biased while the other remains unconnected, or both may be biased, depending upon the desired type of operation. For single-gate operation, the SCS has the same characteristics as an SCR when a negative potential is applied to the anode gate or a positive potential is applied to the cathode gate. Like the GCS, the device can be turned off by applying a potential to the gate whose polarity is opposite that needed to turn it on.

1-7.6 The triac

The triac is a bilateral, or bidirection, thyristor which conducts in either direction. Although the SCR is used as a controlled rectifier, it allows current to flow in one direction only. In some applications it is beneficial to have a device that can be triggered to conduct during either the positive or negative alternations. The triac permits this type of operation. It operates as two SCRs connected in an anode-to-cathode parallel arrangement, as seen in Figure 1-65. Since the triac can conduct equally well in either direction, the device does not have a conventional anode and cathode as the SCR does. Instead, it has two anode terminals, either of which can function as an anode or a cathode. The triac conducts in the forward direction when A_1 is positive with respect to A_2 and a positive potential is applied to the gate.

Figure 1-64

Silicon-controlled
switch (SCS)

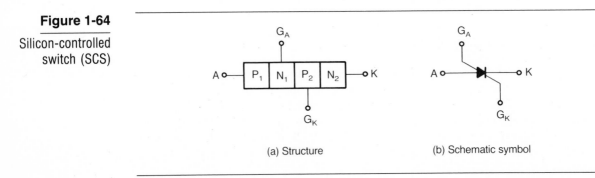

(a) Structure   (b) Schematic symbol

Figure 1-65

Triac

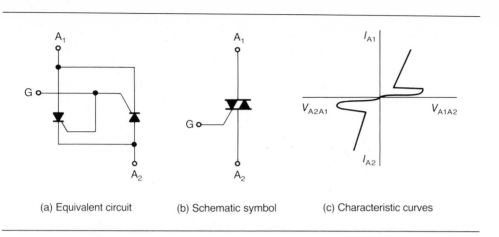

(a) Equivalent circuit   (b) Schematic symbol   (c) Characteristic curves

It conducts in the negative direction when A_1 is negative with respect to A_2 and a negative voltage is applied to the gate. Like the SCR, the triac is turned off by interrupting the anode current or by reducing the anode current below the holding-current value. Although the triac is superior to the SCR because of its ability to conduct in either direction, it does not have the current-carrying capacity of the SCR. SCRs can be manufactured to carry much higher currents and dissipate considerably larger amounts of power than triacs.

1-7.7 The diac

The diac is a bilateral device that conducts equally well in either the forward or reverse directions when the breakover voltage is exceeded, as shown in Figure 1-66. Except for its bilateral characteristics the diac operates much like the Shockley diode. Since it conducts in both directions, the diac does not have conventional anode and cathode terminals. Either terminal can be used as an anode or a cathode depending upon the desired direction of conduction. Applications for the diac and the other thyristor devices are summarized in Table 1-3.

Figure 1-66

Diac

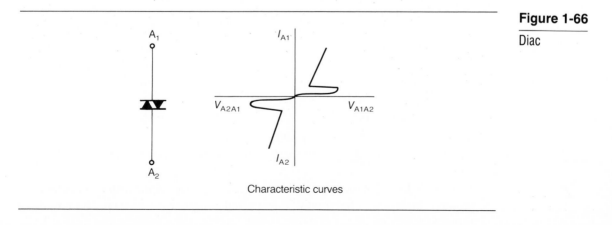

Characteristic curves

Table 1-3

THYRISTOR
APPLICATIONS

DEVICE	APPLICATION
Shockley diode	Major application is that of an electronic switch. It is often connected in series with a capacitor to trigger an SCR.
SCR	Can be used as a high-power switch. It is often used as a high-power controlled rectifier.
GCS	May be used in low-power electronic switch, inverter, and chopper applications.
SCS	This device turns on and off faster than an SCR. It can be used as a fast-acting low-power electronic switch or low-power controlled rectifier. In digital circuits it can be used in binary counters, shift registers, and timers.
Triac	May be used as a low-power electronic switch or a low-power controlled rectifier. The triac can be switched on and off during the positive and negative alternations of the input
Diac	Used as a bidirectional electronic switch.

Summary

1. A switch is a device for making, breaking, or changing connections in a circuit.
2. A circuit breaker is a switching device that opens automatically whenever excessive current flows.
3. A thermal circuit breaker reacts to heat developed when current flows through a bimetallic control element.
4. A magnetic circuit breaker is actuated by the attraction between an electromagnet and a tripping mechanism.
5. Ground Fault Interrupters (GFIs) are used to protect humans if they come in contact with an energized line or load.
6. A relay is a switch that can be actuated from a remote position.
7. Pickup current is the minimum current required to energize a relay.
8. The service-life rating identifies the minimum number of switching operations expected from a relay.
9. A contactor is a relay used in high-power applications.
10. Logic OR and AND gate functions can be accomplished by using switches and relays.
11. A relay-ladder diagram consists of vertical lines (called rails) that represent the voltage source, and horizontal lines (called rungs) that represent the circuit elements.

12. A delay timer may provide a delay between the time a switch is actuated and a load is energized or between the time a switch is opened and the load is deenergized.

13. An interval timer is used whenever it is necessary to operate a load for a specific length of time.

14. A cycle timer is used whenever operations must be performed repeatedly.

15. A transistor performs the function of a closed switch when biased at saturation.

16. When biased at cutoff, a transistor behaves as an open switch.

17. Thyristors are a group of semiconductor devices that conduct whenever their forward breakover potential is reached.

18. A Shockley diode is a unilateral thyristor that operates as an open switch until the forward breakover potential is reached.

19. A silicon-controlled rectifier (SCR) is a unilateral thyristor whose breakover voltage can be varied.

20. SCR forward breakover potential can be reduced by applying a positive potential at the gate terminal.

21. The conduction period of an SCR can be varied by delaying the application of a gate potential.

22. The SCR can be turned off by either interrupting or reducing the anode current.

23. The gate-controlled switch (GCS) is a three-terminal unilateral thyristor that can be turned off by applying a negative potential to its gate.

24. The silicon-controlled switch (SCS) is a two-gate unilateral thyristor.

25. The triac is a three-terminal thyristor that conducts in either direction.

26. The diac is a two-terminal bilateral thyristor.

Chapter Examination

1. Draw the diagram of a TPDT switch.
2. List seven types of switches.
3. List three types of overload devices.
4. Draw the schematic symbols for an SCS, diac, SCR, and triac.
5. List four types of delay timers.
6. Calculate the value of R_B required to cause the silicon NPN transistor shown in Figure 1-47 to be driven into saturation.
7. What is the purpose of the capacitor connected in shunt with the base resistor for the circuit appearing in Figure 1-47?
8. Compute the minimum current rating of a switch that is to be used in a circuit in which 80 amperes of current are flowing.

9. Determine the minimum current rating of a switch used to control a 220-volt AC motor that requires 25 amperes of current when rotating.

10. What is the minimum current rating of a switch used to control a 150-volt DC motor if the motor requires 25 amperes of current while rotating?

11. From Table 1-1, determine the enclosure best suited for a switch used to control a conveyor which carries coal.

12. The minimum current required to overcome the spring tension to actuate a relay is called
    a. surge current
    b. pickup current
    c. inrush current
    d. holding current

13. Which of the following is not a type of delay timer?
    a. Dashpot
    b. Thermal
    c. Electronic
    d. Hydrostatic

14. The _____ is a two-terminal bilateral thyristor which conducts whenever the breakover voltage is exceeded.
    a. SCR
    b. GCS
    c. diac
    d. triac

15. _____ timers utilize the time required for a fluid to flow through an orifice to achieve a delay.
    a. Dashpot
    b. Pneumatic
    c. Hydraulic
    d. Hydrostatic

16. _____ switches are composed of one or more contacts mounted on a shaft that makes connection with multiple stationary contacts attached to a circular disk.
    a. Knife
    b. Slide
    c. Limit
    d. Rotary

17. A (An) _____timer is employed whenever a load is required to operate for a particular length of time.
    a. delay
    b. cycle
    c. interval
    d. hydrostatic

18. Which of the following is not a relay variable?
    a. operate time
    b. release time

    c. dropout current

    d. coil resistance

19. A (An) _____timer is one that provides a delay from the time an event is initiated to the time that the event actually occurs.

    a. delay

    b. cycle

    c. interval

    d. hydrostatic

20. _____ switches are often used to control the motion of machine-tool tables, conveyors, and robot manipulators.

    a. Knife

    b. Slide

    c. Limit

    d. Rotary

21. T F A contactor is a relay used in high-power applications. ___

22. T F Relays and switches can be used to perform NAND and NOR gate functions. ___

23. T F Slide switches are actuated by depressing a shaft that extends from the switch housing. ___

24. T F A GFI is primarily used to provide protection if a short circuit occurs. ___

25. T F The same voltage is used to energize both the coil and load in a relay locking circuit. ___

26. T F Holding current is the minimum coil current required to overcome the spring tension to actuate a relay. ___

27. T F A thermal circuit breaker is actuated by the expansion of a unimetallic expansion column. ___

28. T F An interval timer is used when it is necessary to provide a delay between the time a switch is opened and the time the load is actually deenergized. ___

29. T F The conduction period of an SCR can be varied by delaying the time a potential is applied to the gate terminal. ___

30. T F The expansion timer is a type of delay timer. ___

31. T F The transistor switch is biased for class A operation. ___

32. T F An SCR can be turned off by applying a negative potential at the gate terminal. ___

33. T F A diac cannot be used as a controlled rectifier. ___

34. T F The load is energized whenever the transistor switch is in its conducting state if the load is connected in parallel with the transistor. ___

35. T F Number of throws identifies the number of possible actuated positions associated with a particular switch. ___

36. T F A circuit breaker is a switching device that automatically opens whenever excessive current flows. ___

37. T F Rotary switches are primarily used for switching power circuits. ___

38. T F With the exception of its lower current and power ratings, the GCS operates exactly like the SCR. ___

39. T F The triac is a thyristor that has two anodes and two gate terminals. ___

40. T F Limit switches are composed of two electrodes and a pool of mercury, all sealed in a tube. ___

41. T F The Shockley diode can be used as a controlled rectifier diode in a rectifier power supply. ___

42. T F The diac operates much like the Shockley diode except that it conducts in either the forward or reverse direction. ___

43. T F T_{OFF} is the time required to turn an SCR off. ___

44. T F From the data shown in Table 1–2, a NEMA type 4 switch enclosure should be used on a switch used to control a sump pump submerged in water. ___

45. T F According to Table 1–2, a NEMA type 1 switch enclosure should be used on a switch used to control an air-conditioning condenser that sits outdoors. ___

CHAPTER
2

GENERATORS AND MOTORS

OBJECTIVES

1. Identify the major parts of a basic AC generator.

2. Identify the major parts of a basic DC generator.

3. Identify the four types of DC generators, draw their schematic diagrams, and describe the operating characteristics of each.

4. List the major parts of a DC motor.

5. Describe how torque is produced in a DC motor.

6. Calculate the current flowing through the armature of a rotating DC motor.

7. Identify the three types of DC motors, draw their schematic diagrams, and describe their operating characteristics.

8. Identify the two types of stator winding connections used in three-phase alternators.

9. Describe the operating principles and identify the types of three-phase induction motors.

10. Define the term *slip* as it applies to an induction motor and identify three factors affecting slip.

11. Describe the operating characteristics of a three-phase synchronous motor and discuss the general applications of this type of motor.

12. Discuss the general operating characteristics of the single-phase split-phase and single-phase capacitor-start induc-

tion motors.

13. Define the terms, *varying speed, constant speed, adjustable speed, multispeed*, and *adjustable varying speed* as they pertain to motor speed.

14. Define the terms *general-purpose, definite-purpose*, and *special-purpose* as they relate to the type of intended service of motors.

15. Identify the general applications for the universal, single-phase synchronous, and torque motors.

16. Discuss the principles of operation of the DC stepping motor.

17. List three ways in which DC motors may be started and describe the operating principles of each.

18. List three ways in which the speed of DC motors may be varied and discuss the principles of operation for each.

19. List three methods used to start AC motors and describe the operation of each.

20. Describe two methods used to vary the speed of AC motors.

21. Discuss the four methods used to stop motors.

22. Discuss the purpose of jogging as it pertains to motors.

23. Identify the techniques used to reverse the direction of rotation of DC motors, the capacitor-start single-phase induction motor, and three-phase motors.

2-1 INTRODUCTION

Electric motors and generators are essential to the twentieth-century lifestyle. Motors are used extensively in consumer and industrial applications. Items such as kitchen appliances, vacuum cleaners, refrigerators, air conditioners, washing machines, clothes dryers, sewing machines, cassette players, small electric tools, and computer-memory disk-drive systems are all examples of consumer products that use motors. In the manufacturing industries electric motors are used to power machines, provide machine-tool motion, cause robots to move, and move material from place to place.

Unless the motors are portable and battery powered, generators are needed to provide the electrical energy that drives them. The energy for the motors used in consumer products and in many industrial machines comes from a generator owned and operated by a utility company. This energy is transported to our homes, offices, and industrial plants by means of electric transmission lines. Many motors used in industrial equipment require energy in a form not supplied by the utility company. The voltage amplitude or frequency may not be of the proper type. In many cases this can be changed by the use of rectifier or inverter power supplies. In some situations, however, this is not practical. In these cases, a separate generator is needed to supply the energy to drive the motor.

Motors and generators are both types of transducers. Transducers are devices which convert one form of energy into another. Motors convert electrical (input) energy into mechanical (output) energy whereas generators convert mechanical (input) energy into electrical (output) energy. The many different types of generators and motors can be grouped according to the type of operating voltage produced or required—DC and AC. The principles associated with these machines are discussed in the following sections.

2-2 DC GENERATORS

2-2.1 Introduction

Both DC and AC generators operate on the principle of magnetic induction as expressed by Faraday's law. Faraday found that a voltage is induced across a conductor whenever the conductor is moved through a magnetic field. The magnitude of the induced voltage is proportional to the intensity of the magnetic field, the angle at which the field is cut, and the rate at which the conductor moves through the magnetic field. AC-generator principles must be understood before DC-generator action can be presented.

A basic AC-generator is shown in Figure 2-1. The device consists of a folded conductor, called an armature, which is made to revolve in the air gap between the poles of a field magnet. A mechanical force, called a prime mover, is needed to turn the armature. The prime mover may be an electric motor, internal-combustion engine, or a turbine.

Figure 2-1

Basic AC generator

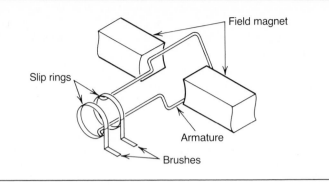

A sinusoidal voltage is induced across the armature as it turns between the field-magnet poles. Two cylindrical bands, called slip rings, are attached to the armature to allow voltage to be taken from it. One slip ring is electrically connected to one side of the armature and the other is connected to the opposite side. Carbon brushes make contact with the rotating slip rings. One revolution of the armature produces one cycle of sinusoidal voltage, as illustrated in Figure 2-2. The armature is shown in five different positions in this diagram. The two brushes and two slip rings are identified as A and B. In Figure 2-2(a) brush A is making contact with slip ring A and brush B makes contact with slip ring B. As the armature begins to rotate, it moves parallel to the magnetic flux lines and the induced voltage is zero. As the armature rotates from 0 to 90 degrees in Figure 2-2(b), the half of the armature

Figure 2-2

Basic AC-generator action

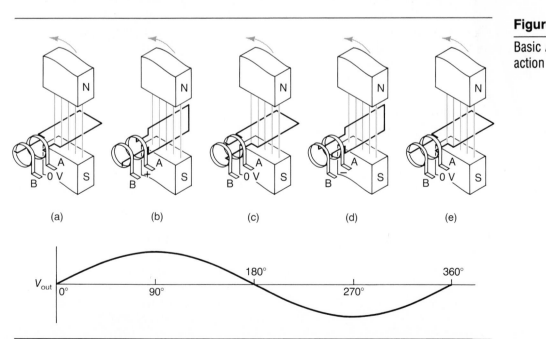

connected to slip ring A moves upward while the half connected to slip ring B moves downward, causing a voltage to be induced across the armature. Assuming that an upward cutting of magnetic lines creates a positive induced voltage, brush A will be positive with respect to brush B. The armature has rotated an additional 90 degrees in Figure 2-2(c) and is again moving parallel to the magnetic lines of force. At this point there is no voltage being induced across the armature. The armature has rotated 180 degrees in Figure 2-2 (a)-(c) and has produced one positive sinusoidal alternation.

The side of the armature connected to slip ring B is moving upward for the next half revolution, as shown in Figure 2-2(d), causing the voltage at brush A to be negative with respect to that at brush B. In Figure 2-2(e), the armature is again moving parallel with the magnetic flux lines, completing the negative alternation of the cycle. During the entire time, brush A has made contact with slip ring A and brush B has maintained contact with slip ring B. When a circuit is connected to the brushes, sinusoidal current will flow through the armature winding and the externally connected circuit.

If the slip rings are replaced with two symmetrical metal segments called a commutator, a DC generator is formed. The commutator segments are insulated from one another and attached to opposite ends of the armature, as illustrated in Figure 2-3. Commutator segment A is connected to one end of the armature and segment B to the other end, as shown in Figure 2-4. When the armature is caused to rotate, a voltage is induced across it, as shown in Figure 2-4(a)-(c). During this time brush A has been making contact with commutator segment A, and the polarity of the induced voltage at brush A is positive with respect to brush B. The brushes switch commutator segments at 180 degrees. During the second half of the revolution, the position of the armature is such that commutator segment B is making contact with brush A. Since that segment of the armature is moving upward, the voltage appearing at brush A is again positive with respect to brush B, as indicated in Figure 2-4(c)-(e). Therefore, the potential at brush A is always positive with respect to brush B. The commutator acts like a switch causing the upward- moving half of the armature to stay in contact with brush A.

One revolution of the armature produced two cycles of voltage. The output is a pulsating DC voltage that has a ripple amplitude ranging from zero to a peak value.

Figure 2-3

Basic DC generator

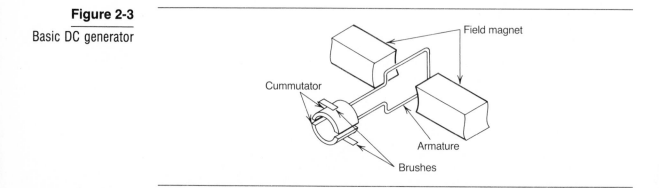

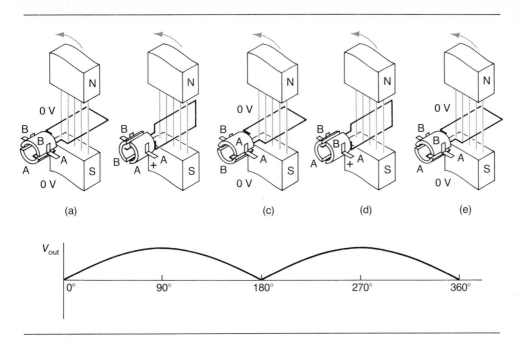

(a)   (c)   (d)   (e)

Figure 2-4

Basic DC-generator action

The amplitude of the ripple voltage can be reduced, as indicated in Figure 2-5, by using two armature windings and commutator segments.

The generators just described have limited applications because of their low output power. The output voltage and current capacity can be increased by strengthening their magnetic fields. This is accomplished by employing an electromagnet to produce the magnetic field, as shown in Figure 2-6. The electromagnet is called a field coil or field winding. The current flowing through the field coil is called field current I_F. The strength of the magnetic field produced by the field coil is controlled by varying the intensity of the field current. For a constant speed, the generator's output voltage is proportional to the strength of the magnetic field produced by the field coil.

Whenever a circuit is connected to a generator, current flows through the armature winding, creating a circular magnetic field about the winding, as depicted in Figure 2-7. The interaction of the field winding and armature-winding magnetic fields creates a torque that is opposite in direction to the torque of the rotating armature. The prime mover has to expend more energy to overcome the counter torque when the generator is loaded. The higher the intensity of the armature current, the greater the torque produced and the more work the prime mover has to

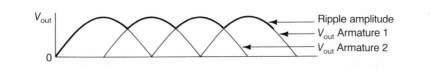

Figure 2-6

Generator
employing an
electromagnet

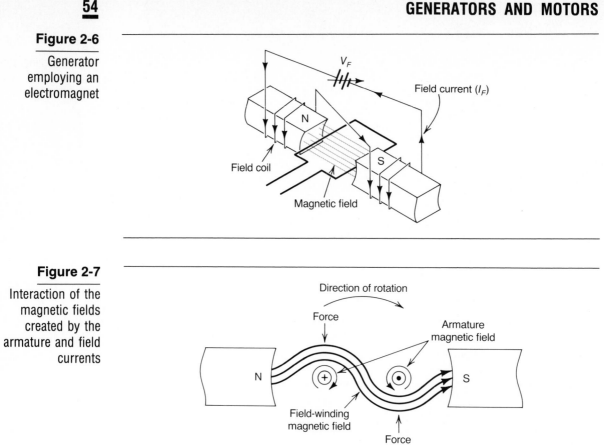

Figure 2-7

Interaction of the
magnetic fields
created by the
armature and field
currents

perform. The loaded generator tends to act like a motor turning against the prime mover.

Four different types of DC generators are utilized for industrial applications—separately excited, shunt wound, series wound, and compound wound.

2-2.2 Separately excited generators

The diagram of a separately excited DC generator appears in Figure 2-8. A voltage source V_F, called an exciter voltage, provides the field current required to produce the magnetic field that causes voltage to be induced across the rotating armature. The field current and output voltage are controlled by a field rheostat R_F.

When the generator is first started, R_F is adjusted for minimum resistance. This allows maximum field current to flow through the field winding, creating a magnetic field about the field coil, as shown in Figure 2-9(a). As the prime mover begins to turn, voltage is induced across the armature. When the armature reaches its rated speed, the rheostat is adjusted for the desired output voltage.

The separately excited generator is characterized by its excellent load regulation. A constant output voltage is maintained over a wide range of load-current values, as

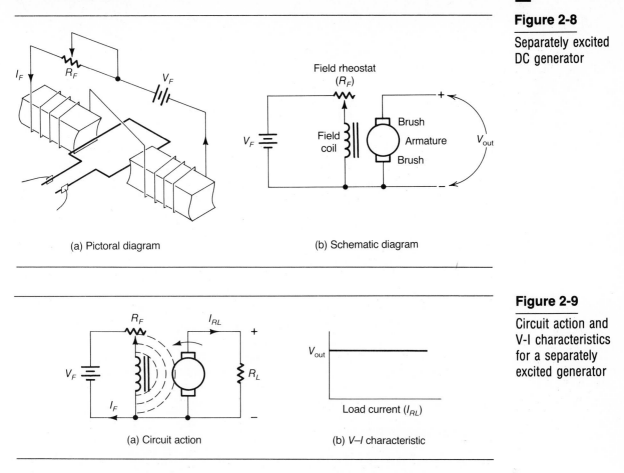

(a) Pictoral diagram   (b) Schematic diagram

Figure 2-8

Separately excited
DC generator

(a) Circuit action   (b) V–I characteristic

Figure 2-9

Circuit action and
V-I characteristics
for a separately
excited generator

shown in Figure 2-9(b). This type of generator has the disadvantage of requiring an exciter voltage supply. A DC source is required to produce DC voltage.

2-2.3 Shunt-wound generators

The shunt-wound generator is one of three self-excited generators, the other two being the series and compound wound. Self-excited generators do not require a separate DC exciter supply.

The diagram of a shunt-wound generator is shown in Figure 2-10. The generator gets its name from the method in which the field coil is connected to the armature. The field winding is placed in shunt (parallel) with the armature. The voltage buildup on this machine depends upon residual magnetism being present in the core of the field winding. The field rheostat R_F is adjusted for minimum resistance when the generator is initially started. As the armature begins turning, it cuts the very weak residual magnetic flux lines about the field coil. This creates a small induced voltage across the armature, which causes an armature current I_A to flow. With the load switch open, the armature current flows to node A and through the

Figure 2-10

Shunt-wound DC
generator

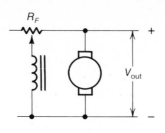

field winding, as seen in Figure 2-11(a). This creates a stronger magnetic field, which causes a larger voltage to be induced across the armature, which in turn causes a higher armature and field current to flow. This process continues until maximum voltage buildup is reached, as shown in Figure 2-11(b). The load is energized by closing the switch and R_F is adjusted for the desired output voltage.

After the load switch has been actuated, the armature current I_A is the sum of the field current I_F and the load current I_{RL}, as indicated in Figure 2-12. The resistance of the field winding is much greater than that of the armature, causing the field current to be considerably less than the armature current.

The shunt generator has the advantage of not requiring a separate DC exciter source. It is characterized by rather poor voltage regulation, however. If the load resistance decreases, the load current increases, causing additional armature current to flow through the load. This decreases the field current, causing the output voltage to decrease as shown by the volt–ampere curve in Figure 2-12(b).

2-2.4 Series-wound generators

This is another type of self-excited generator. In this type of machine, the field coil is connected in series with the armature winding, as illustrated in Figure 2-13(a). The voltage buildup depends upon residual magnetism being present in the core of the field winding. When the prime mover begins to turn the generator, the armature cuts the weak residual magnetic field, inducing a small voltage across the windings. The resulting current I_T flowing through the armature, load, and field

Figure 2-11

Voltage buildup in
a shunt-wound
generator

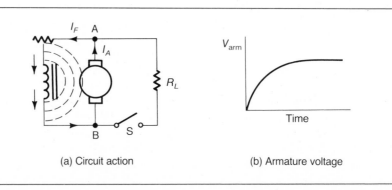

(a) Circuit action   (b) Armature voltage

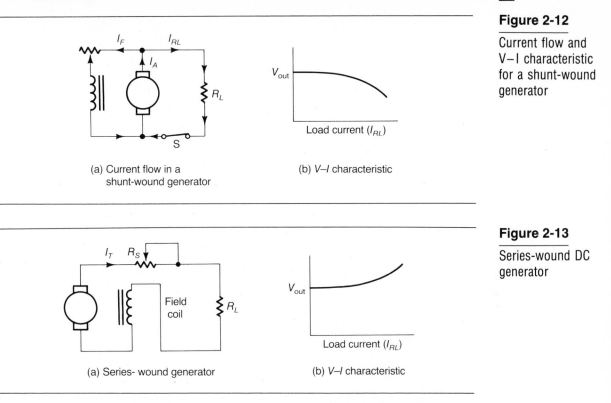

Figure 2-12

Current flow and
V–I characteristic
for a shunt-wound
generator

(a) Current flow in a
shunt-wound generator

(b) V–I characteristic

Figure 2-13

Series-wound DC
generator

(a) Series- wound generator

(b) V–I characteristic

winding causes the magnetic field about the field winding to become stronger. This creates a larger voltage across the armature, which causes the field, armature, and load current to increase. At the same time, the prime mover is increasing the speed of the armature, bringing it up to the desired speed of rotation. The increase in armature speed also helps the output voltage to increase. This process continues until the operating speed and output voltage are constant. The series rheostat R_S can be adjusted for the desired output voltage.

Like the shunt-wound generator, the series-wound generator has the advantage of not requiring a separate DC exciter source. It, too, is characterized by poor voltage regulation, however. If the load resistance decreases, the current flowing through the field winding increases, causing the output voltage to increase as seen in Figure 2-13(b).

2-2.5 Compound-wound generators

The shunt- and series-wound generators react differently for changes in load current. As previously noted, an increase in load current causes the output voltage to decrease for a shunt-wound generator, whereas the output voltage increases for a series-wound generator when the load current increases.

The compound-wound generator utilizes the undesirable effects of both the shunt- and series-wound machines to produce an output voltage that is relatively constant for a wide range of load currents. The compound-wound generator is a type of self-excited generator that utilizes two field coils, as depicted in Figure 2-14(a). Field coil L_1 is in parallel with the armature while field coil L_2 is in series with the armature winding. The shunt and series field coils interact with one another to maintain a constant output voltage, as shown in Figure 2-14(b). The output voltage can be varied by adjusting the shunt field rheostat.

2-3 DC MOTORS

2-3.1 Introduction

If, instead of a load, a DC voltage is connected across the brushes of any of the generators just discussed, the machine will operate as a motor. Reciprocity exists in DC generators and motors—that is, these machines may be operated as either generators or motors. The diagram of a basic DC motor appears in Figure 2-15. Two DC voltage sources are used—a field supply V_F and an armature source V_S. Current flowing through the armature winding creates a circular magnetic field about the winding, which distorts the magnetic flux lines created by the field winding. This creates a torque that causes the armature to rotate, as seen in Figure 2-16. The amount of torque developed depends upon the intensity of the current flowing through the armature and field windings. The direction of the rotation created by the torque depends upon the direction of current flow through either winding. The motor appearing in Figure 2-16 is rotating in the C.C.W. direction.

As the armature begins to rotate, it cuts the field-winding flux lines, causing a voltage to be induced across the armature. A rotating motor has a generator effect, just as a rotating generator has a motor effect. The polarity of the induced voltage opposes the armature source voltage. The net difference of potential V_A appearing across the armature brushes is the difference between the armature source voltage V_S and the counter induced voltage V_{CEMF}, as shown by Equation 2-1.

$$V_A = V_S - V_{CEMF}$$ *Eq. 2-1*

The intensity of the current I_A flowing through the armature depends upon the speed of the armature and its resistance. If the armature is not rotating, the arma-

Figure 2-14

Compound-wound
DC generator

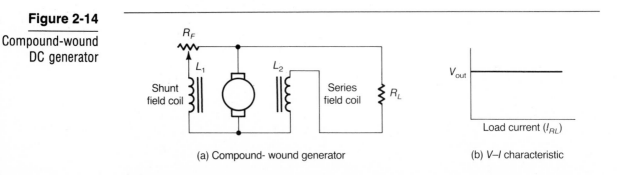

(a) Compound- wound generator   (b) V–I characteristic

Figure 2-15

Basic DC motor

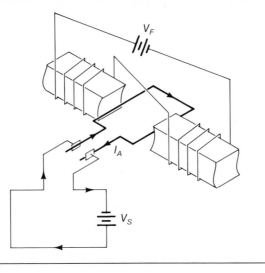
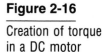

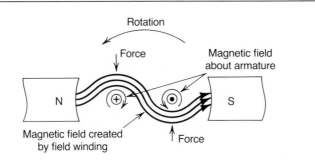

Figure 2-16

Creation of torque
in a DC motor

ture current is limited only by its resistance and is quite high, as indicated by Equation 2-2.

$$I_A = \frac{V_S}{R_A}$$

Eq. 2-2

A motor that requires 150 volts and has an armature resistance of 5 ohms will have 30 amperes of armature current when the motor is not turning, as seen in Example 2-1.

Example 2-1

CALCULATING THE ARMATURE CURRENT IN A STATIONARY MOTOR

Problem: Determine the current flowing through the armature of a stationary motor if the armature requires 150 volts and has a resistance of 5 ohms.

Solution: $I_A = \dfrac{V_S}{R_A} = \dfrac{150\ V}{5} = 30\ A$

Figure 2-17

Rheostat to limit
armature current

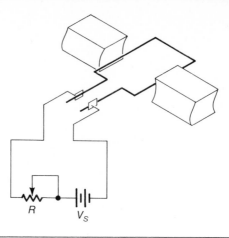

Equation 2-3 is used to calculate armature current in a motor that is turning.

$$I_A = \frac{V_S - V_{CEMF}}{R_A} \qquad\qquad Eq.\ 2\text{-}3$$

If the motor just described produces 135 volts of CEMF when rotating at 3600 rpm its armature current is reduced to 3 amperes as indicated in Example 2-2.

To limit the current flowing through the armature when voltage is initially applied, a rheostat is often connected in series with the motor supply voltage and armature, as shown in Figure 2-17. This is discussed further in Section 2-7.2.

The motor just discussed is similar to the separately excited generator. To develop the magnetic flux about the field winding, a separate DC source is required. Actual DC motors utilize a self-excitation method to develop this field. Just as there are three types of self-excited generators, there are three types of DC motors. These include the shunt, series, and compound.

2-3.2 Shunt DC motors

The diagram of a shunt DC motor appears in Figure 2-18. The field coil is connected in parallel with the armature. A rheostate R is connected in series with the

Example 2-2

CALCULATING THE
ARMATURE
CURRENT IN A
ROTATING MOTOR

Problem: The motor described in Example 2-1 produces 135 volts of CEMF while turning at a speed of 3600 rpm. Calculate the current flowing through the armature.

Solution:
$$I_A = \frac{V_S - V_{CEMF}}{R_A} = \frac{150\ V - 135\ V}{5} = 3\ A$$

Figure 2-18

Shunt DC motor

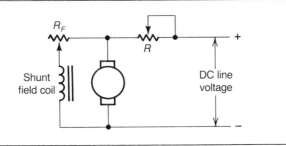

motor and DC line voltage V_S to limit the armature current while starting the machine. After the motor has been started and brought up to operating speed, its speed can be changed by varying the resistance of the field rheostat R_F. The intensity of the field current controls the amplitude of the counter EMF induced across the armature, which determines the speed of the motor.

It is extremely important that an open circuit not occur in the field-coil circuit. If one does occur, no counter EMF will be developed across the armature and the motor will turn faster and faster until destroyed by centrifugal action.

The shunt motor is characterized by its low starting torque. This is due to the weak magnetic field about the field winding during the starting process. The motor depends upon residual magnetism in the field-coil core to initially start the motor turning. The motor has excellent speed regulation after it has reached its operating speed and maintains a rather constant speed for varying mechanical-load conditions.

2-3.3 Series DC motors

The schematic diagram of a series DC motor is shown in Figure 2-19. The DC line voltage applied to the series-connected armature and field windings creates a tremendously high starting torque; the speed of the motor is controlled by the mechanical load connected to its armature shaft. It is very important that this motor never be operated without a mechanical load coupled to its shaft. If the motor is operated without a load, its speed will continue to increase until the motor is destroyed by the centrifugal force of the armature.

Series DC motors are characterized by high starting torque and poor speed reg-

Figure 2-19

Series DC motor

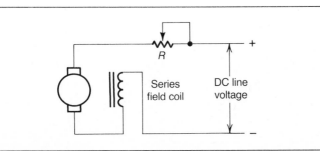

ulation. They are usually used in applications where the mechanical load is continually applied.

2-3.4 Compound DC motors

Like the compound-wound generator, the compound DC motor utilizes both series and shunt field coils, as seen in Figure 2-20. Although the starting torque is not as great as that developed by the series motor, it is considerably greater than that of the shunt machine. Speed regulation is relatively constant for a wide range of mechanical-load variations. A summary of the different types of DC machines and their characteristics appears in Table 2-1.

2-4 AC GENERATORS

The principles of operation of the AC generator were presented in Section 2-2.1. As illustrated in Figure 2-1, the generator consists of an armature rotating between the poles of a magnet. Slip rings allow the induced voltage to be taken from the armature.

Slip rings operate satisfactorily for low-power generators but there is considerable energy loss between the slip rings and brushes for higher power generators where the armature current is high. For this reason, most generators utilize a rotating field, called a rotor, and a stationary armature, called a stator. As long as magnetic lines of flux are cut, it does not matter whether the field is stationary and the armature rotates or the field winding rotates and the armature winding remains stationary. Since the field current is usually much smaller than the armature current, the energy loss between the slip rings and brushes is much less when a rotating field is used. A generator that utilizes a stator and a rotor is called an alternator. An alternator uses an exciter to produce a separately excited field. This is often a compound-wound generator whose armature is mounted on the rotor shaft of the alternator.

The rotor is composed of wire wound around a metallic core. Stator winding sets are wound on metallic cores, which form poles that are attached to the motor housing. Although a single stator winding set could be utilized, most alternators utilize multiple stator-pole sets connected in series, as depicted in Figure 2-21. Two sets

Figure 2-20

Compound DC
motor

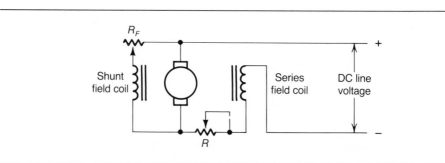

AC GENERATORS

Table 2-1

SUMMARY OF
DC-MACHINE
CHARACTERISTICS

MACHINE	CHARACTERISTICS
Generators:	
Separately Excited	Requires an exciter voltage. Output voltage is constant regardless of load-current variations.
Shunt Wound	Output voltage varies inversely with load current.
Series Wound	Output voltage varies proportionally with load current.
Compound Wound	Output voltage is relatively constant as load current changes.
Motors:	
Series	High starting torque, poor speed regulation. Cannot be operated without a load.
Shunt	Low starting torque, good speed regulation. Field-coil circuit cannot be allowed to open.
Compound	Moderate starting torque, relatively good speed regulation.

of stator windings forming four poles are utilized in this alternator. In a single-phase alternator, the output voltage at any instant is the sum of the voltages induced across each stator winding.

Alternators may be designed to produce single- and polyphase voltage. Single-phase alternators are primarily small units designed for supplying temporary power to operate farm and construction-site equipment; to supply emergency power for hospitals and other important institutions during a major power outage; and to provide power at remote installations, such as microwave relay stations.

Most polyphase alternators produce three-phase voltage. Three separate voltages, each separated by 120 degrees, are produced by these machines, as shown in Figure 2-22. The stator windings are connected in either a delta or wye configura-

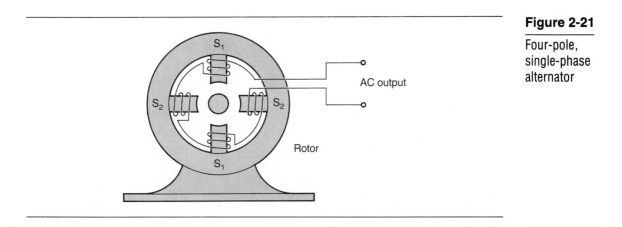

Figure 2-21

Four-pole,
single-phase
alternator

Figure 2-22

Three-phase
voltage

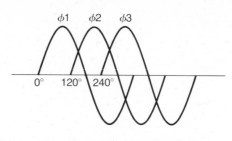

tion, as seen in Figure 2-23. For an identical number of stator winding turns, the
delta connection provides a higher current capacity than does the wye connection,
whereas the wye configuration supplies a higher output voltage. The wye configuration can be connected for either three- or four-wire operation, with the neutral
wire being optional. Either single- or three-phase voltage may be obtained when
the neutral wire is utilized.

The frequency of the AC voltage produced by an alternator is proportional to the
product of the number of stator poles P and the speed (in rpm) at which the rotor is
turning, as indicated by Equation 2-4. As seen in Example 2-3, an eight-pole alternator turning at 900 rpm generates a 60-hertz voltage.

$$f = \frac{P \, (\text{rpm})}{120} \text{ in Hz} \qquad\qquad Eq.\ 2\text{-}4$$

2-5 AC MOTORS

2-5.1 Introduction

AC motors have many applications in industry. Because electrical energy is generated and transmitted in AC form, this is the type of motor most often used to drive
production machines. AC motors can be grouped into two categories—polyphase
and single-phase. The former is primarily of the three-phase type.

2-5.2 Three-phase motors

The major components forming a three-phase motor are a stator and a rotor. Like
their DC counterparts, AC motors create their turning torque by the interaction of

Example 2-3

CALCULATING
ALTERNATOR
FREQUENCY

Problem: Calculate the frequency of the voltage produced by an alternator
containing eight poles and turning at 900 rpm.

Solution: $\quad f = \dfrac{(P)\,(\text{rpm})}{120} = \dfrac{(8)(900)}{120} = 60 \text{ Hz}$

Figure 2-23

Three-phase
alternator stator
connections

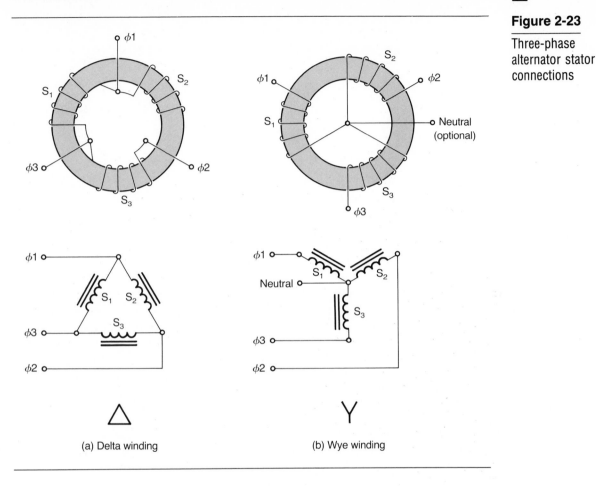

(a) Delta winding

(b) Wye winding

two magnetic fields. In a DC motor, torque is created by the interaction of the fields associated with the stationary electromagnetic-field winding and the rotating armature winding. Torque is created in a three-phase motor by the interaction of a rotating field created by three stationary electromagnetic stator windings and the magnetic field of a rotor mounted in the air gap between the stator windings. Three-phase voltage, one phase per stator, is applied to the stator windings. A basic three-phase motor utilizing a permanent-magnet rotor appears in Figure 2-24(a). The current flowing in the stator windings is continually changing in amplitude and direction, causing the stator-pole polarity to change at the line-frequency rate. For the purpose of discussion, assume that a south pole is established in stator winding S_3 when the motor, shown in Figure 2-24, is actuated. The north pole of the rotor is attracted to the stator, causing the rotor to turn in the clockwise direction. As the rotor approaches S_3 the polarity of the voltage applied to S_1 causes a south pole to be developed in that stator and the rotor moves past S_3 because of its inertia and continues to turn toward S_1. Just as it approaches the stator a south pole is established in S_2 and the rotor turns past S_1 and moves toward that stator. In this fashion

Figure 2-24

Basic three-phase
motor

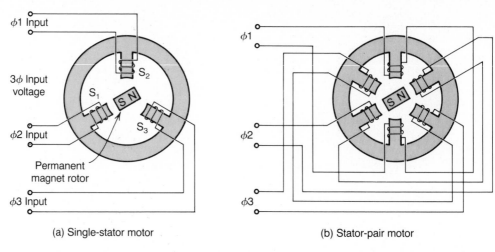

(a) Single-stator motor   (b) Stator-pair motor

the phase displacement of the stator currents creates a circular magnetic field that continually rotates and pulls the rotor along with it. Instead of using a single stator winding, stators are wound in pairs, as shown in Figure 2-24(b). This creates additional torque by allowing both ends of the rotor to be attracted to the stators. Additional stator poles may be used by connecting multiple windings in series.

The speed of the rotating field established by the stator windings is referred to as synchronous speed and can be calculated using Equation 2-5.

$$\text{Synchronous speed} = \frac{120\,f}{P}$$

f = Line voltage frequency in hertz
P = Number of poles

Eq. 2-5

As seen in Example 2-4, an eight-pole motor connected to a 60-hertz supply line has a synchronous speed of 900 rpm. The number of poles inversely affects the speed of the motor. The greater the number of poles, the slower the motor turns. Speed can be changed in discrete steps by switching in or out the number of stator poles used for a particular motor.

Permanent-magnet rotors are used only in fractional-horsepower motors. The magnetic field produced by a permanent magnet is not strong enough to create the torque required in larger motors. Instead, electromagnet rotors are employed in

Example 2-4

CALCULATING
SYNCHRONOUS
SPEED

Problem: Determine the synchronous speed of a motor connected to a 220-volt three-phase 60-hertz line if the motor has eight stator windings.

Solution:
$$\text{Synchronous speed} = \frac{120\,f}{P} = \frac{(120)(60)}{8} = 900 \text{ rpm}$$

these motors. Two different types of rotors are used resulting in two kinds of three-phase motors. These include the induction motor and the synchronous motor.

Induction motor: This is the most common type of motor. It has stator windings like those just discussed and a slotted rotor core. Two types of rotors are utilized—the squirrel cage and the wound rotor. The squirrel-cage rotor has heavy bars embedded in the slots of the core, while wire is wound in the slots in a wound-rotor motor. The operation for either type is basically the same. The induction motor operates much like a transformer that has a stationary primary winding and a rotating secondary winding. When voltage is applied to the stator windings the resulting rotating magnetic field causes a voltage to be induced across the rotor. This produces current in, and a magnetic field about, the rotor. The interaction of this field and the rotating stator field creates the torque that causes the rotor to turn. Because the rotor obtains its energy through mutual induction, there is no physical connection between the stator and rotor windings. Unlike most other motors, there are no brushes to wear out.

The rotor will never turn as rapidly as the synchronous speed of the stator field. The difference between the synchonous speed and the rotor speed is called slip. Slip may be expressed as a percentage or in rpm, as shown in Equation 2-6.

$$\text{Percent slip} = \frac{\text{synchronous speed} - \text{rotor speed}}{\text{synchronous speed}} \times 100$$

Eq. 2-6

$$\text{Slip in rpm} = \text{synchronous speed} - \text{rotor speed}$$

A motor having a synchronous speed of 1800 rpm and a rotor speed of 1710 rpm has a slip of 5 percent, as seen in Example 2-5.

Slip is affected by load, rotor resistance, and stator voltage. It is proportional to both load and rotor resistance and may be as low as 1 percent for a nonloaded motor and as high as 10 percent for a fully loaded motor. Slip will double if the rotor resistance is doubled. Slip varies inversely as the square of the voltage applied to the stator windings. The torque developed at any instant in an induction motor is affected by slip, rotor resistance, and the applied voltage.

Synchronous motors: The three-phase synchronous motor is constructed much like the induction motor except that two rotor windings are used, an AC and a DC.

Example 2-5

CALCULATING MOTOR SLIP

Problem: Calculate the slip of a motor in both percent and rpm if the synchronous speed is 1800 rpm and the rotor speed is 1710 rpm.

Solution:

(a) $\text{Percent slip} = \dfrac{\text{synchronous speed} - \text{rotor speed}}{\text{synchronous speed}} \times 100$

$\qquad = \dfrac{1800 - 1710}{1800} \times 100 = 5\%$

(b) $\text{Slip in rpm} = \text{synchronous speed} - \text{rotor speed}$

$\qquad = 1800 - 1710 = 90 \text{ rpm}$

The AC winding is similar to that used in induction motors and causes the rotor to begin turning when voltage is applied to the stator windings. As the rotor speed approaches the synchronous speed of the stator, the DC winding is excited creating nonchanging north and south poles on the rotor. These poles are attracted to, and become synchronized with, the opposite poles of the stator, causing the rotor to turn at the same rate as the synchronous speed of the stator field. Since two rotor windings are used, this motor is more complex than the induction motor. Although the AC winding is excited by electromagnetic induction, brushes and a commutator are required to excite the DC winding.

The synchronous motor has two characteristics that are important in some applications. It operates at a constant speed from no-load to full-load conditions and its power factor can be varied. The power factor of a synchronous motor depends upon the current flowing in the DC rotor winding. This current can be adjusted to produce a unity, leading, or lagging power factor. The power factor of the main electrical-transmission line in manufacturing plants having production machines powered by induction motors may be low. If some of the induction motors are replaced with synchronous motors, the synchronous motors can be used to correct the low power factor of the electrical system in addition to delivering mechanical power to a load. Synchronous motors cannot be used in heavy load applications. If the motor slows under load conditions, the rotor falls out of synchronization with the rotating stator field and stops. Synchronous motors are usually employed to drive loads which require constant speeds and do not require frequent starting and stopping.

2-5.3 Single-phase induction motors

Single-phase motors are used in light-duty industrial applications and are the type used in most consumer applications in the home. The induction motor is the most popular type of single-phase motor. Like the three-phase induction motor, this machine is composed of a rotor and stator and has no mechanical connection between these windings. Unlike the three-phase motor, which has three separate stator windings and a rotating stator field, the basic single-phase induction motor utilizes one stator winding and has no rotating field. A basic single-phase induction motor appears in Figure 2-25(a). When voltage is first applied to the stator at t_0, no magnetic field is developed across the stator poles and the rotor does not turn. As time progresses and reaches t_1, the stator becomes magnetized and the rotor turns in the clockwise direction because of the attraction between the rotor and stator, as shown in Figure 2-25(b). At t_2, the magnetic field has disappeared and the rotor continues its movement by inertia, as illustrated in Figure 2-25(c). The magnetic field is reversed at t_3 and the rotor continues to move as seen in Figure 2-25(d). Inertia causes the rotor to continue to turn when the stator field decreases to zero, and seen in Figure 2-25(e).

The motor action just described depended upon the rotor being initially offset from the stator pole, as illustrated in Figure 2-25(a). If the rotor was aligned with the stator when the voltage was initially applied, as seen in Figure 2-25(b) and (d), the motor would not have started. The rotor would have revolved in the opposite

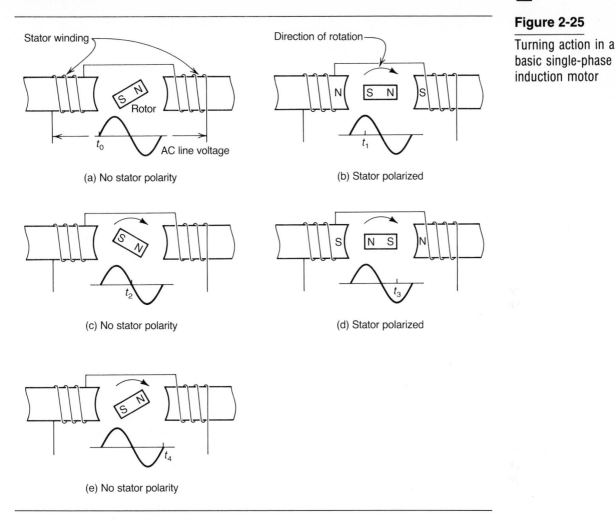

Figure 2-25

Turning action in a basic single-phase induction motor

(a) No stator polarity

(b) Stator polarized

(c) No stator polarity

(d) Stator polarized

(e) No stator polarity

direction if it was reversed 180 degrees in Figure 2-25(a) when the stator was first energized.

Split-phase motors: Since the single-phase induction motor does not generate a rotating stator field it has little or no starting torque, as just discussed. A number of different methods have been developed to make the motor self-starting and to control its direction of rotation. One technique is to connect an auxillary, or starting, winding in parallel with the main, or running, stator winding, as depicted in Figure 2-26. This creates a phase-splitting effect that develops a rotating stator field— causing the rotor to begin turning when voltage is applied. This is called a split-phase induction motor. The main stator winding has high inductance and low resistance, while the auxiliary winding has low inductance and high resistance. For maximum starting torque, the current flowing through the main winding should lag the

Figure 2-26

Split-phase motor

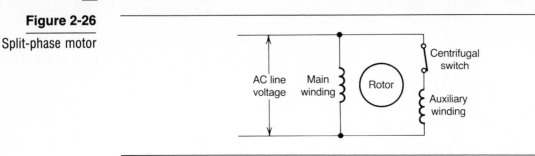

current flowing through the auxiliary winding by 90 degrees. After the motor reaches approximately 75 percent of its rated speed, a centrifugal switch located in the auxiliary winding opens, causing the motor to operate with just the main stator winding. Although the motor would continue to turn with the auxiliary winding energized, that winding is disconnected to conserve energy. Split-phase induction motors have limited starting torque. As previously mentioned, maximum torque occurs when the currents in the main and auxiliary windings are displaced 90 degrees. Because of the resistance in the main winding, the phase angle is approximately 70 degrees, however. For this reason, the split-phase motor is primarily used in applications where the motor is not started under loaded conditions.

Capacitor-start motors: Maximum torque is developed in a split-phase motor when the current flowing through the main winding lags the current flowing through the auxiliary winding by 90 degrees. This angle can be created by connecting an AC electrolytic capacitor of proper size in series with the auxiliary winding, as shown in Figure 2-27. This forms a capacitor-start motor. The capacitor causes the current to lead the applied voltage by approximately 20 degrees in the auxiliary winding. Since the current is lagging the applied voltage by approximately 70 degrees in the main stator, a 90-degree differential is created, as seen in Figure 2-28. A centrifugal switch disconnects the auxiliary winding at approximately 75 percent of the rated speed. Capacitor-start motors are the type usually used in washing machines, clothes dryers, refrigerators, air conditioners, and other consumer appliances and equipment.

Figure 2-27

Capacitor-start motor

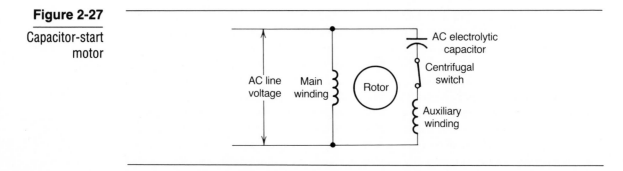

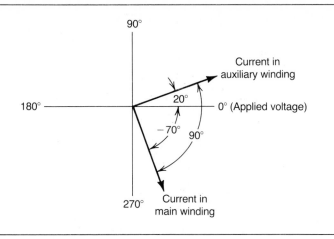

Figure 2-28

Current displacement in a capacitor-start motor

2-5.4 Motor classification

In general, three-phase motors are less expensive to manufacture and more economical to operate than equivalent-size single-phase motors. Single-phase motors require relatively high starting currents and are less efficient than their three-phase counterparts. Both single- and three-phase motors are manufactured in standard horsepower ratings which range from a low of about 1/250th to a high of approximately 5000 horsepower.

The National Electrical Manufacturers Association (NEMA) has grouped motors into two broad categories which include speed classification and type of intended service.

Speed classification: Motors can be grouped into five speed categories—varying speed, constant speed, adjustable speed, multispeed, and adjustable varying speed. A varying-speed motor is one whose speed is dependent upon the load. The speed usually decreases as the mechanical load is increased. A constant-speed motor is a machine whose speed remains constant as it goes from a nonloaded to a fully loaded condition. The speed of an adjustable-speed motor can be varied, but the speed remains constant, once adjusted, regardless of the load conditions. A multispeed motor can be operated at any one of several specific speeds. After the speed has been selected, it remains constant whatever the load conditions. The speed of an adjustable-varying-speed motor can be continuously adjusted over a wide range. After a specific speed has been selected, the motor's speed will vary considerably, depending upon the load. These motor types are summarized in Table 2-2.

Type of intended service. This classification method deals with motor application and includes general-purpose, definite-purpose, and special-purpose. General-purpose motors are designed and manufactured for general use. Their usage is not restricted to any particular application. Definite-purpose motors have the same standard operating characteristics and ratings as general-purpose motors but are de-

Table 2-2
NEMA
MOTOR-SPEED
CLASSIFICATION

SPEED CLASSIFICATION	CHARACTERISTICS
Varying speed	Speed is inversely proportional to load.
Constant speed	Speed remains constant as the motor goes from a nonloaded to a fully loaded condition.
Adjustable speed	Speed can be continuously varied, but once adjusted remains constant regardless of load conditions.
Multispeed	Motor may be operated at any one of several discrete speeds. Once adjusted, the speed remains constant whatever the load conditions.
Adjustable varying speed	Speed may be continuously varied over a wide range, but after a speed has been selected it will vary depending upon the load conditions.

signed for a particular application. Special-purpose motors have unique operating characteristics and are designed for a particular application.

2-6 OTHER TYPES OF MOTORS

2-6.1 Introduction

In addition to the DC and AC motors just discussed, there are several other types of motors. These are usually smaller devices used for special applications. The more common of these are discussed in the following sections.

2-6.2 Universal motor

This is a fractional-horsepower motor used in many household appliances. This motor can be operated with either AC or DC voltage and is used in appliances such as fans, vacuum cleaners, kitchen appliances, and small electric tools.

2-6.3 Single-phase synchronous motor

The single-phase synchronous motor is a small, low-torque, constant-speed motor used in electric clocks and timers. The motor is made up of a single-pair stator winding and a soft iron rotor that revolves in the air gap between the stator poles. The polarity of the magnetic field created when AC voltage is applied to the stator winding alternates at the frequency of the line voltage. The magnetic lines of force passing through the air gap between the stator poles magnetizes the rotor, causing it to turn at the synchronous speed of the stator field.

2-6.4 Torque motor

This type of motor is often used to open and close fluid power valves, open and close building ventilation dampers, and turn machine-tool screw mechanisms. Torque motors are available for either single- or three-phase operation and are designed to produce a high amount of torque. They can be operated in a stalled condition for short periods of time without becoming overheated. These motors are not designed for continuous rotation activity.

2-6.5 Permanent-magnet motors

A permanent-field magnet is employed to produce a constant magnetic field and a conventional electromagnet is used as the armature in this type of motor. This motor is available with either a stationary permanent-magnet field and a rotating electromagnetic armature or a stationary electromagnetic armature (stator) and a rotating permanent-field magnet (rotor). Because of its constant-field magnet characteristics, the permanent-magnet motor develops almost 50 percent more torque and has approximately twice the acceleration capabilities of other motors of similar size. This motor is used in many of the same applications as the torque motor.

2-6.6 DC stepping motor

The DC stepping motor converts binary pulses into angular, or rotary, shaft movement. These motors are driven by binary pulses and turn in discrete steps rather than turning continuously. The stepping motor, or stepper motor as it is sometimes called, has the advantage, when compared with equivalent-sized DC and AC motors used in digital systems, of not requiring a digital-to-analog (D/A) converter to change the input binary voltage into an equivalent analog voltage to cause the rotor to turn. The stepping motor has characteristics similar to both the permanent-magnet and three-phase synchronous motors. Like the permanent-magnet motor, it has a permanent-field-magnet rotor. It uses multiple angular-displaced stator windings like the three-phase synchronous motor, as seen in Figure 2-29. Rotor movement occurs when binary pulses are applied to the stator windings in a particular sequence to produce a rotating magnetic field. The action of a basic stepping motor is illustrated in Figure 2-30. Although not shown, initially the rotor is aligned with stator winding S_1 with no binary excitation pulses applied to any of the stator windings. When a pulse is applied to stator winding S_2, that stator becomes a south pole, the north pole of the rotor is attracted to and aligns itself with that stator, as depicted in Figure 2-30(a). The rotor advances 90 degrees and aligns itself with stator winding S_3 when a binary pulse is applied to that stator, as shown in Figure 2-30(b). In Figure 2-30(c) the rotor steps an additional 90 degrees and aligns itself with stator winding S_4 when a binary pulse is applied to that stator. Finally, the rotor realigns itself with stator winding S_1 when a pulse is applied to that winding, as illustrated in Figure 2-30(d). Four binary pulses were required to cause the rotor to turn 360 degrees. The sequence of stator excitation caused the rotor to turn in

Figure 2-29

DC stepping motor

Figure 2-30

Creating rotation in
a stepping motor

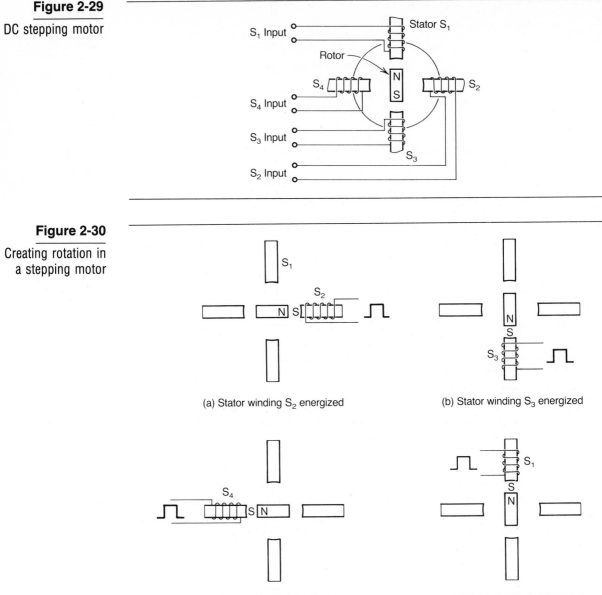

(a) Stator winding S_2 energized

(b) Stator winding S_3 energized

(c) Stator winding S_4 energized

(d) Stator winding S_1 energized

the clockwise direction. If the sequence is reversed, the rotor will turn in the counterclockwise direction.

Energizing the stators individually as just described results in rotor motion called half-step positions. Two stators can be energized simultaneously, with the result that the rotor moves and aligns itself halfway between the two energized stators. This is called full-step rotor-position operation. Instead of applying full current to the two stator windings during binary-pulse application, we can reduce the current in one stator winding. This allows for smaller or larger rotor movement than that obtained by the conventional full-step rotor operation, and is called microstep operation.

The term step-angle sensitivity is used to describe the angular movement of the rotor as pulses are applied to the stator windings. The step-angle sensitivity of the motor just discussed for half-step operation is 90 degrees (360°/4 pulses). Step-angle sensitivity can be increased by using full and microstep operation. This variable can be increased even further by using a circular gear-shaped rotor. Each gear tooth represents a separate field magnet. This decreases rotor movement as successive binary pulses are applied to the stator windings. Each pulse causes the rotor to move one-fourth of a tooth distance instead of 90 degrees for single-stator excitation. Four pulses are needed to make the rotor move one tooth distance. If a 60-tooth rotor is utilized, the rotor turns 360 degrees when 240 pulses (4×60) are applied to the stator windings.

In a stepping motor, the total number of pulses applied to the stator windings determines the distance the rotor is turned. The pulse rate determines the velocity at which the rotor turns. Rotor acceleration and deacceleration is controlled by the pulse rate of change while the direction of rotation is controlled by the sequence in which the stator windings are energized.

Stepping motors may be used to position small robot manipulators, turn machine-tool feed screws, and perform similar activities where accurate shaft positioning is required and the operations are digitally controlled. Rotor displacement can be very accurately repeated with each succeeding pulse. Compared with other small electric motors used in similar applications, stepping motors can often be used without a feedback circuit, simplifying the controller considerably. They are limited to relatively small load applications and will slip if overloaded. Angular-displacement error is introduced when slip occurs, and the error may not be detected.

Binary pulses used to drive stepping motors usually come from a controller. This may be hardwired or programmable depending upon the application. Binary-pulse power, or current-driving capability, may not be great enough to fully magnetize the stepping-motor stator cores. Buffer amplifiers are often connected between the controller and the stepping motor, as illustrated in Figure 2-31, to provide the current required to magnetize the stator cores. These are discussed further in Chapter 6.

2-7 MOTOR CONTROLLERS

2-7.1 Introduction

Motor controllers are devices or circuits that control the electrical power applied to a motor. Usually, a motor is an integral part of the machine to which it is connected and the controller is designed to meet the requirements of the machine. Motor controllers range from the very basic, consisting of a simple start–stop switch, to the more complex—made up of switches, contactors, resistors, and semiconductor

Figure 2-31

Stepping-motor
controller

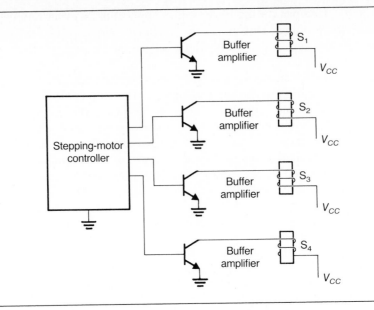

circuits to start and stop the motor, change its speed, and reverse its direction of rotation. Many motor controllers include current-overload and under-voltage protection.

2-7.2 DC-motor starters

Depending upon the size of the motor, several different techniques are used to start DC motors. The more common of these include across-the-line, the starting rheostat, and the drum switch.

Full line voltage can often be applied to DC motors approximately one-fourth horsepower or less in size. The weight of the armature is small in these motors and the period required for the motor to reach its rated speed and develop the CEMF needed to limit the armature current is short. A DPST switch is often used to connect electrical power to the motor, as shown in Figure 2-32.

Because of the greater inertia of their armatures, larger DC motors require that a resistance be connected in series with the armature, to limit armature current until sufficient speed is reached to develop an adequate CEMF. A common method of achieving this is by using a starting rheostat, as depicted in Figure 2-33(a). The rheostat is usually mounted in a rectangular case that is mounted near the motor, as seen in Figure 2-33(b), and is referred to as a starting box. The starting rheostat is connected in series with the motor armature. Contact with the rheostat is made when the operator moves the handle from the OFF position and slowly turns it to the ON position, stopping at each contact for one or two seconds. This causes the motor speed to increase; at the same time the CEMF becomes greater. A holding coil located near the last rheostat contact holds the handle in the ON position. In

Figure 2-32

Across-the-line DC
motor starter

this position, full line voltage appears across the armature. The motor is stopped by means of a stop switch. When actuated, the holding coil is deenergized and the handle, which is spring loaded, snaps back to its original position.

   Drum-switch controllers consist of an inner movable drum on which is attached a set of heavy-duty switch contacts and a set of stationary contact fingers connected to a tapped starting resistance, as shown in Figure 2-34(a). The starting resistance is connected in series with the armature of the motor. The armature is connected to the starting resistance by the first set of movable switch contacts coming in contact with the set of stationary fingers, as the handle is moved from the OFF position. As the handle is slowly moved toward the ON position, the starting resistance is decreased by the succeeding movable drum contacts making contact with the stationary fingers, as seen in Figure 2-34(b).

Figure 2-33

Starting rheostat

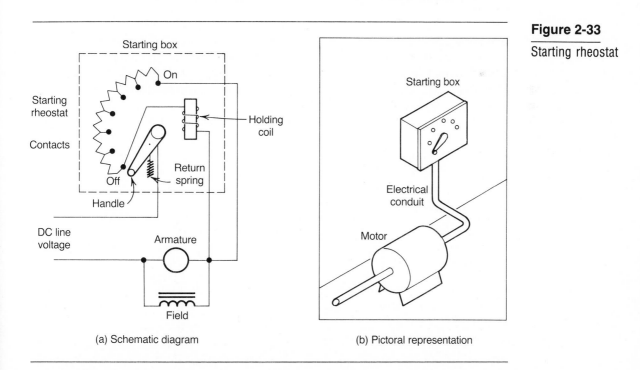

(a) Schematic diagram   (b) Pictoral representation

Figure 2-34

Drum switch

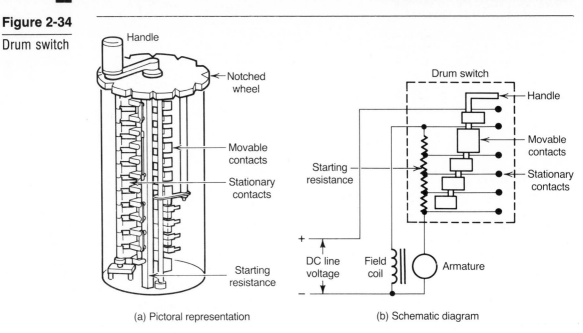

(a) Pictoral representation   (b) Schematic diagram

2-7.3 DC-motor speed control

The speed of a DC motor can be changed by varying field current, armature current, or armature (input) voltage. Methods used to change these variables include the use of field rheostats, armature rheostats, and controlled-rectifier circuits.

Field-rheostat speed control: Shaft speed in shunt and compound motors can be varied by connecting a rheostat in series with the shunt-field winding, as illustrated in Figure 2-35. This motor employs two rheostats—a starting rheostat and one for controlling speed. The motor is started with the shunt-field rheostat set for minimum resistance. This allows maximum field current to flow and causes the motor to reach its minimum running speed. The shunt-field rheostat is adjusted to increase the speed to a desired value. Increasing the resistance decreases the field current, causing the speed to increase.

Armature-rheostat control.   Since the field winding is connected in series with the armature in a series-wound motor, speed can be varied by connecting a rheostat in series with the armature and field windings, as seen in Figure 2-36. A starting rheostat and speed-control rheostat are both utilized. Motor speed increases when the resistance of the speed-control rheostat is decreased and vice versa. The speed-control rheostat is a much heavier device than that used in Figure 2-35. Since the full armature current must flow through this rheostat, it must be capable of dissipating considerably more power than does the shunt-field rheostat.

Armature-voltage speed control.   The DC voltage used to drive DC motors may come from DC generators or from controlled-rectifier power supplies. The controlled-rectifier power supply is the most common method of controlling motor

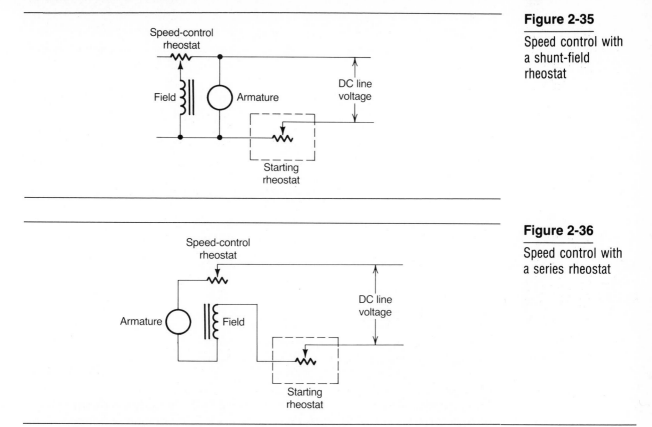

Figure 2-35

Speed control with
a shunt-field
rheostat

Figure 2-36

Speed control with
a series rheostat

speed, as it eliminates the need for heavy rheostats. Thyristors, such as SCRs, may
be used as the rectifier diodes, or conventional PN-junction diodes may be used
and a thyristor employed to control the period in which current flows through the
armature of the motor. The latter technique is used in Figure 2-37. In this circuit a
bridge rectifier converts AC line voltage into a pulsating DC voltage. The rectifier
is connected to a shunt-wound motor. An SCR is connected in series with the armature. The motor will not rotate until the SCR is turned on. The triggering angle
and conduction period of the SCR is controlled by an RC triggering circuit (R_1 and

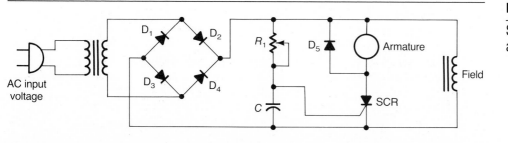

Figure 2-37

Speed control with
a thyrsitor

C) connected to the rectifier power supply. Current flowing through the rheostat causes the capacitor to charge. This provides the gate energy required to trigger the SCR. When the voltage across the capacitor reaches a predetermined amplitude the SCR begins conducting. This completes the path in the armature circuit, causing the motor to begin turning.

The armature voltage, and turning speed, is proportional to the conduction period of the SCR. This is varied by adjusting the rheostat in the triggering circuit. Since the SCR anode–cathode voltage is pulsating, its conduction period is terminated at the end of each cycle. To sustain conduction, the SCR must be retriggered for each succeeding cycle. The capacitor in the triggering circuit discharges through the rheostat, armature, and SCR when the SCR is conducting, and is ready to resume charging when the SCR is cut off, to provide the gate potential required to retrigger the device.

A reverse-biased diode is connected across the armature winding to prevent the CEMF developed across the armature when the SCR stops conducting from appearing across the anode–cathode terminals of the SCR, as this voltage may exceed the PRV rating of the SCR.

2-7.4 AC-motor starters

Several methods are employed to start AC motors. The more common of these include across-the-line, the autotransformer, and series-resistor starters.

Most single- and three-phase motors can be started by applying full voltage to the motor. Across-the-line AC motor starters are usually made up of a start–stop switch, a relay, an overload protective device, and a low-voltage protective device. Although full line voltage can be applied to a nonturning AC motor without harming it, extra current is required by the power source until the rotor reaches its rated or desired speed. Because of this, three-phase motors are often started with reduced voltage. An autotransformer or a series resistor is used to reduce the voltage. If an autotransformer starter is used, the motor is connected to the secondary of the transformer when initially started, as depicted in Figure 2-38(a). As the motor begins turning and approaches the desired speed, full line voltage is applied to the motor by actuating a switch or relay, as shown in Figure 2-38(b).

A resistor is connected in series with the supply line and motor, as depicted in

Figure 2-38

Autotransformer
AC motor starter

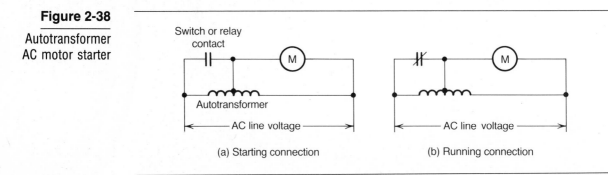

(a) Starting connection   (b) Running connection

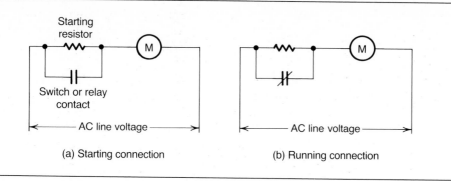

Figure 2-39

Series-resistor AC
motor starter

(a) Starting connection

(b) Running connection

Figure 2-39, when a series-resistor starter is used. Current flowing through the resistor reduces the voltage applied to the motor when initially started. As the motor speed develops, the resistor is bypassed by the closed switch or relay contact—allowing full line voltage to be applied to the motor, as seen in Figure 2-39(b). Motor-starting techniques are summarized in Table 2-3.

TYPE OF MOTOR	METHOD OF STARTING
Three-phase AC:	
Induction motors *Squirrel cage* *Wound rotor* *Synchronous*	Full line voltage can be applied without motor damage. All may be started with reduced voltage with the use of auto-transformers or series resistors.
Single-phase AC:	
Induction motors *Split-phase* *Capacitor-start* *Synchronous* *Torque* *Permanent magnet*	All are started with full line voltage.
DC:	
Series *Shunt* *Compound*	Full line voltage can often be applied to motors of ¼ HP or less. All others require that a variable resistance, in the form of a rheostat or drum switch, be connected in series with the armature.
Other types:	
Universal *Stepping*	Started with full line voltage. Utilizes binary pulses.

Table 2-3

METHODS OF
STARTING
MOTORS

2-7.5 AC-motor speed control

From Equation 2-5 in Section 2-5.2 it was found that the speed of AC motors may be varied by changing either the number of stator poles or the frequency of the voltage applied to the stator windings.

The speed of multispeed motors is varied by changing the number of stator poles. This is accomplished by using tapped stator windings, which can be switched to change the stator-winding configuration. For example, a stator winding may provide six poles for one type of connection, nine poles when connected in another configuration, and 12 poles when connected in yet a third configuration. This allows for three separate operating speeds. Push-button switches and relays or a drum switch may be used to accomplish the switching activity.

The shaft speed of adjustable-speed motors is varied by changing the frequency of the voltage applied to the stator windings. This is accomplished by using a rectifier power supply to convert the AC line voltage into a DC voltage and an inverter power supply to convert the DC voltage into an AC squarewave voltage whose frequency can be varied, as shown in Figure 2-40. This is called an adjustable-frequency AC-motor drive. Two different inverter techniques are employed—the variable-voltage inverter (VVI) and the pulse-width modulated inverter (PWM). Pulse-width modulation is the more commonly used technique. Energy is applied to the motor stator windings in the form of pulses whose width can be varied. An SCR bridge circuit is used as an inverter. Pulse width is varied by controlling the conduction period of the SCRs. These may be switched on and off several times during each cycle. The techniques used to vary motor speed are summarized in Table 2-4.

2-7.6 Commercial electric-motor drives

One of the most significant advances in industrial electronics has been the ability to accurately control the speed of electric motors. A wide variety of motor-speed controllers, often referred to as adjustable-speed motor drives, are available for both

Figure 2-40

A basic adjustable-frequency AC motor drive

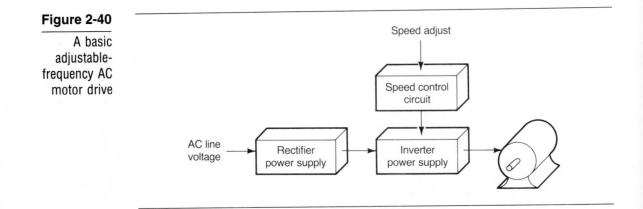

TYPE OF MOTOR	SPEED-CONTROL TECHNIQUE
Three-phase AC:	Speed is changed on multispeed motors by varying the number of stator poles. Speed is changed on adjustable-speed and adjustable-varying-speed motors by changing the frequency of the AC line voltage.
DC:	Field rheostat may be used on shunt and compound motors. Armature rheostat may be used on series motors. The speed of all may be changed by varying the amplitude of the voltage applied to the armature.
Stepping:	Speed is changed by varying the rate of the pulses applied to the stator windings.

Table 2-4

METHODS OF CHANGING MOTOR SPEED

DC- and AC-motor applications. Models have been designed to control motors ranging in size from one-fourth to 600 horsepower.

Adjustable speed-motor drives used to control DC motors of one-fourth to five horsepower usually employ a circuit similar to that appearing in Figure 2-37 and operate off of single-phase line voltage. Many models utilize prepackaged rectifier power modules and LSI ICs, which provide feedback and allow for very accurate speed control. DC motors in excess of five horsepower usually operate from three-phase voltage. Phase-control rectification is used to control motor speed, just as it is in the smaller DC-motor drives.

Most AC adjustable-speed motor drives operate from three-phase voltage, except those designed for use in consumer appliances. AC-motor drives used for industrial applications are designed to use three-phase voltage of 200, 208, 230, and 460 volts. Both VVI and PWM drives are available. VVI drives employ power transistors to develop an adjustable-frequency step voltage that is applied to the stator windings of the motor. An SCR is used to control the DC voltage applied to the inverter. Three-phase voltage is rectified and a fixed DC voltage is applied to the inverter when PWM is employed. High-speed switching transistors deliver narrow voltage pulses to the motor.

The data sheet for a typical adjustable-speed motor drive appears in Illustration 2-1 on page 91. This controller is designed to control the speed of a 10-horsepower three-phase squirrel-cage induction motor. Input voltage to the system is three-phase, 60 hertz, at 460 volts. The output frequency is variable from 0 to 200 hertz.

2-7.7 Stopping motors

Motors may be stopped in a number of ways. In some applications the motor is allowed to coast to a stop, while in others rotation must cease the moment the stop switch is actuated. Three methods are used to stop a motor rapidly—dynamic braking, plugging, and friction braking.

When dynamic braking is employed, the motor is connected as a generator when the stop switch is actuated. This creates a counter torque that rapidly slows and stops the motor.

Plugging is accomplished by causing the motor to attempt to reverse direction when the stop switch is actuated. As the motor very rapidly comes to a halt, the line voltage is removed, preventing the motor from rotating in the reverse direction.

Friction braking is a mechanical technique that utilizes brake shoes and a break drum attached to the rotor. An energized relay keeps the brake shoes open and away from the drum while the motor is running. When the stop switch is actuated the relay is deenergized, allowing the brake shoes to clamp around the brake drum, thereby stopping the motor.

2-7.8 Jogging and inching

Some mechanisms driven by motors, such as machine tools, must be precisely positioned manually. This can be accomplished with a jogging or inching switch. The terms jogging and inching are often used interchangeably to refer to the rapid and repeated closing of a switch to achieve small increments of machine motion. Although conventional start and stop switches can be used for jogging, they do not usually provide the precision required. Whenever the start switch is depressed, the motor remains running until the stop switch is actuated. When a jogging switch is employed, the motor runs only as long as the jogging switch is depressed.

2-7.9 Reversing motor direction

The direction in which a DC motor turns depends upon both the polarity of the magnetic-field poles and the direction of current flowing through the armature. Rotation direction may be changed by interchanging either the field-coil or armature winding leads. Reversing the polarity of the line voltage will not change the direction of rotation, as this changes both the field-pole polarity and the direction of armature current.

The technique employed to reverse AC motors depends upon the type of motor. The direction of rotor movement for a three-phase motor depends upon its phase sequence. Interchanging any two of the three-line voltage leads causes the direction of rotation to reverse. Single-phase, split-phase, and capacitive-start induction motors may be reversed by interchanging the connections of either the main or auxiliary stator winding leads. The methods used to reverse motor direction are summarized in Table 2-5.

2-7.10 Overload current and undervoltage protection

Overload devices such as line fuses and circuit breakers must be used to protect the motor from current overloads whenever excessive mechanical loads are connected to the motor.

Many motors will overheat and even burn out if operated at low voltage. Under-

Table 2-5

METHODS OF
REVERSING
MOTORS

TYPE OF MOTOR	REVERSING TECHNIQUE
Three-phase AC: *Induction motors*	
Squirrel cage *Wound rotor* *Synchronous*	Interchange any two of the three AC input lines.
Single-phase AC: *Induction motors*	
Split-phase *Capacitor-start*	Interchange either the main or auxiliary winding connections.
DC: *Series* *Shunt* *Compound*	Interchange either the field or armature winding connections.
Stepping motor:	Reverse the sequence of the binary pulses applied to the stator windings.

voltage protective devices cause the motor to shut down during low- and no-voltage conditions. Many of these will allow for automatic restarting when the voltage amplitude returns to normal or is restored.

Summary

1. Generators operate on the principle of electromagnetic induction as expressed by Faraday's law.
2. An AC generator uses slip rings to transfer the energy developed in the armature to the external circuit.
3. A DC generator employs a commutator to transfer energy from the armature to the external circuit.
4. An exciter voltage provides the field current needed to produce the magnetic field in a separately excited DC generator.
5. The field coil is connected in shunt with the armature in a shunt-wound DC generator.
6. The field coil is connected in series with the armature in a series-wound DC generator.
7. A compound-wound generator utilizes both series and shunt field windings.
8. A shunt DC motor is characterized by low starting torque and good speed regulation.

9. A series DC motor has high starting torque and poor speed regulation.

10. Alternators utilize a rotating field, called a rotor, and a stationary armature, called a stator.

11. The stator windings in a three-phase alternator can be connected in either a delta or wye configuration.

12. A three-phase induction motor is essentially a transformer whose stator forms the primary winding and rotor forms the secondary winding.

13. The synchronous speed is the rate at which the stator field revolves in a three-phase induction motor.

14. Slip is the difference between the synchronous and rotor speed in a three-phase induction motor.

15. A three-phase synchronous motor can be used to correct a low power factor in addition to providing mechanical power.

16. Single-phase split-phase and capacitor-start motors employ an auxiliary stator winding.

17. A varying-speed motor is one whose speed depends upon the load.

18. The speed of an adjustable-speed motor can be varied, but its speed remains constant once adjusted, regardless of the load conditions.

19. Definite-purpose motors are designed with standard operating features but are manufactured for particular applications.

20. The universal motor is a fractional-horsepower motor that will operate with either AC or DC voltage.

21. The torque motor produces a high amount of torque and is used to operate devices such as fluid power valves and furnace dampers.

22. DC pulses are used to turn, or step, the rotor in a stepping motor. This motor employs stator windings, which are angularly displaced, and a permanent-magnet rotor, which has toothlike poles.

23. Starting rheostats and drum switches are used to limit armature current when starting large DC motors.

24. The speed of DC motors may be varied by changing the field current, armature current, or armature voltage. The most common technique is to vary the armature voltage with a thyristor or controlled-rectifier circuit.

25. Across-the-line starters can be used to start most three-phase motors.

26. The speed of most three-phase motors can be varied by changing either the number of stator poles or the frequency of the line voltage.

27. Dynamic braking, friction braking, and plugging are methods used to rapidly stop motors.

28. Jogging refers to the rapid and repeated closing of a switch to achieve small movement of the rotor.

29. The direction of DC-motor rotation may be reversed by interchanging either the field-coil or armature-winding connections.

30. The direction of rotation of a three-phase motor may be reversed by interchanging any two of the three line-voltage leads.

31. The direction of rotation of a split-phase or capacitor-start motor may be reversed by interchanging either the main or auxiliary stator-winding connections.

Chapter Examination

1. An induction motor is connected to a 220-volt three-phase 60-hertz line. Calculate the synchronous speed if the motor has six stator windings.

2. If the rotor in the motor described in question 1 is turning at 1085 rpm, determine the percentage of slip.

3. An alternator that has six poles is turning at 2000 rpm. What is the frequency of the output voltage?

4. Calculate the current flowing through a nonturning armature in a DC motor that has a resistance of 7 ohms and requires 125 volts.

5. If the motor identified in question 4 produces a 112-volt CEMF while turning at 1800 rpm, determine the current flowing through the armature.

6. Describe how torque is produced in a DC motor.

7. What is the purpose of undervoltage protection for a motor?

8. Draw the schematic diagrams for the following DC machines.
   a. series-wound generator
   b. shunt-wound generator
   c. compound-wound generator
   d. shunt DC motor
   e. series DC motor
   f. compound DC motor

9. Which of the following is not a method used to rapidly stop a rotating motor?
   a. plugging
   b. static braking
   c. dynamic braking
   d. friction braking

10. The output voltage of a _____ wound DC generator decreases as the load current increases.
    a. rotor-
    b. shunt-
    c. series-
    d. compound-

11. The speed of a _____ DC motor increases as the mechanical load is reduced.
    a. rotor
    b. shunt
    c. series
    d. compound

12. The energy required in the rotor of a (an) _____ motor is obtained by mutual inductance.
    a. torque
    b. stepping
    c. induction
    d. universal

13. Which of the following motors employ(s) both DC and AC rotor windings?
    a. torque
    b. split-phase
    c. three-phase induction
    d. three-phase synchronous

14. Slip is present in a _____ motor.
    a. torque
    b. split-phase
    c. three-phase induction
    d. three-phase synchronous

15. Which one of the following motors requires a starting rheostat?
    a. shunt
    b. torque
    c. three-phase induction
    d. three-phase synchronous

16. The output voltage of the _____ generator increases when the load resistance decreases.
    a. shunt-wound
    b. series-wound
    c. compound-wound
    d. separately excited

17. Which one of the following DC generators has a single field coil and provides a constant output voltage regardless of load current?
    a. shunt-wound
    b. series-wound
    c. compound-wound
    d. separately excited

18. The speed of a(n) _____ motor can be varied; it remains constant after it has been adjusted, regardless of the load conditions.
    a. multispeed
    b. varying-speed
    c. constant-speed
    d. adjustable-speed

19. _____ motors have special operating characteristics and are designed for specific applications.
    a. Exact-purpose
    b. General-purpose
    c. Special-purpose
    d. Definite-purpose

20. The _____ motor may be used to operate fluid power valves and furnace dampers.
   a. torque
   b. universal
   c. split-phase
   d. capacitor-start

21. The stator windings of a _____ motor are energized by binary pulses.
   a. torque
   b. stepping
   c. universal
   d. capacitor-start

22. Which one of the following is *not* a method used to start three-phase motors?
   a. autotransformer
   b. across-the-line
   c. series resistor
   d. starting rheostat

23. Which one of the following motors will race, or run away, if operated without a load?
   a. shunt
   b. series
   c. induction
   d. synchronous

24. Thyristors may be used to control the speed of _____ motors.
   a. DC
   b. torque
   c. split-phase
   d. capacitor-start

25. T F The speed of a multispeed motor may be varied continuously over   ___
a wide operating range.

26. T F Plugging is accomplished by causing the motor to attempt to re-   ___
verse direction when the stop switch is actuated.

27. T F The speed of three-phase synchronous motors can be varied by   ___
changing the frequency of the line voltage.

28. T F The split-phase and capacitor-start motors are types of three-phase   ___
induction motors.

29. T F The speed of a three-phase synchronous motor is constant as it   ___
goes from a nonloaded to a fully loaded condition.

30. T F The speed of a shunt motor will increase if its field current is de-   ___
creased.

31. T F Alternators have a rotating field coil and a stationary armature.   ___

32. T F The direction of rotation of three-phase motors can be changed by   ___
interchanging the rotor lead connections.

33. T F When jogging is utilized, the motor remains running only as long   ___
as the jogging switch is actuated.

34. T F Both the armature- and field-winding connections must be inter-   __
changed to reverse the direction of rotation of a DC motor.

35. T F A stepping motor is a type of single-phase induction motor.   __

36. T F Compound motors require two stator windings and a capacitor to   __
cause them to be self-starting.

37. T F Single-phase induction motors require a starting rheostat to limit   __
the armature current when the motor is first started.

38. T F Three-phase induction motors can be used to correct the line-   __
voltage power factor.

39. T F The speed of three-phase motors may be changed by varying the   __
amplitude of the line voltage.

40. T F Inverter power supplies are used to convert DC voltage into a vari-   __
able-frequency AC voltage that can be used to vary the speed of
AC motors.

ALLEN-BRADLEY ADJUSTABLE FREQUENCY
AC MOTOR DRIVES

10 HP Bulletin 1334 NEMA Type 12 Enclosure

DESCRIPTION – The Bulletin 1334 AC drive converts 460V, 3 phase, 60 Hz input power to an adjustable AC frequency and voltage source for controlling the speed of AC squirrel cage motors. The 1334 is available in horsepower ratings from 5 to 50 HP.

The output voltage varies proportionally with the output frequency to maintain a typical constant VOLTS/HZ value from 0 to 60 Hz.

The 1334 may be operated above 60 Hz with the output voltage staying at a constant 460V. The maximum frequency is 200 Hz.

The standard controller includes power conversion components and control devices and regulator circuitry. A custom LSI circuit is used to generate the control signals for the power transistor modules which in turn generates a pulse width modulated (PWM) output waveform.

The drive is available as an open chassis or with a NEMA Type 1 or 12 enclosure, sized to dissipate the heat generated by the drive within the limits of the specified environmental and service conditions.

Illustration 2-1

CHAPTER
3

FLUID-POWER PRINCIPLES AND DEVICES

OBJECTIVES

1. State Pascal's law.

2. Calculate the pressure exerted on a confined liquid by a force.

3. Compute the force exerted on a piston by a pressurized liquid.

4. Identify the components forming a typical hydraulic system.

5. Identify the components forming a hydraulic power unit and discuss the purpose of each.

6. List the three types of hydraulic pumps and discuss the operating principles of each.

7. Describe the construction of the hydraulic reservoir.

8. Describe the operation of, and the applications for, two-, three-, and four-way direction-control valves.

9. Discuss the principles of operation and applications for pressure-relief, pressure-reducing, flow-control, and check valves.

10. List four techniques used to operate direction- and flow-control valves.

11. Compare the operating characteristics and applications for hydraulic and pneumatic motors.

12. Identify three types of hydraulic motors and discuss the operating characteristics of each.

13. Identify the major components forming a hydraulic cylinder; list the two types of cylinders; and describe how motion is controlled in each.

14. List the functions of the fluid used in a hydraulic system.

15. List the different types of transmission lines used in hydraulic systems.

16. Discuss the purpose of the heat exchanger, accumulator, and intensifier when used in a hydraulic system.

17. Identify the components forming the air compressor unit used in a pneumatic system and discuss the purpose of each.

18. List the two types of air compressors and discuss the operating characteristics of each.

19. Identify the devices used for treating the air in a pneumatic system and describe the purpose of each.

20. Identify the types of valves commonly used in a pneumatic system and discuss the operation and applications of each.

21. Discuss the operating principles of the rotary-vane air motor.

22. Identify the two types of pneumatic cylinders and describe how motion and speed is controlled in each.

23. Draw the schematic symbols for the more commonly used fluid power-devices.

3-1 INTRODUCTION

Although much of the power required to drive machines, move material, and control manufacturing processes is accomplished through the use of electrical energy, many of these activities are performed by means of fluid power.

Fluid power is the technology concerned with the transmission and control of energy through a medium of a pressurized liquid (usually oil) or a compressed gas (usually air). The transmission of energy by a pressurized liquid is called hydraulics and that by a compressed gas is called pneumatics.

Many of the principles of fluid power apply equally well to both hydraulics and pneumatics. However, some applications are best suited for hydraulics, while others are best implemented through pneumatics. Hydraulic power is usually used in applications where very accurate control is needed, high pressure is required, or cold temperatures are encountered. Hydraulic power is frequently the choice in those applications requiring power in excess of one horsepower. In applications involving elevated temperatures, low-to-moderate pressure, and low power, pneumatic power is often the better medium. In addition, pneumatic power systems usually are simpler, operate faster, and are cleaner than their hydraulic counterparts. Although a hydraulic system is designed and built not to leak, the possibility exists that oil may leak from seals and packings or hydraulic lines may break. In addition to the resulting uncleanliness, leaking fluid cannot be tolerated where products such as food, chemicals, pharmaceuticals, or textiles are processed. Pneumatic power systems are usually less expensive and less accurate than are the hydraulic types. Because of the compressability of air, they are usually not employed where rigidity is important—such as in hoists, lifts, and jacks.

3-2 PRINCIPLES OF HYDRAULICS

Individuals such as Euler, Pascal, and Torricelli, living in the 17th century, discovered the laws and developed many of the principles that made hydraulic technology possible. Widespread practical applications were not attainable until the industrial revolution made possible the production of precision interchangeable parts in large quantities.

The major operating principles of hydraulics are based on Pascal's law. This law states that whenever a force is exerted on a confined liquid, the resulting pressure is distributed equally in all directions, as illustrated in Figure 3-1. Because of this, a pressure can be transmitted through a confined liquid, as shown in Figure 3-2. In this diagram, two pistons, separated by a liquid, are placed at opposite ends of a cylinder. When a force of 100 pounds is exerted on one piston, it is transmitted through the liquid to the other, exerting a force of 100 pounds on that piston. A movement of one piston causes the other to move even though no mechanical linkage exists between them.

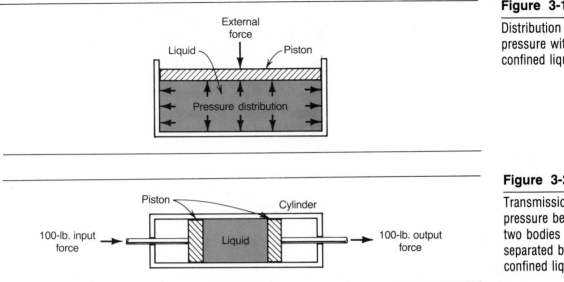

Figure 3-1

Distribution of a
pressure within a
confined liquid

Figure 3-2

Transmission of a
pressure between
two bodies
separated by a
confined liquid

If the single cylinder appearing in Figure 3-2 is replaced by two separate cylinders connected by a pipe containing a liquid, pressure is developed in the confined liquid whenever a force is exerted on one of the pistons. This pressure is proportional to the force exerted on the piston and inversely related to its area, as expressed by Equation 3-1.

$$P = \frac{F}{A} \text{ in psi} \qquad \qquad Eq.\ 3\text{-}1$$

where P = pressure in pounds per square inch (psi)
F = force in pounds (lbs)
A = area in square inches (in^2)

When a 200-pound force acts on a piston having an area of 4 square inches, a pressure of 50 pounds per square inch is developed, as seen in Example 3-1. Equation 3-1 can be rearranged to determine the force exerted by a pressurized liquid on a body, such as a piston, as expressed by Equation 3-2. A liquid having a pressure of

$$F = (P)(A) \text{ in lbs} \qquad \qquad Eq.\ 3\text{-}2$$

50 pounds per square inch exerts 1000 pounds of force on a 20-square-inch piston,

Example 3-1

CALCULATING THE
PRESSURE
EXERTED ON A
CONFINED LIQUID

Problem: Determine the pressure exerted on the liquid in the pipe in Figure 3-3.

Solution:
$$P = \frac{F}{A} = \frac{200 \text{ lbs}}{4 \text{ in}^2} = 50 \text{ psi}$$

Figure 3-3

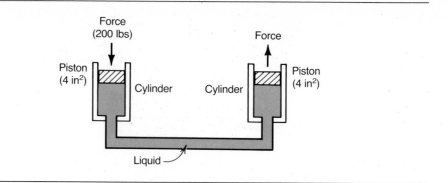

as illustrated in Example 3-2. Equation 3-2 expresses a major principle governing the application of hydraulics. A small force applied to a small body, such as a piston, results in a much larger force acting on a larger body if the two bodies are connected together by a pressurized liquid. A multiplication effect is produced. This effect is at the expense of speed, however. The velocity of the larger piston is less than that of the smaller piston. This is illustrated by Equation 3-3.

$$V = \frac{Q}{A} \text{ in in/s} \qquad\qquad Eq. \ 3\text{-}3$$

Where V = piston speed in inches per second (in/s)
  Q = fluid flow rate in cubic inches per second (in^3/s)
  A = piston cross-sectional area in square inches (in^2)

Referred to as the flow continuity equation, this equation illustrates that piston velocity is inversely proportional to the cross-sectional area of the piston.

Figure 3-4

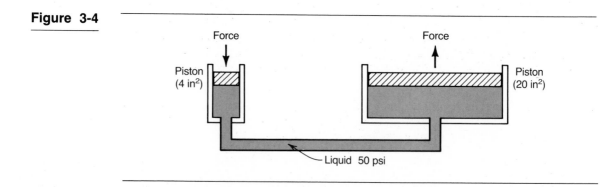

Example 3-2

CALCULATING THE
FORCE EXERTED
ON A PISTON BY
A PRESSURIZED
LIQUID

Problem: Calculate the force exerted on the 20-in^2 piston appearing in Figure 3-4.

Solution: $F = (P)(A) = (50 \text{ psi})(20 \text{ in}^2) = 1000$ lbs

Components forming a typical hydraulic system include a power unit, a pressure-control valve, a direction-control valve, a flow-control valve, and an actuator (hydraulic cylinder or motor)—all connected together by a transmission line made up of piping, tubing, or hoses, as illustrated in Figure 3-5. Hydraulic systems are designed to form a closed circuit. Oil is pumped from the reservoir to the actuator and is returned to the reservoir.

3-3 THE POWER UNIT

3-3.1 Introduction

The hydraulic power unit produces the flow required to transmit power to the actuator. It performs the same function in the hydraulic circuit that the voltage source performs in an electric circuit. As illustrated in Figure 3-6, the power unit consists of a hydraulic pump, a prime mover, and an oil reservoir.

3-3.2 The prime mover

The prime mover provides the power to drive the hydraulic pump. Either internal-combustion engines or electric motors may be employed. Electric-motor drive is, by far, the more commonly used. Although almost any type of motor of suitable size may be used, the single-phase capacitor-start induction motor is often employed in small systems, while the three-phase squirrel-cage induction motor is the type most commonly used in larger systems. Both are chosen because of their high torque characteristics.

The size of the prime mover is proportional to the flow rate and pressure of the system oil. For a specific pressure, if the flow rate is to be doubled, the motor size has to be doubled. In a similar manner, for a specified flow rate a motor having twice the horsepower rating will have to be used, if the pressure is to be doubled.

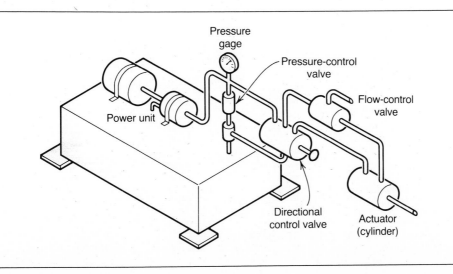

Figure 3-5

Basic hydraulic system

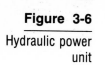

Figure 3-6

Hydraulic power
unit

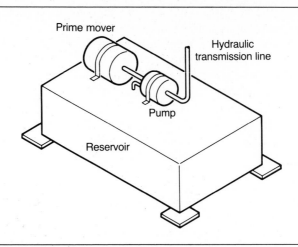

3-3.3 The hydraulic pump

The hydraulic pump forms the heart of the power unit. This is the device that causes fluid to flow through the hydraulic system. Three different types of pumps are available—the gear, vane, and piston. Although each is constructed differently, they all operate on the principle that a partial vacuum is formed as the pump cycles. Atmospheric pressure acting on the oil in the reservoir forces oil from the reservoir into the pump through the input port. The oil cycles through the pump and exits under pressure through the discharge port.

The gear pump: This is a type of rotary pump made up of a pair of meshed gears, as seen in Figure 3-7. The gears are mounted on shafts coupled to the prime mover and rotate whenever the prime mover is actuated. A partial vacuum is created by the disengaging of the rotating teeth, causing oil to be pulled from the reservoir

Figure 3-7

Gear pump

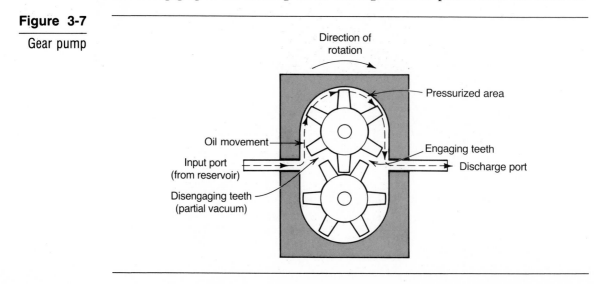

Figure 3-8

Vane pump

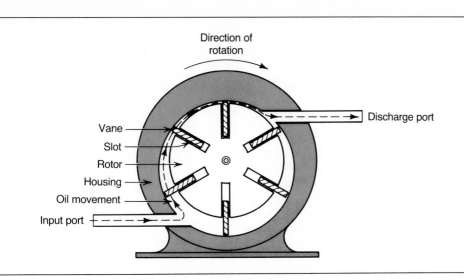

into the pump housing through the input port. Oil fills the evacuated space left by the disengaging teeth and is carried along with the rotating teeth until they again mesh. This causes the oil to be forced out of the pump, under pressure, through the discharge port.

This is the most commonly used pump. It is simple, durable, and is the least sensitive to dirt and contaminants in the fluid. It is a medium-pressure pump capable of producing pressures in the range of 1000 to 1500 psi at gear speeds of 1500 to 1800 rpm. The volume of fluid pumped is dependent upon the size of the gears. Models are available that pump as little as a few gallons per hour to as much as several hundred gallons per minute.

The vane pump: The vane pump is a rotary pump in which a cylindrical rotor having movable vanes set in radial slots is made to rotate in an offset circular housing, as shown in Figure 3-8. When actuated, the prime mover causes the rotor to turn and the vanes move in and out in their slots, following the offset path of the housing. Oil is pulled into the housing through the input port, located where the distance between the vanes and housing is maximal, and is forced out the discharge port, which is located where the distance between the vanes and housing is minimal.

This is a medium-pressure pump producing pressures of between 500 to 1500 psi with liquid-flow capacity in the 5 to 40 gpm range. It has a low internal mass and is often connected to a high-speed prime mover. Rotor speeds as high as 4000 rpm are common.

The piston pump: There are two types of piston pumps—the radial, and the axial, or in-line. The radial pump is made up of an even number of cylinders placed radially about a centrally located cam-shaped rotor. Pistons positioned in the cylinders move inward, pulling in oil through an intake valve as the piston rod travels along the flat side of the rotor, as depicted by the partial diagram appearing in Figure 3-9. The piston moves outward, causing the oil to be ejected when the piston

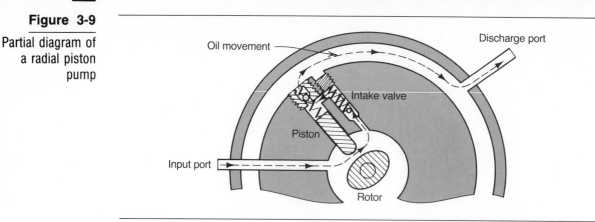

rod moves along the lobe part of the rotating cam. The axial pump operates in much the same fashion, except that the cylinders are located in-line.

These are high-pressure pumps, with most operating in the 2500 to 10,000 psi range. Most are designed to be driven by a prime mover that rotates at approximately 1800 rpm.

3-3.4 The oil reservoir

Although the reservoir primarily serves as a storage and supply source for the system oil, it has other uses as well. It is here that the oil having performed work in the system is cooled, the returning oil is purged of air, and contaminants such as dust and small metallic particles are removed from the oil.

The reservoir is an air-tight tank with a baffle located in its center, as illustrated

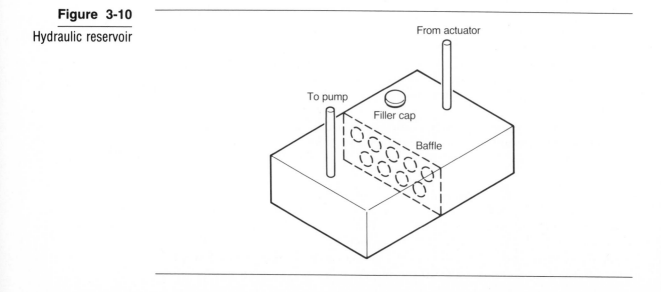

in Figure 3-10. The returning oil has to flow through the baffle before being pumped back into the system. This provides time for the oil to cool, air to be purged, and contaminants to settle. A filler cap is located at the top of the tank to allow oil to be added when required.

The size of the reservoir is important. For small and light-duty systems, reservoir capacity should be large enough that all the cylinders in the system can be fully extended at the same time. In large and heavy-duty systems, the capacity should be enough to hold two to four times the pump flow capacity—to provide for adequate cooling.

It is important that the oil remain clean. Approximately 75 percent of all problems occuring in hydraulic systems are caused by contaminated oil. A filter is usually mounted on the return side of the line while a strainer is mounted on the pump inlet line, as seen in Figure 3-11. These help remove very small abrasive particles which may repeatedly circulate through the system causing accumulated wear on pumps, motors, cylinders, and valves. Both should be cleaned on a scheduled basis. An oil drain plug is located on the bottom of the reservoir. The oil should be routinely drained and replaced.

3-4 HYDRAULIC VALVES

3-4.1 Introduction

Hydraulic valves are devices used to control the direction, pressure, and velocity of the oil flowing through the hydraulic system and to protect operating personnel, equipment, and property. Hydraulic valves used in industrial applications are manufactured for use in several pressure ranges, which include 0 to 1500 psi, 1500 to 2500 psi, and 1000 to 5000 psi. Most applications are in the 0 to 1500 psi range. Valves used in hydraulic power systems include direction-control, pressure-relief, pressure-reducing, flow-control, and check valves.

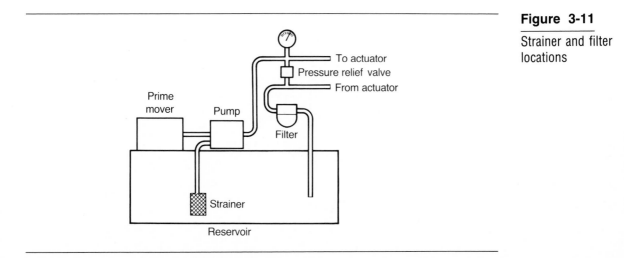

Figure 3-11

Strainer and filter locations

3-4.2 Direction-control valves

Several different designs are employed in the manufacture of direction-control valves. Most include an input port, an output port, and one or more control ports. The spool valve is a popular type of direction-control valve. One or more movable pistons, called spools, mounted on a stem, determine the direction of oil flow. As the spool is moved, internal chambers are opened and closed, directing the oil where required. These devices may be grouped into the categories appearing below.

Two-way valves: Two-way direction-control valves are used to open and close hydraulic lines to the flow of oil. They are used to turn hydraulic motors and cylinders on and off and are sometimes called on–off valves. As illustrated in Figure 3-12(a), in the open position oil flows in through the input port, passes through the chamber, and exits through the output port. In the closed position the upper spool is located between the input and output ports, preventing oil from flowing through the output port, as seen in Figure 3-12(b).

Three-way valves: These valves have three ports and can be used for selection, diversion, or direction-control purposes. When the valve is used as a selector, two different fluid sources are connected to two ports while the third port serves as an output. Depending upon the position of the valve stem, fluid exiting the output port can come from either input port, as shown in Figure 3-13(a).

When the valve is used as a diverter, one port serves as an input while the other two are connected as outputs, as depicted in Figure 3-13(b). Fluid may exit either output depending upon the position of the valve stem. Three-way valves may be used to control the movement of a single-acting cylinder. This application is discussed in Section 3-6.3.

Four-way valves: Four-way direction-control valves have four ports. These valves are usually used to control the direction of rotation of a hydraulic motor or the movement of a double-acting cylinder, as described in Section 3-6.3. As depicted in Figure 3-14, this valve has three spools and five chamber openings. Two openings

Figure 3-12

Two-way valve

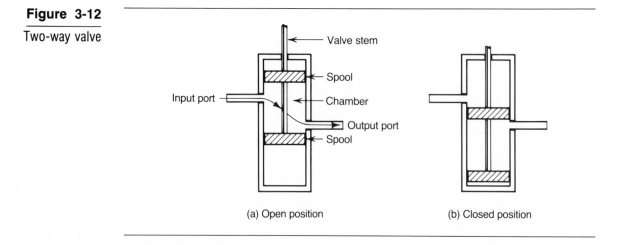

(a) Open position   (b) Closed position

Figure 3-13

Three-way valves

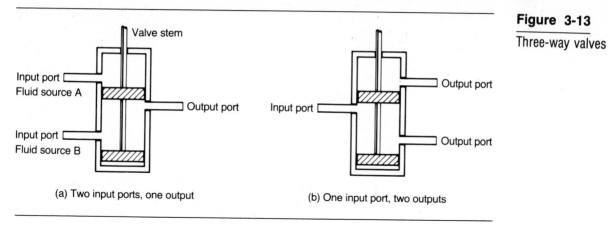

(a) Two input ports, one output   (b) One input port, two outputs

are connected together to form a total of four ports. These can be configured a number of different ways. For the valve appearing in Figure 3-14, two ports serve as inputs while the other two act as outputs.

3-4.3 Pressure-relief valves

Sometimes called relief valves, or safety valves, these devices are used to ensure that the pressure in the system does not rise above a predetermined level. They are designed to operate in a specific pressure range and must be large enough to handle the maximum output pressure of the pump.

A variety of designs are employed in the production of relief valves. A spring-loaded type appears in Figure 3-15. Spring action acting on a ball, or poppet, blocks the flow of fluid through the valve. If the line pressure rises so that the spring action is overcome, a small amount of fluid flows through the valve and back to the reservoir as the potential energy, or pressure, is converted into kinetic energy, or fluid flow. The pressure can be set with the adjustment screw.

Figure 3-14

Four-way valve

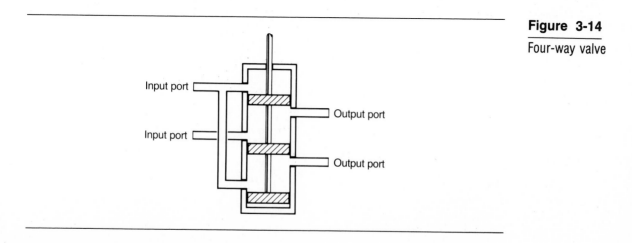

Figure 3-15

Pressure-relief
valve

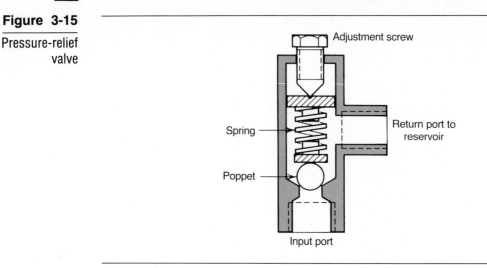

3-4.4 Pressure-reducing valves

These valves are used for reducing line pressure. Although they can be used for reducing main-line pressure, pressure-reducing valves are more commonly used for supplying fluid to low-pressure, or secondary, lines from a high-pressure line. In a typical application, the main high-pressure line provides oil to drive hydraulic motors and cylinders, while a secondary line supplies oil to devices such as hydraulic clamps and low-pressure actuated valves. A pressure-reducing valve appears in Figure 3-16. Maximum secondary-line pressure is set by the adjustment screw, which controls the tension of the spring. When the pressure in the secondary line exceeds that set by the adjustment screw the spool is forced upward, reducing both the volume and pressure of the oil supplied to the secondary line.

Figure 3-16

Pressure-reducing
valve

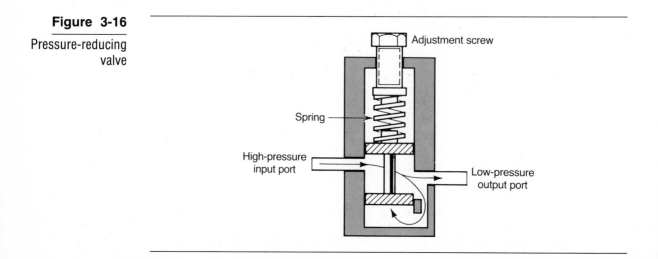

Figure 3-17

Flow-control valve

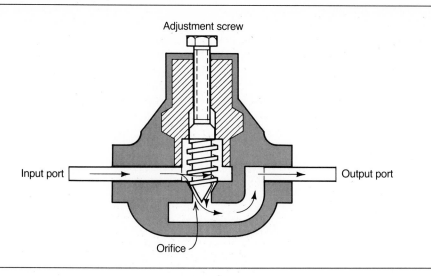

Adjustment screw

Input port

Output port

Orifice

3-4.5 Flow-control valves

Flow-control valves control the volume of oil flowing in a transmission line. A typical flow-control vavle appears in Figure 3-17. Oil enters the valve through the input port, flows through an orifice, and exits through the output port. The volume of fluid leaving the valve is governed by the size of the orifice. Orifice size, and flow control, can be changed by varying the adjustment screw. Flow-control valves are used to vary the speed of fluid-power motors and cylinders and are often called speed-control valves.

Figure 3-18

Check valve

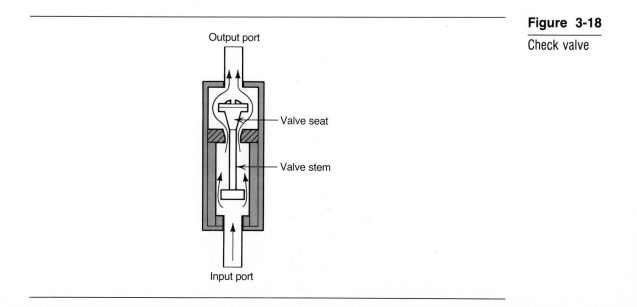

Output port

Valve seat

Valve stem

Input port

3-4.6 Check valves

These are one-way valves that let oil flow in one direction but not the other. In installations, such as a basement, where the power unit is located below the other system components, a check valve is used to prevent oil from draining back into the reservoir whenever the pump is turned off. A number of different designs are employed in the construction of check valves. The pressure-relief valve appearing in Figure 3-15 can be used for this application. Another typical design is illustrated in Figure 3-18. Pressure from the input port forces the valve stem upward, allowing oil to flow through the chamber and out the output port. If oil should attempt to enter the valve through the output port, the stem is forced downward, causing the valve to close.

3-5 METHODS OF OPERATING VALVES

3-5.1 Introduction

Several different techniques are used to operate direction-control and flow-rate valves. The technique employed depends upon the frequency at which the valves must be actuated, fluid flow rate, pressure, and the degree of automation required. Valves may be operated manually; mechanically, by a low-pressure fluid called a pilot pressure; or electrically, by solenoids and torque motors.

Figure 3-19

Methods of operating a valve manually

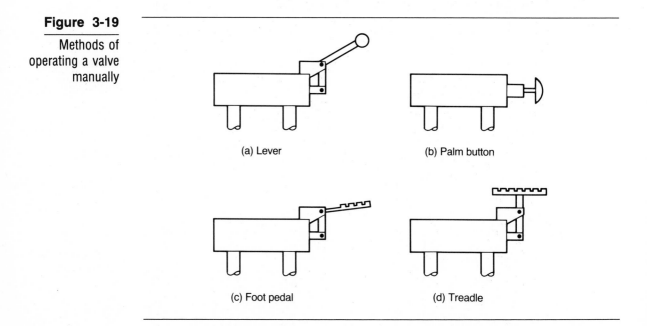

(a) Lever

(b) Palm button

(c) Foot pedal

(d) Treadle

3-5.2 Manual operation

Manually operated valves are controlled either by hand or by foot, as illustrated in Figure 3-19. The valve shown in Figure 3-19(a) is operated by a lever, while that shown in Figure 3-19(b) utilizes a knob attached to the end of the valve stem. This is called a palm-button control. Both of these valves are operated by hand. A foot-pedal control is seen in Figure 3-19(c). This valve is operated by the toe of the operator. Either the toe or heel of the operator can be used to operate the foot treadle control appearing in Figure 3-19(d). Foot-operated valves are usually less fatiguing to operate than the hand-operated types and allow the operator to have both hands free. This is especially important for many machine-tool operators who must have both hands free to handle and position workpieces.

3-5.3 Mechanical operation

An almost unlimited number of techniques are available to operate valves mechanically. Two typical methods are depicted in Figure 3-20. In Figure 3-20(a), the valve is actuated by a cam. A cam roller is connected to the valve stem. Whenever the cam lobe comes in contact with the roller, the valve is actuated. This method is often used when valves must be operated in a sequential or cyclical fashion. A clevis is attached to the stem of the valve appearing in Figure 3-20(b). A mechanical linkage, such as a rod, is connected between the clevis and the operating force.

3-5.4 Pilot operation

Low-pressure oil, called a pilot pressure, can be used to operate a valve. This technique is beneficial when the valve is located at a considerable distance from the operator workstation, as it eliminates the need for long runs of high-pressure transmission line. Low-pressure and smaller diameter line can be used instead. The low-pressure oil is controlled by a smaller valve called a pilot valve, which may be manually, mechanically, or electrically operated.

Other types of pilot devices are also available. Called fluidic circuits, these devices are composed of a sealed block of metal having a series of internal chambers. Fluidic circuits are available to perform such functions as AND, OR, NOR, and NAND gate-logic operations in addition to flip-flop and counter activities.

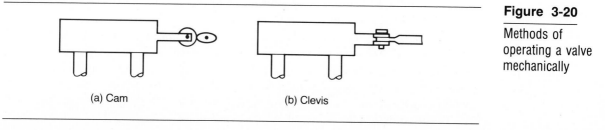

(a) Cam   (b) Clevis

Figure 3-20

Methods of operating a valve mechanically

3-5.5 Electric operation

Solenoids and torque motors both may be used to operate valves. Both are fast acting and can be used with electronic controllers. The solenoid is by far the more commonly used and is directly coupled to the valve stem. Depending upon the design, either push or pull types may be utilized.

3-6 HYDRAULIC ACTUATORS

3-6.1 Introduction

Hydraulic actuators are devices which convert hydraulic power into rotary or linear mechanical power. They can be grouped into two categories—hydraulic motors and hydraulic cylinders. A motor is employed whenever a rotational force is required. A cylinder is used when a linear force is needed as in pushing, pulling, lifting, and clamping.

3-6.2 Hydraulic motors

Hydraulic motors produce more torque and horsepower from less bulk than any other type of motor. They can be operated submerged in a liquid; they produce no sparks, which makes them explosion-proof; they may be operated in a stalled condition indefinitely without damage; and they have a continuously variable speed when used with a flow-control valve.

Hydraulic pumps may be operated as hydraulic motors in the same sense that DC generators can be connected to operate as DC motors. To operate a pump as a motor, a mechanical load is substituted for the prime mover. Pressurized oil is injected into the input port, causing the shaft to turn. Just as there are three types of hydraulic pumps, there are three kinds of hydraulic motors—the gear, vane, and piston.

Gear motors: Pressurized oil enters the motor through the intake port, travels around the periphery of the gears, exits through the exhaust port, and flows back to the reservoir. The pressure exerted on the teeth as the oil travels from the intake to the exhaust ports creates a torque, causing the motor to turn. Shaft speed is determined by the volume of oil pumped through the motor.

Vane motors: Torque is produced when pressurized oil entering the intake port strikes the internal vanes, causing the rotor to turn. Rotor speed is proportional to the volume of oil traveling through the motor.

Piston motors: Piston motors operate on the principle that pressurized oil acting on a piston creates a force, which causes the piston to move in a cylinder. This movement causes a cam to move, which in turn creates shaft torque. Shaft speed is determined by the volume of oil acting on the piston. Since the speed of all three motors is proportional to the volume of oil flowing into the input port, motor speed can be varied with a flow-control valve.

Figure 3-21

Single-acting
cylinder

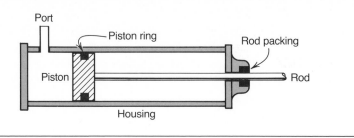

Port

Piston ring

Rod packing

Piston

Rod

Housing

3-6.3 Hydraulic cylinders

Hydraulic cylinders, sometimes called rams, are available in various sizes for different applications. Diameters range from approximately 1/2 inch to 20 inches. Stroke length may vary from less than one inch in some cylinders to several feet in others. Hydraulic cylinders are designed to operate in specific pressure ranges, such as 0 to 250 psi, 0 to 500 psi, 0 to 2500 psi, and 0 to 10,000 psi.

There are two types of cylinders—single-acting and double-acting. The single-acting cylinder develops a force in one direction only. After it is extended by pressurized oil, it is returned to its resting position either by gravity acting on the load or by a spring. As illustrated in Figure 3-21, the device consists of a housing, piston, and piston rod. The piston is forced outward whenever pressurized oil is injected into the port. Automobile-type piston rings or O rings are mounted on the outer circumference of the piston to prevent oil from escaping.

The force created by a moving piston is proportional to the pressure of the oil and the size of the piston, as expressed by Equation 3-2. A cylinder having a diameter, or bore, of three inches will exert a force of 3532.5 pounds if the oil pressure is 500 pounds per square inch, as shown in Example 3-3.

The diagram of a double-acting cylinder appears in Figure 3-22. This is the more commonly used type of cylinder. The construction of this device is similar to that of the single-acting cylinder except that a second port is included. Pressurized oil is injected into either the piston (blind) or rod ends of the cylinder. The force developed when oil is injected into the blind end is the same as that developed by a single-acting cylinder. When oil is injected into the rod end, the resultant force is slightly less because the area of the cylinder has been reduced by the cross-sectional area of the rod. Equation 3-4 is used to determine the force when oil is injected at the rod end.

Example 3-3

DETERMINING
THE FORCE
DEVELOPED BY A
SINGLE-ACTING
CYLINDER

Problem: Calculate the force produced by a cylinder having a bore of 3 in. if the oil pressure is 500 psi.

Solution: $F = (P)(A) = (P)(\pi)(r^2)$
$= (500)(3.14)(1.5^2) = 3532.5$ lbs

Figure 3-22

Double-acting
cylinder

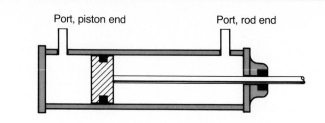

$$F = (P)(A - A_1) \text{ in lbs} \qquad Eq. \text{ } 3\text{-}4$$
Where A = area of piston
A_1 = area of piston rod

As shown in Example 3-4, a force of 3434.38 pounds is developed for a cylinder having a bore of 3 inches and a connecting rod diameter of 1/2 inch when the oil pressure is 500 pounds per square inch. This is approximately three percent less force than that acting on the blind end, as calculated in Example 3-3.

Occasionally, erratic motion may be encountered in clyinder operation. This is usually caused by air that is trapped in the system. This may be especially noticeable in new systems or systems which have been recently repaired. The system will usually purge itself of air after a few hours of operation. If the problem persists, the transmission-line fittings to the cylinder should be loosened slightly for a short period of time to provide an escape path for the air.

Piston motion is controlled by direction-control valves. A three-way valve may be used to control the movement of a single-acting cylinder, as illustrated in Figure 3-23. During the extension stroke, the reservoir return port is closed and oil is injected into the cylinder, causing it to extend as seen in Figure 3-23(a). The cylinder retracts when the input port is closed and the reservoir port is open, as shown in Figure 3-23(b). In this position the oil flows from the cylinder back to the reservoir. Some three-way direction-control valves have a mid-position that closes both ports. This type of valve allows the cylinder to be maintained in a stationary position for applications in lifts, hoists, jacks, and presses.

A four-way valve can be used to control the movement of a double-acting cylinder, as seen in Figure 3-24. When the valve stem is in the extended position, oil flows into the blind end of the cylinder while oil in the rod end flows out the reservoir port. This causes the piston to be forced outward, as shown in Figure 3-24(a).

Example 3-4

CALCULATING THE
FORCE EXERTED
ON THE ROD END
OF A
DOUBLE-ACTING
CYLINDER

Problem: Calculate the force created at the rod end of a double-acting cylinder having an internal diameter of 3 in. and a rod diameter of 1/2 in. Oil pressure equals 500 psi.

Solution: $F = (P)(A - A_1)$
$= (500)[(3.14)(1.5^2) - (3.14)(0.25^2)]$
$= 3434.38$ lbs

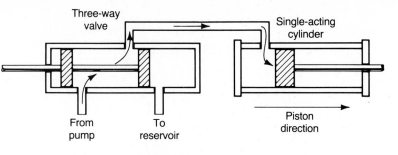

(a) Cylinder extending (outstroke)

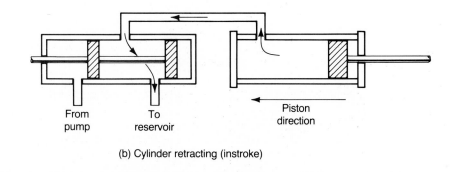

(b) Cylinder retracting (instroke)

Figure 3-23

Three-way valve used to control a single-acting cylinder

When the stem is depressed and is in the actuated position, oil flows into the rod end while that in the blind end exits through the reservoir port, causing the piston to move inward as illustrated in Figure 3-24(b). Some four-way valves have a mid-position to allow stationary positioning of the cylinder.

3-7 MISCELLANEOUS HYDRAULIC DEVICES

3-7.1 Hydraulic fluid

The fluid used in hydraulic systems performs three functions—transmission of power, lubrication, and sealant. A number of different grades of petroleum-base oils and synthetic fluids have been developed, with the former being the most frequently used medium.

3-7.2 Hydraulic transmission lines

Transmission lines are used to carry the oil from the pump to the other components in the system and to return it to the reservoir. Special piping, tubing, and hoses have been developed for this purpose. Heavy steel piping is employed in high pressure systems. Aluminum, seamless steel, and annealed steel tubing are utilized for medium- and low-pressure applications. Where movement of devices occurs or

Figure 3-24

Four-way valve
used to control a
double-acting
cylinder

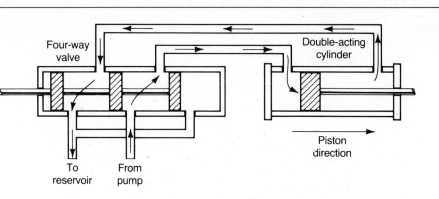

(a) Cylinder extending

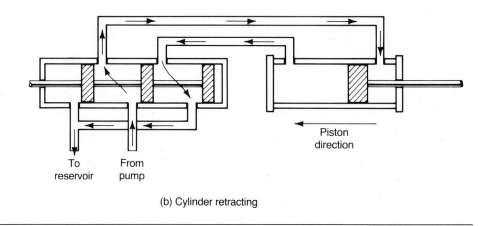

(b) Cylinder retracting

where excessive vibration is present, hose is used. This is a specially constructed hose composed of an inner tube, a reinforcement lining, and an outer protective covering, as illustrated in Figure 3-25.

Hydraulic transmission lines must have a sufficiently large cross-sectional area to avoid unwanted pressure drops along the line. Pressure drops occur in hydraulic lines that are too small, just as voltage drops occur along electrical conductors of insufficient size. The relationship of pressure and transmission-line size is illus-

Figure 3-25

Hydraulic hose

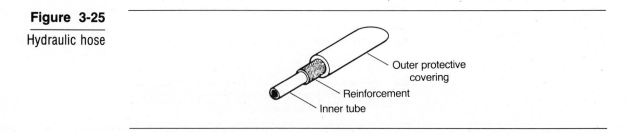

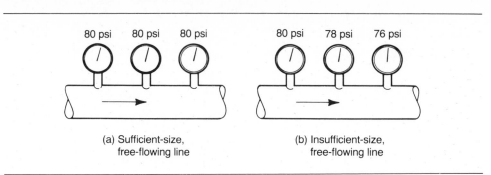

trated in Figure 3-26. Pressure drops may also be developed by roughness of the inner surface area of the line, length of the line, and the type of fluid.

3-7.3 Heat exchanger

Heat exchangers are an important component in hydraulic systems where extreme heat or cold becomes a factor in system performance. Heat may come from either external or internal sources and has several adverse affects on the performance of the system. Heat reduces the efficiency of the power unit and causes the hydraulic fluid to break down, valves to malfunction, and packings to leak. The heat in the oil must be removed if the environmental temperature or work performed by the system causes the oil temperature to exceed approximately 140° F. On the other hand, if the system is to be operated in an extremely cold environment, heat may have to be added to the oil to decrease its viscosity. The heat exchanger may be used either to cool or to warm the oil.

3-7.4 Accumulator

The accumulator is a reservoir that stores fluid during nonpeak work activities and returns it to the system whenever the work demand is greater than the pump supply. An accumulator is used where a high pressure must be sustained for a prolonged period of time while the power unit is diverted to other work, to supplement the energy supplied by the power unit, and to absorb shocks caused by sudden changes in the load.

3-7.5 Intensifier

Sometimes called a pressure booster, this device is used to increase the fluid pressure available from the power unit by converting a high-volume, low-pressure oil supply into a proportionally low-volume, high-pressure output. An intensifier is used in applications such as punches and presses where the pressure from the pump may be insufficient to produce the force needed to perform the work.

3-8 THE PNEUMATIC SYSTEM

3-8.1 Introduction

Fluid power is a bifurcated technology. As such, it contains two components—hydraulics and pneumatics. This section deals with the components that form the pneumatic power system. Many of the devices used in pneumatics are similar to those found in hydraulic systems. System circuitry is different, however. In hydraulics, a closed circuit is employed. Pressurized fluid flows from the pump through the system back to the reservoir where it is recirculated. This is not the case in a pneumatic system. Here, a compressed gas, usually air, is sent to the actuator. After it transfers its energy to the actuator, the air is expelled into the atmosphere. This is an open circuit. The air is not returned to the air compressor.

The diagram of a basic pneumatic system appears in Figure 3-27. Major components include an air-compressor unit, air-treatment devices, control valves, an actuator, and air lines to connect the components together.

3-8.2 Air-compressor unit

The air-compressor unit performs the compression activity. It performs the same function in a pneumatic system that the power unit provides in a hydraulic system. The air-compressor unit is made up of a prime mover, an air compressor, and a storage tank. The electric motor is the most commonly used type of prime mover, although gasoline, diesel, and steam engines; turbines; and hydraulic motors may also be utilized.

There are two types of compressors. These include the rotating and the reciprocating. The rotating air compressor is used in low- to medium-pressure systems.

Figure 3-27

Basic pneumatic system

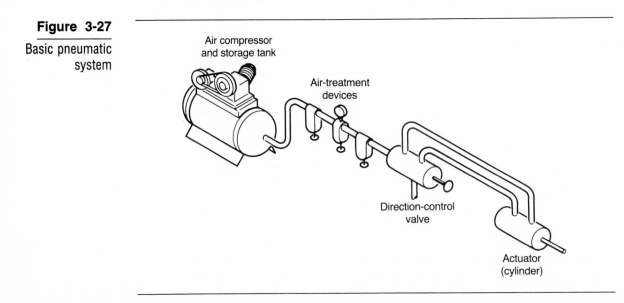

Figure 3-28

Rotating air compressor

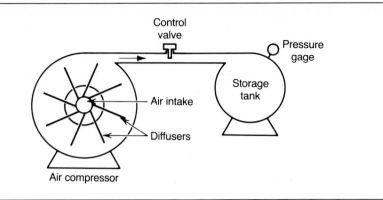

This device consists of a bladed wheel that rotates within an enclosed circular housing, as shown in Figure 3-28. Air is drawn in at the center of the wheel and is accelerated by the centrifugal action of the revolving wheel. The energy contained in the moving air is compressed by the diffusers and is forced into the storage tank.

Air is compressed by the action of a piston in a cylinder in a reciprocating air compressor. As illustrated in Figure 3-29, air is drawn into the cylinder through the intake valve on the down stroke of the piston. During the upstroke of the piston the intake valve is closed and the air is compressed and forced into the storage tank through the open exhaust valve. A single-cylinder compressor is limited in the amount of pressure that can be developed. Multistage compressors are used in high-pressure applications. These air compressors are made up of two or more cylinders. Air pressure is successively increased by connecting the output of one cylinder to the input of the next.

3-8.3 Air-treatment devices

Air filters, lubricators, and regulators form part of the air-compressor unit and are referred to as air-treatment devices. They are essential to the proper operation of

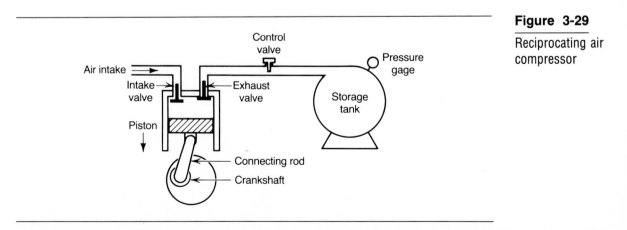

Figure 3-29

Reciprocating air compressor

the system. The air must be clean, free of moisture, and be maintained at a specified pressure.

Air taken in from the atmosphere and compressed by the air compressor contains varying amounts of foreign particles. The air filter is used to remove these. This device is usually connected to the air-compressor intake line. Air drawn in by the compressor also contains water vapor, which condenses in the compressor storage tank and air lines, forming water. A reservoir is included as part of the filter mechanism, as shown in Figure 3-30, to collect the water to prevent rust and corrosion from forming in the system.

A light lubricating oil must be injected into the air supply to prevent metal-to-metal contact from occurring in actuators and to prevent packings and seals from drying out. As shown in Figure 3-31, an air-line lubricator is made up of an oil reservoir, a needle valve to meter the oil, and a Venturi tube in which air atomizes the oil.

3-8.4 Pneumatic valves

Pneumatic valves are used to control the pressure, direction, and flow rate of air moving in a pneumatic system.

There are several different types of pressure valves. Two commonly used types include the pressure-relief valve and the pressure regulator. A pressure-relief valve, often referred to as a relief valve, protects both the air compressor and the components in the system by setting the maximum safe system operating pressure. If this pressure is exceeded, the valve is automatically actuated. Although a number of different factors may cause an increase in the compression of the air, this condition is often a result of an actuator encountering a heavy work resistance. A simple relief valve appears in Figure 3-32. The valve is connected in shunt with the air-supply line and contains a pressure port, relief port, a poppet, and a spring. The

Figure 3-30

Air-line filter

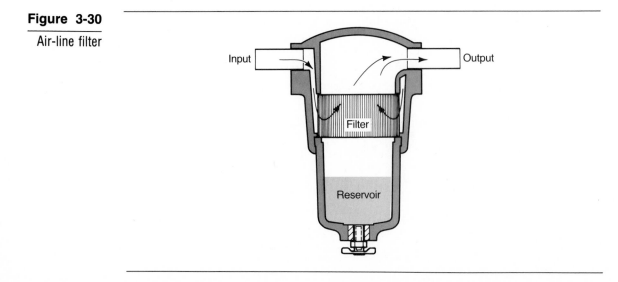

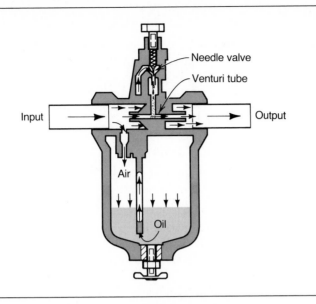

Figure 3-31

Air-line lubricator

relieving pressure can be preset by varying the spring tension with a setscrew. The poppet blocks the passage of the air as long as the air pressure is equal to or less than the spring tension. If the air-line pressure rises above the preset value, the spring tension is overcome, the poppet is forced upward, and part of the air in the line is expelled through the relief port.

The pressure-regulating valve is used to provide the proper pressure to the pneumatic system. It is usually mounted near the air-filter and lubricator units, as

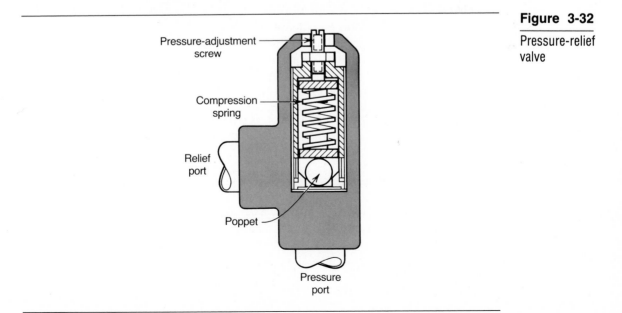

Figure 3-32

Pressure-relief valve

Figure 3-33

Location of
pressure-regulating
valve

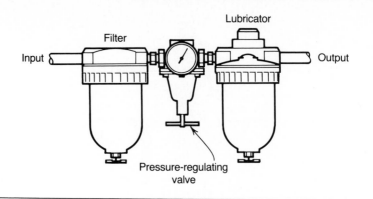

seen in Figure 3-33, and is usually the first control device the air flow passes through as it enters the system. This valve is adjustable, allowing the air pressure to be varied continuously from the maximum line pressure downward to zero pressure.

Direction-control valves used in pneumatic systems are similar in appearance to those used in hydraulic systems and have the same applications. As such, they are used to cause cylinders to extend and retract and control the direction of rotation of pneumatic motors. Unlike hydraulic systems, where the oil is returned to the reservoir, in a pneumatic system the spent air is usually exhausted into the atmosphere. A return line is not required.

Flow- or speed-control valves regulate the volume of air flowing through the system. These devices may be placed in the air line in series with either the input or output ports of a cylinder or motor to vary the speed of operation of these devices, as illustrated in Figure 3-34. In this circuit, a four-way direction-control valve is used to control the motion of a double-acting cylinder. The flow-control valve is connected between the cylinder and direction-control valve to control the outstroke speed of the cylinder. Air flow is controlled as it leaves the cylinder. A second flow-control valve is needed if the speed of both the outstroke and instroke of the piston is to be controlled, as seen in Figure 3-35.

Figure 3-34

Location of
flow-control valve,
outstroke control

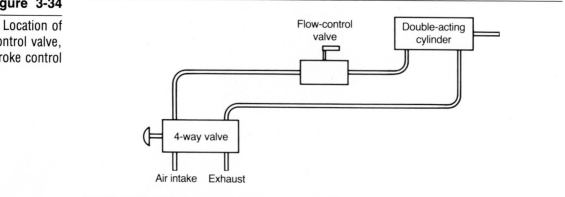

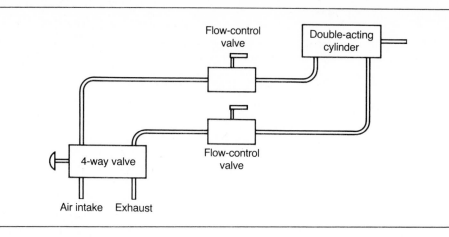

Figure 3-35

Location of
flow-control valves
for instroke and
outstroke control

3-8.5 Pneumatic motors

Air motors have advantages over hydraulic and electric types in some applications. Like hydraulic motors, these devices can be stalled indefinitely without damage, have a continuously variable speed, can be operated submerged in a liquid, and are explosion-proof. In addition they are clean. However, they do not develop the power or torque that hydraulic and electric motors produce.

The rotary vane is the most commonly used type of air motor. As illustrated in Figure 3-36, the motor consists of a slotted rotor mounted on a shaft which is offset in a cylinder. Rectangular vanes, which are free to slide, are mounted in the rotor slots. Centrifugal force causes the vanes to move outward when the rotor turns. Compressed air is applied at the intake port. Because of the unequal distribution of the cylinder area about the rotor, a higher pressure is exerted on the vanes located in the confined area of the cylinder, causing the rotor to turn. Rotor speed is proportional to the air pressure applied at the intake port and inversely proportional to the load connected to the shaft. The spent air exits the motor through the exhaust port.

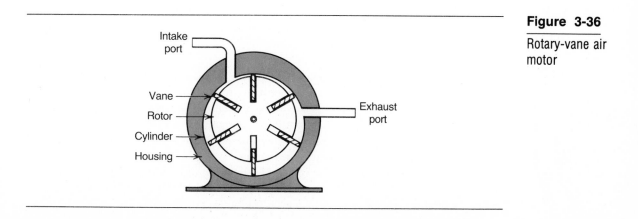

Figure 3-36

Rotary-vane air
motor

3-8.6 Pneumatic cylinders

These devices are used to convert pneumatic energy into linear mechanical energy. Although some are designed to be used with pressures as high as 250 psi, most operate at pressures of 90 psi or less. The basic air cylinder is constructed much like its hydraulic counterpart. Since it is usually used in lower pressure and lighter load applications, it is usually fabricated from lighter materials.

Like hydraulic cylinders, the pneumatic types may be either single- or double-acting. A three-way direction-control valve can be used to control the movement of a single-acting cylinder. Compressed air is applied at the input port to make the piston extend. The piston is returned to its resting position either by gravity acting on the load or by a spring.

Air is applied to either end of a double-acting cylinder to produce power in both directions of movement. A four-way valve is used to create back-and-forth motion. Flow-control valves connected to each air line allow for cylinder-speed control.

Because of the compressibility of air, a pneumatic cylinder normally cannot be stopped and held in an intermediate stationary position and maintain a reasonable degree of accuracy. When intermediate stopping is required, two three-way valves can be used. Each valve controls a single direction of movement of the piston, just as when connected to a single-acting cylinder. Compressed air applied to both ends of the piston causes it to remain stationary. A pressure regulator is used to reduce the pressure applied to the piston end to allow the pressure at both ends of the cylinder to be equal.

3-9 SCHEMATIC SYMBOLS AND DIAGRAMS

Schematic diagrams are often used to display fluid-power circuits and systems rather than pictoral representations, just as they are for electronic circuits and systems. Each fluid-power device is assigned a unique schematic symbol. Symbols for the more commonly used devices appear in Figure 3-37. For the most part, fluid-power devices are represented symbolically by circles, squares, rectangles, and diagonal shapes. Like many electronic symbols, the symbols used to represent fluid-power devices are often similar in appearance to either the function or fabrication of the device. Symbols are connected together to form fluid-power schematic diagrams.

The schematic diagram of a hydraulic power unit is shown in Figure 3-38. The unit illustrated is made up of an oil reservoir, strainer, pump, electric motor, pressure-relief valve, and pressure gage. Since the reservoir symbol forms part of the pressure-relief valve and strainer symbols, a separate reservoir symbol is not required. This simplifies the diagram as lines connecting the pressure-relief valve and strainer to a common reservoir are eliminated.

The diagram of a pneumatic system used to drive a double-acting air cylinder is seen in Figure 3-39. This system is composed of an air compressor, electric motor, filter, regulator, lubricator, four-way direction-control valve, two flow-control

Figure 3-37

Selected
fluid-power
schematic symbols

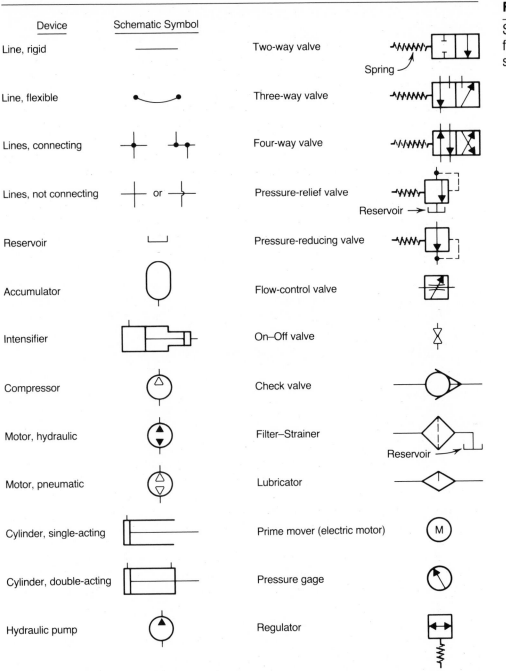

Device	Schematic Symbol
Line, rigid	
Line, flexible	
Lines, connecting	
Lines, not connecting	or
Reservoir	
Accumulator	
Intensifier	
Compressor	
Motor, hydraulic	
Motor, pneumatic	
Cylinder, single-acting	
Cylinder, double-acting	
Hydraulic pump	
Two-way valve	Spring
Three-way valve	
Four-way valve	
Pressure-relief valve	Reservoir
Pressure-reducing valve	
Flow-control valve	
On–Off valve	
Check valve	
Filter–Strainer	Reservoir
Lubricator	
Prime mover (electric motor)	M
Pressure gage	
Regulator	

Figure 3-38

Hydraulic power
unit

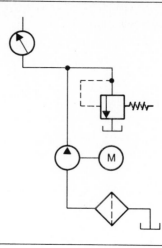

Figure 3-39

Pneumatic system
employing a
double-acting
cylinder

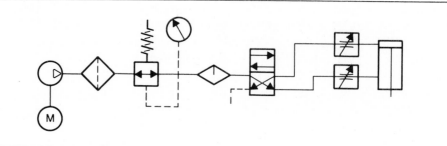

Figure 3-40

Hydraulic system
employing a
double-acting
cylinder

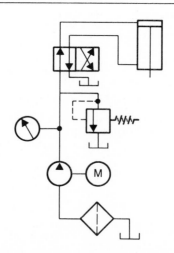

valves, and a double-acting cylinder. The electric motor is mechanically coupled to the air compressor. The filter, pressure regulator, and lubricator treat the air and maintain the proper line pressure. The four-way direction-control valve allows the piston in the cylinder housing to move back and forth. The speed of the back-and-forth motion is controlled by the two flow-control valves.

A double-acting cylinder is connected to the hydraulic system appearing in Figure 3-40. This system includes a reservoir, strainer, pump, electric motor, pressure gage, pressure-relief valve, four-way direction-control valve, and double-acting cylinder. Although three reservoir symbols are used, they all represent the same reservoir. Reservoir symbols are included as part of the strainer and pressure-relief symbols. A third reservoir symbol has been added in the fluid-return path of the four-way valve.

Summary

1. Fluid power is the technology dealing with the transmission and control of energy through a medium of pressurized liquid or a compressed gas.
2. Hydraulic power is often used in high-power, high-pressure, and high-precision applications.
3. Pneumatic power is usually used for low-power, low-pressure, and clean environment applications.
4. The application of hydraulic power is based on Pascal's law, which states that whenever a force is exerted on a confined liquid the resulting pressure is distributed equally in all directions.
5. The hydraulic pump causes fluid to flow in a hydraulic system. Three different types of pumps are employed. These are the gear, vane, and piston.
6. The reservoir serves as a storage and supply source for hydraulic fluid. After performing work, the oil is cooled here, purged of air, and contaminants are removed.
7. It is important that the hydraulic fluid be clean. Approximately 75 percent of all problems encountered in hydraulic systems are caused by dirty or contaminated oil.
8. Filters and strainers mounted on hydraulic lines entering and leaving the reservoir help keep the fluid clean.
9. Valves are used to control the direction, pressure, and flowrate of oil in a hydraulic system.
10. Two-way valves serve as gates to open and close hydraulic lines and to turn actuators on and off.
11. Three-way valves are used for selection, diversion, and direction-control applications.

12. Whenever a three-way valve is used as a selector, two different fluid sources are connected to two ports while the third port serves as an output.

13. When a three-way valve is utilized as a diverter, one port serves as an input while the other two ports are connected as outputs.

14. Three-way valves are used to control the movement of a single-acting cylinder.

15. The back-and-forth movement of a double-acting cylinder is controlled by a four-way valve.

16. Pressure-relief valves ensure that the pressure in a hydraulic system does not rise above a predetermined level.

17. Pressure-reducing valves may be used to supply fluid to a secondary hydraulic line.

18. Flow-control valves are used to control the volume of hydraulic fluid flowing in a line.

19. A check valve allows oil to flow in one direction but not the other.

20. Hydraulic motors produce more torque and horsepower from less bulk than any other type of motor. There are three types of hydraulic motors. They are the gear, vane, and piston.

21. Hydraulic cylinders convert hydraulic power into linear mechanical power. There are two types of cylinders— single- and double-acting.

22. The functions performed by the fluid in a hydraulic system include power transmission, lubrication, and sealant.

23. A heat exchanger may be used to either cool or heat hydraulic fluid.

24. An accumulator can be used to supplement the regular hydraulic oil supply.

25. An intensifier may be used to increase the pressure produced by the power unit.

26. The energy transmitted in a pneumatic system is in the form of compressed gas, usually air.

27. In a pneumatic system, air is compressed by an air compressor.

28. Pneumatic air-treatment devices are used to filter, lubricate, and regulate the compressed air.

29. The pneumatic system is composed of the same basic types of valves and actuators as are used in hydraulic systems.

30. Schematic symbols are used to represent graphically the devices connected together to form a fluid-power system.

Chapter Examination

1. Determine the force acting on a piston in a single-acting cylinder if the piston has a diameter of 3 inches and the oil pressure is 1700 pounds per square inch.

2. Calculate the pressure exerted on the fluid in the pipe appearing in Figure 3-3 if the piston diameter is 1.5 inches and the force remains the same.

3. List the three types of hydraulic pumps.

4. List the pressure ranges associated with each of the pumps identified in question 3.

5. List the major components forming a hydraulic cylinder.

6. List four techniques used to operate direction- and flow-control valves.

7. Discuss the basic principles of operation for the gear pump.

8. Describe the operating characteristics of a three-way valve.

9. What is the purpose of the baffle used in a hydraulic reservoir?

10. Compare the operating characteristics of hydraulic and pneumatic motors.

11. What is the purpose of the lubricator used in a hydraulic system?

12. What type of valve is used to produce the back-and-forth motion of a double-acting cylinder?

13. What type of valve allows oil to flow in one direction but not the other?

14. Draw the schematic symbols for the following devices:

a. check valve   **d.** double-acting cylinder
b. pressure-relief valve   **e.** hydraulic pump
c. four-way valve   **f.** reservoir

15. T F The reservoir should be large enough to hold at least two to four   —
times the pump-flow capacity in heavy-duty hydraulic systems.

16. T F Two-way valves are used to control the back-and-forth motion of a   —
double-acting cylinder.

17. T F Erratic hydraulic-cylinder motion is usually caused by air being   —
trapped in the system.

18. T F The pneumatic system forms a closed circuit. The air is recirculated back to the compressor after it has completed its work.   —

19. T F Pressure-relief valves are used to maintain the proper pressure in a   —
pneumatic system.

20. T F Flow-control valves are used to vary cylinder speed.   —

21. T F An intensifier is a reservoir of pressurized fluid used to supplement   —
the regular hydraulic-fluid supply.

22. T F Pneumatic motors develop more torque and horsepower than do   —
hydraulic motors of similar size.

23. T F The speed of fluid-power motors can be governed with a flow-control valve.   —

24. T F According to Pascal's law, whenever a pressure is exerted on a confined liquid, the pressure is distributed at right angles to the direction of liquid propagation.   —

25. T F The rotating air compressor is mainly used in medium- to high-pressure pneumatic systems.   —

26. T   F   Automobile-type or "O" rings are used to prevent oil from leaking out the rod end of a hydraulic cylinder.   __

27. T   F   Pressure-relief and check valves perform similar functions. They both prevent fluid from exceeding a maximum predetermined pressure.   __

28. T   F   Three-way direction-control valves may be used for selection and diversion purposes in addition to direction control.   __

29. T   F   Flow-control valves are sometimes referred to as speed-control valves.   __

30. T   F   A pressure-regulating valve is used to protect the air compressor and components in a pneumatic system.   __

CHAPTER
4

TRANSDUCERS AND SENSORS

OBJECTIVES

1. Define the terms *transducer* and *sensor*.

2. Define the term *temperature* and identify the three scales used in temperature measurement.

3. Identify four methods used to measure temperature in industrial applications and discuss the operating principles of each.

4. Define the terms *humidity* and *relative humidity*, identify two methods used to measure relative humidity, and describe the principles of operation of each.

5. Define the term *strain*, list three techniques used to measure this variable, and discuss the operating principles of each.

6. Define the term *acceleration*, identify the sensor used to measure this variable, and describe the sensor's principle of operation.

7. Define the term *linear displacement* and discuss the operating principles of the linear variable-differential transformer (LVDT).

8. Define the term *angular displacement*, identify three methods used to measure this variable, and discuss the operating principles of each.

9. Identify two methods used to measure shaft speed and discuss the principles of operation for each.

10. Define the term *fluid pressure*, list three methods used to measure this variable, and describe the operating characteristics of each.

11. Define the term *fluid flow measurement*, list five ways of measuring this variable, and discuss the operating principles of each.

12. List four ways of measuring liquid level and discuss the principles of operation for each.

13. Discuss the principles of light, list three methods of generating light, and identify the frequency and wavelength of visible light.

14. Describe the principles of operation of the light-emitting diode (LED).

15. Discuss the principles of operation of the gas, synthetic ruby, and semiconductor lasers.

16. Describe the operating principles of the photovoltaic cell.

17. Describe the operation of the photodiode and phototransistor.

18. Discuss the operating principles of the inductive and photoelectric proximity sensors.

4-1 INTRODUCTION

A transducer is a device which converts one form of energy into another. Batteries, lamps, motors, and internal combustion engines are examples of transducers encountered in day-to-day living. Almost all areas of electronics make extensive use of transducers. In industrial electronics much of the physical phenomena regulated by a control system has to be measured. Transducers convert phenomena such as temperature, humidity, strain, acceleration, linear and angular displacement, shaft speed, fluid pressure, fluid flow, and liquid level into a quantifiable variable. Transducers used for this purpose are referred to as sensors because they sense the presence of a phenomenon and change it into a variable that can be measured.

Sensors are employed in almost all types of manufacturing—including the food, chemical, petroleum, pharmaceutical, textile, paper, and metal-process manufacturing industries and in almost all discrete product manufacturing. In addition, they are used in water treatment, hazardous waste disposal, weather forecasting, sewage treatment, heating, ventilation, air conditioning, air filtration, telecommunications, and consumer products.

Approximately 900 companies manufacture over 300 different types of transducer and sensor devices. This chapter deals with some of the more common types of sensors and transducers used in industrial measurement applications.

4-2 TEMPERATURE MEASUREMENT

4-2.1 Introduction

Temperature is one of the most closely monitored variables in manufacturing and other industrial applications. It is a measurement of the relative warmness or coldness of a material. Three different scales are employed in temperature measurement—Fahrenheit (F), Celsius (C), and Kelvin (K). The Fahrenheit and Celsius (or centigrade) scales are both referenced against the freezing and boiling temperatures of water. At standard atmospheric pressure (sea level), the freezing and boiling temperatures of water are 32 and 212° F respectively and 0 and 100° C. The Kelvin scale uses absolute zero as a reference. Absolute zero is the temperature at which all molecular activity ceases. Zero degrees Kelvin (0° K) is approximately equal to $-273.16°$ C. The degree intervals used on the Kelvin scale are the same as those on the Celsius scale.

The thermometer is used to measure temperature. This instrument is composed of a temperature sensor and an indicator. A variety of sensors are available. Those commonly used in industrial applications include the thermocouple, the resistance temperature detector (RTD), the thermistor, and the solid-state temperature sensor.

4-2.2 The thermocouple

The thermocouple is a sensor that converts thermal energy into electrical energy. It operates on the Seebeck effect. This phenomenon is observed when two dissimilar metals are joined together at one end. When heat is applied to this junction, a voltage, proportional to the intensity of the heat, is developed across the opposite open-ended terminals, as shown in Figure 4-1. A voltmeter connected across the open terminals, either directly or through an amplifier, and calibrated in degrees, is used to quantify the temperature.

The metals composing the thermocouple should be linear over their entire operating range, be sensitive to small changes in temperature, and resist corrosion. Common thermocouple metal pairs and their temperature include nickel–chromium and constantan (−100 to 1000° C), iron and constantan (0 to 760° C), nickel–chromium and nickel (0 to 1370° C), platinum–rhodium and platinum (0 to 1750° C), and copper and constantan (−160 to 400° C).

A hollow tube or probe is generally used to protect the element and provide a method of mounting. Thermocouples are mounted in pipes, air ducts, and furnace walls, and on manufacturing machinery. Often the thermocouple is located a considerable distance from the indicating unit. Extension wires are used to connect the thermocouple element to the indicator. These should be made from the same material as the thermocouple element. The only exceptions are the wires used with the platinum–rhodium and platinum thermocouple element. Because of the high cost of those materials, copper and copper-nickel extension wires are used.

4-2.3 The resistance temperature detector

The resistance temperature detector (RTD) operates on the principle that the resistance of some pure and alloy metals is proportional to the temperature of their environment. These metals have a positive temperature coefficient. Whenever the ambient temperature rises the resistance of the metals increases.

The metal used in an RTD element should exhibit a linear resistance change over its entire temperature measuring range and be sensitive to small changes in temperature. Copper, nickel, nickel–iron, gold, silver, iron, tungsten, and platinum are all used as RTD elements. RTD resistance at 0° C ranges from 8.8 ohms for silver to approximately 2000 ohms for nickel-iron. The metal-film RTD is one of the newest RTDs. This device utilizes a platinum or metal-glass slurry film deposited

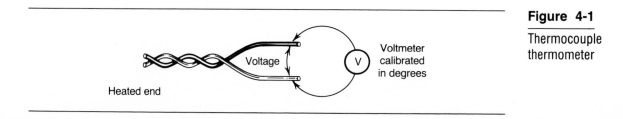

Figure 4-1

Thermocouple thermometer

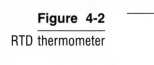

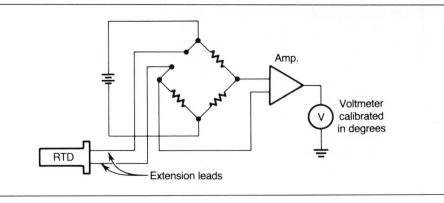

onto a ceramic substrate. Metal-film RTDs are smaller and more sensitive than the wire devices. However, at present they are not as stable as the wire types.

The RTD is usually connected in a bridge circuit for increased measurement sensitivity. As illustrated in Figure 4-2, the RTD forms one leg of the bridge. The bridge resistors are normally located in the indicating unit. Extension leads are used to connect the RTD to the bridge. The additional resistance introduced by the extension leads must be compensated for to eliminate measurement error. This can be accomplished by using a three-lead RTD, as seen in Figure 4-3. Conductors X and Z (made of the same material as the RTD) have identical length and the same cross-sectional area. These conductors are connected to opposite ends of the leg and cancel one another. The resistance of conductor Y does not affect bridge balance as it is a sensing wire and is connected to the amplifier. The temperature indicator is a voltmeter calibrated in degrees and connected across the amplifier output terminals.

4-2.4 The thermistor

In its basic form, a thermistor is a temperature-sensitive resistor. The device is fabricated from oxides of cobalt, copper, iron, nickel, manganese, tantanium, or from an alloy containing those metals. The materials used have a negative temperature

Figure 4-3

Three-wire RTD

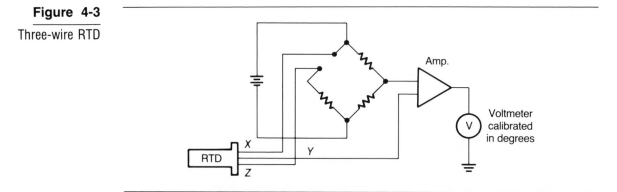

Figure 4-4

Thermistor
structures

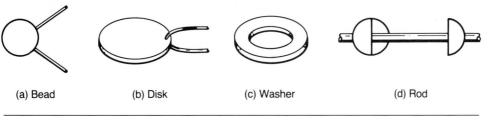

(a) Bead   (b) Disk   (c) Washer   (d) Rod

coefficient. An increase in ambient temperature causes the resistance of the metal to decrease, and vice versa. The type of metal chosen for a particular application is determined by the characteristics required of the thermistor, which include nominal resistance, linearity, sensitivity, and temperature range. Thermistors are manufactured in numerous shapes—including beads, disks, washers, glass rods, and glass probes, as illustrated in Figure 4-4.

Nominal resistances range from a low of less than one ohm to a high of several megaohms, with most having resistances in the hundreds or thousands of ohms. Thermistors are highly sensitive. A very small change in temperature creates a considerable change in resistance. Because of their small mass, they rapidly respond to changes in temperature.

The thermistor is often connected as a leg in a bridge circuit, as shown in Figure 4-5. Since the resistance of the thermistor is high, the resistance of the extension leads connecting it to the bridge usually does not create any appreciable measuring error.

4-2.5 Solid-state temperature sensors

With the increased applications of microminiature electronics in manufacturing equipment, industrial processes, automobiles, and consumer appliances, the need for smaller temperature sensors has arisen. Two types of solid state sensors have been developed to fill this demand. They include the thin-film detector (TFD) and the thin-film RTD.

Figure 4-5

Thermistor
thermometer

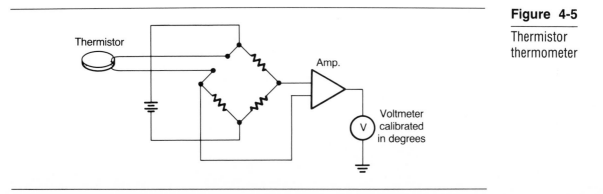

Thermistor

Amp.

Voltmeter
calibrated
in degrees

The TFD consists of a temperature-sensitive platinum layer deposited on a ceramic substrate which is encapsulated in glass. Leads made of platinum wire are bonded to the layer. Available models will measure temperatures ranging from −200 to 500° C.

The thin-film RTD is made from a platinum or metal−glass slurry film which is deposited or screened onto a ceramic substrate. The desired resistance of the RTD is obtained by laser-trimming the deposited material. Thin-film RTDs are available with 100, 200, 400, 500, 1000, and 2000 ohms of resistance. These sensors are capable of measuring temperatures ranging from −50 to 850° C.

The thermocouple is noted for its versatility; the RTD is characterized by its stability; the thermistor is known for its sensitivity; and solid-state temperature sensors are characterized by their small size and sensitivity.

4-3 RELATIVE-HUMIDITY MEASUREMENT

4-3.1 Introduction

Humidity is the term used to quantify the amount of water vapor, or moisture, present in the atmosphere. The maximum amount of moisture the air can contain is referred to as saturation and is proportional to air temperature. For example, five pounds of water vapor saturates 1000 pounds of air at 40° F whereas it takes 41 pounds of vapor to saturate 1000 pounds of air at 100° F.

Humidity can be expressed in either absolute or relative terms. Absolute humidity identifies the weight of water vapor contained in a volume of air. Relative humidity is the ratio of the actual amount of water vapor in the air at a particular temperature to the maximum amount that it can hold. Relative humidity is usually expressed as a percentage.

Closely associated with humidity is dew point. Whenever air containing water vapor is cooled and drops below the temperature at which it is saturated, the excessive water vapor condenses and forms droplets of water called dew. The dew point is the temperature at which dew begins to form from air containing a particular quantity of water vapor. Frost forms when the dew point is below the freezing temperature of water.

Humidity is important not only for human comfort and work efficiency but also for the environment in many manufacturing industries. The atmospheric moisture content in industries such as semiconductor manufacturing, paper, and textiles is important in the production of these goods. Traditionally, two types of instruments have been used to measure relative humidity. They include the psychrometer and the hygrometer.

4-3.2 The psychrometer

The psychrometer is made up of two RTDs (called bulbs), a water reservoir, a fan, and a housing, as illustrated in Figure 4-6. One of the RTDs is used to measure the temperature of the surrounding air and is called the dry bulb. The other measures

Figure 4-6

Psychrometer

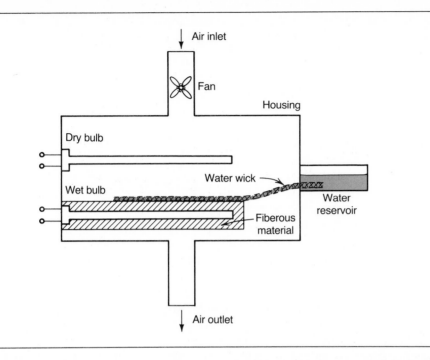

the temperature of the air surrounding a saturable porous fiber material and is called the wet bulb. The fan is used to force air over the fibrous material, causing the water to evaporate and lowering the air temperature measured by the wet bulb. The amount of the temperature decrease depends upon the rate of evaporation, which is dependent upon the humidity of the air forced across the bulbs. When the air is dry the evaporation rate is high, and the temperature measured by the wet bulb is much cooler than that measured by the dry bulb. When the air is moist the rate of evaporation is low, and the temperature measured by the wet bulb is only slightly less than that measured by the dry bulb. The dry-bulb temperature remains constant regardless of the air's moisture content. After the wet- and dry-bulb temperatures have been measured, the difference between the two is calculated and the relative humidity obtained from a psychromatic table similar to Table 4-1. Notice the inverse relationship between the difference temperature and the relative humidity. The greater the difference temperature, the smaller the percentage of relative humidity. As seen in Table 4-1, the relative humidity is 27 percent if the temperature is 60° F and the difference temperature is 15° F.

4-3.3 The hygrometer

Although highly accurate, the psychrometer does not provide a continuous relative-humidity indication. Whenever a continuous measurement is required and less accuracy is acceptable, a hygrometer is employed. There are several different types of hygrometers. Two of the more commonly used types are the fiber and resistive. The fiber hygrometer operates on the principle that certain organic fibers change

Table 4-1

SAMPLE PSYCHROMATIC TABLE

		DRY BULB TEMPERATURE (°F)							
		30	40	50	60	70	80	90	
DIFFERENCE TEMPERATURE (°F)	1	87	89	90	91	92	92	93	PERCENTAGE OF RELATIVE HUMIDITY
	5	55	58	60	63	65	66	67	
	10		41	43	44	45	47	48	
	15		18	24	27	29	30	31	
	20			9	13	16	18	19	
	25				5	9	12	14	

length when they absorb moisture. These materials lengthen in damp air and compress when the air is dry. A mechanical pointer is attached to the fiber, as illustrated in Figure 4-7. As the moisture content of the air changes the length of the fiber, the pointer is moved across the indicator. A single blond human hair is frequently used as the fiber.

A resistive hygrometer utilizes an element whose resistivity varies with the moisture content of the air. The element consists of two metal-foil electrodes separated from each other by a dielectric. The element is coated with a solution of lithium chloride, which provides continuity between the electrodes. Lithium chloride can absorb moisture very readily. As it absorbs moisture, its resistance decreases. When a voltage is applied to the electrodes, the current flowing through the element is inversely proportional to the resistance of the element. An ammeter calibrated in percentage of relative humidity is used as an indicator.

The fiber hygrometer is the simpler and less expensive of the two instruments. Its range is limited, however. Typically, this type of hygrometer is limited to relative-humidity measurements extending from approximately 15 to 90 percent. The resistive hygrometer will measure the full range of relative humidity.

Another and more recently developed humidity detector is the optical hygrome-

Figure 4-7

Hygrometer

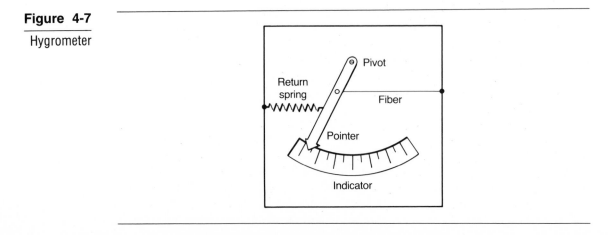

ter. Optical hygrometers measure the attenuation of light shined through the atmosphere for a specified length. The relative-humidity measurement is proportional to the amount of attenuation. This measurement technique utilizes the principle that light rays are absorbed as they travel through water vapor. The absorption is most significant for ultraviolet and infrared light. Because of this, two different types of optical hygrometers are available. One type utilizes ultraviolet rays while the other employs infrared rays.

Optical hygrometers provide more accurate relative-humidity measurements than do the others, especially at high humidities. They can be used in almost any environment, including high altitudes and extreme temperatures.

4-4 STRAIN MEASUREMENT

4-4.1 Introduction

Stress is created by a force acting on an object. Strain is the resulting change in the shape of the object. Stress causes the object to distort, or strain, by stretching or compressing it—depending upon the direction of the force.

The strain gage is used to measure strain. Three commonly used types are the bonded-wire, semiconductor, and thin-film. All can be used for sensing stress caused by torque, pressure, weight, and displacement acting on an object. Strain gages can be used to measure forces as small as a few ounces or as large as several tons.

4-4.2 The bonded-wire strain gage

The bonded-wire strain gage is by far the one most commonly used. This gage is formed from wire looped back and forth and sandwiched between thin sheets of paper or other insulating material, as shown in Figure 4-8(a). It may also be constructed from a thin metallic foil mounted on a dielectric backing, much like a printed circuit, as seen in Figure 4-8(b). Although both are bonded-wire gages and operate on the same principle, the latter is sometimes referred to as a bonded-foil or metal-film strain gage. The bonded-wire strain gage utilizes the principle that the resistance of wire or a conductor foil changes as its length and cross-sectional area are varied. Approximately one inch square or less in size, the strain gage is securely cemented (bonded) to the surface of the object whose strain is to be measured. This allows the strain gage to stretch or compress the same amount as the object being subjected to stress. If the object stretches, the wire or foil stretches the same amount, causing its length to increase and its cross-sectional area to decrease. This increases the resistance of the strain gage. The wire or foil becomes shorter and its cross-sectional area increases when the material undergoing stress compresses. This causes the wire resistance to decrease. In this fashion the strain gage converts displacement or size changes into resistance changes.

The strain gage should provide a linear resistance over its entire strain range and exhibit a relatively high change in resistance when stress is applied. This latter

Figure 4-8

Bonded-wire strain gage

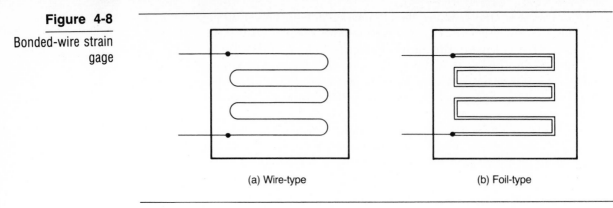

(a) Wire-type   (b) Foil-type

characteristic, called the gage factor (GF), determines the sensitivity of the strain gage. The gage factor of constantan and nickel–chromium, two common strain-gage materials, is two. To increase the measurement sensitivity, the strain gage is normally connected to a balanced Wheatstone bridge with the gage forming one leg of the bridge circuit. The output of the bridge is connected to a voltmeter calibrated in micro-strain units.

4-4.3 The semiconductor strain gage

This is a newer type of strain sensor that replaces the bonded-wire types for some applications. This sensor operates on the principle of the piezoresistive effect present in semiconductor materials such as germanium and silicon. Piezoresistive devices are those whose resistance changes when subjected to stress. They can be fabricated to produce either increases or decreases in resistance when stressed. This type of gage is much more sensitive than the bonded-wire types, with gage factors in the 1000 to 10,000 range being typical.

Although the semiconductor strain gage is extremely sensitive, its resistance change when stressed is nonlinear and its output is temperature dependent. These features have limited its widespread use as a strain sensor.

Figure 4-9

Accelerometer

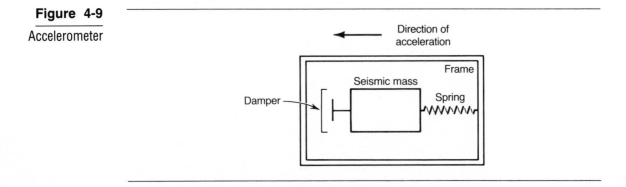

4-4.4 The thin-film strain gage

This is the newest type of strain gage. It has the sensitivity of the semiconductor type but has a more linear output. This is the type of strain sensor used in many robotic and materials-handling-equipment applications. Thin-film strain gages are available that measure forces ranging from 1 ounce to 100 pounds.

4-5 ACCELERATION AND VIBRATION MEASUREMENT

4-5.1 Introduction

Acceleration is the rate at which the velocity of an object increases per unit of time. The rate at which velocity decreases is called negative acceleration or deceleration. Vibration is the back-and-forth acceleration and deceleration of an object.

4-5.2 The accelerometer

Acceleration and vibration are both measured with an accelerometer. This device operates on the principle that an object at rest tends to remain at rest; if it is in motion it tends to maintain movement in a straight line, because of its inertia.

The accelerometer is composed of a small material bulk, called a seismic mass, that moves along an axis against a restraining spring attached to a frame, as illustrated in Figure 4-9. A mechanical damper is attached to the opposite end of the seismic mass to prevent oscillations from occuring. The accelerometer is attached to the object whose acceleration or vibration is being measured. When the object is accelerated the accelerometer is accelerated at the same rate, causing the seismic mass to move in the opposite direction. The seismic mass returns to its original position when the accelerating force is removed. The distance the seismic mass travels

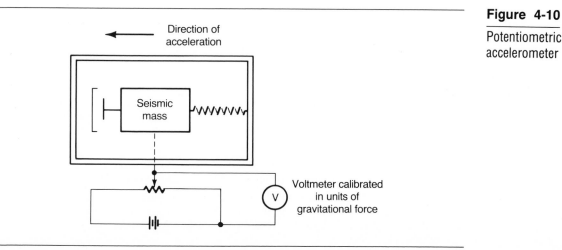

Figure 4-10

Potentiometric accelerometer

Figure 4-11

Piezoelectric
accelerometer

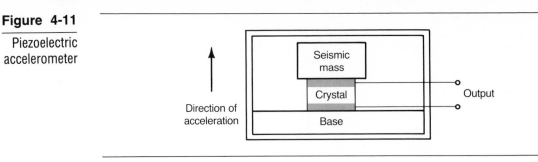

is proportional to the force of acceleration. The accelerometer is a transducer which converts acceleration into linear displacement.

Several different methods are employed to convert the displacement into an equivalent electrical signal that can be quantified. One method utilizes a potentiometer whose shaft is coupled to the seismic mass as shown in Figure 4-10. The movement of the mass causes a resistance change creating a change in output voltage. A voltmeter calibrated in units of gravitational force is used as the indicator.

A piezoelectric crystal is used as the sensing element in the accelerometer appearing in Figure 4-11. A piezoelectric crystal is one that generates an EMF when a pressure is applied across opposite sides. Quartz is commonly used as the piezoelectric crystal. The crystal is placed between the seismic mass and base when used in an accelerometer. When an accelerating force is applied, the seismic mass moves toward the crystal causing it to compress slightly. This creates a small voltage that is proportional to the magnitude of the accelerating force. The voltage is amplified, and a voltmeter calibrated in gravity units is utilized as an indicator.

The accelerometer appearing in Figure 4-12 employs a strain gage. The seismic mass is attached to the frame with an elastic band on which a strain gage is bonded. When an accelerating force is applied, the reactive force of the seismic mass causes the elastic band to stretch, changing the resistance of the strain gage.

4-6 LINEAR-DISPLACEMENT MEASUREMENT

4-6.1 Introduction

Displacement is a very important variable in manufacturing applications. The exact location of a machine cutting tool with respect to a reference must be known so that operations can be performed on a workpiece. There are two types of displacement—linear and angular. Angular-displacement measurement is discussed in the next section. Linear displacement refers to the location of an object, such as a workpiece or cutting tool, in a straight line from a reference point and is measured in units of length. The linear variable-differential transformer (LVDT) is the sensor most often used to quantify this type of movement.

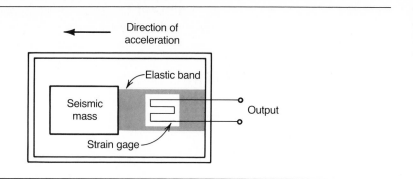

Figure 4-12

Strain-gage
accelerometer

4-6.2 The LVDT

The linear variable-differential transformer is a specially designed transformer with one primary winding and two secondary windings. The secondary windings are linked together by a movable core as shown in Figure 4-13. The core is attached to a shaft coupled to the object whose movement is to be measured. When an AC voltage is applied to the primary winding, voltages are induced across the two secondary windings. The secondary windings are connected in series opposition so that they are 180 degrees out of phase. When the core is centered, the two secondary voltages are equal in amplitude and the output voltage is zero. As the core is moved by the motion of the shaft, the mutual inductance of each secondary winding changes. The voltage induced across one secondary winding increases while the other decreases—creating a differential, or difference, voltage. The polarity of this potential may be either positive or negative, depending upon the direction of the movement of the shaft and core. The amplitude of the differential voltage is linearly related to the amount of core displacement.

LVDT sensitivity is measured in millivolts per 0.001 inch of core displacement ($mV_{out}/0.001$ in) and is somewhat dependent upon input frequency, so the frequency of the primary voltage is often specified whenever a sensitivity value is identified. Primary-voltage frequencies ranging from 50 to 15,000 hertz are used. LVDTs are very sensitive; some models can detect shaft movement of a few hun-

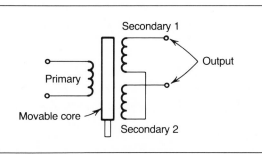

Figure 4-13

Linear variable-differential transformer (LVDT)

dred microinches. A voltmeter (calibrated in inches or centimeters) connected across the secondary windings, either directly or through an amplifier, may be used as an indicator.

4-7 ANGULAR DISPLACEMENT MEASUREMENT

4-7.1 Introduction

Angular displacement refers to the angular position of an object with respect to a reference point. This type of measurement is important in three-dimension machining operations and in positioning the manipulator of a robot. Three methods are employed to measure angular displacement. They include the potentiometer, the angular digital encoder, and the synchro.

4-7.2 The potentiometric angular-displacement sensor

This is the simplest and least accurate method of measuring angular displacement. It is sufficiently accurate for many applications, however. The shaft or rotating object whose angular displacement is to be measured is mechanically coupled to the shaft of a linear potentiometer, as illustrated in Figure 4-14. Movement of the shaft causes the resistance to increase or decrease depending upon the direction of rotation. An ohmmeter calibrated in degrees or radians can be used as an indicator, or a voltage source can be connected across the potentiometer and a voltmeter utilized as an indicator.

4-7.3 The angular digital encoder

A more accurate method of measuring angular displacement is by the use of an angular digital encoder. This device employs an encoder disk that is mechanically coupled to the shaft whose angular displacement is to be measured. The rotating disk generates a binary word whose identity, or weight, is proportional to the angular position of the disk. Two encoder techniques are employed. In one, the encoder disk is constructed from a nonconductive medium and is divided into circular tracks. A metal-oxide layer is deposited at intervals on the tracks, as shown in Figure 4-15. The inside track is completely metallized and is used to complete the cir-

Figure 4-14

Potentiometric angular-displacement sensor

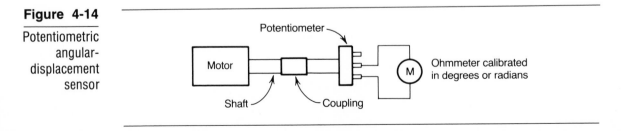

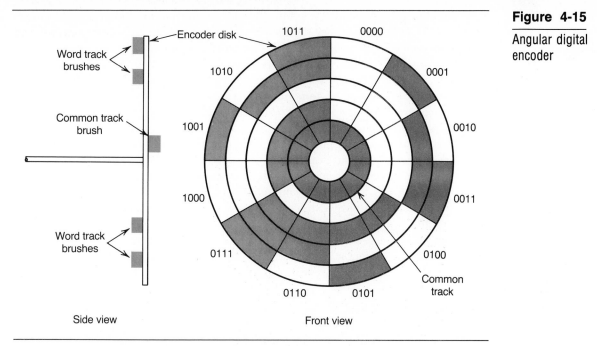

Figure 4-15

Angular digital encoder

cuit. Brushes are used for detection with each track having its own brush. When a potential is connected across the brushes, a pulse representing a binary one is detected everytime a track brush encounters a metallic area on a track. No pulses appear across the output while the track brushes are in contact with the nonmetallic areas. The length of the binary word is determined by the number of tracks. A separate track is required for each bit in the binary word.

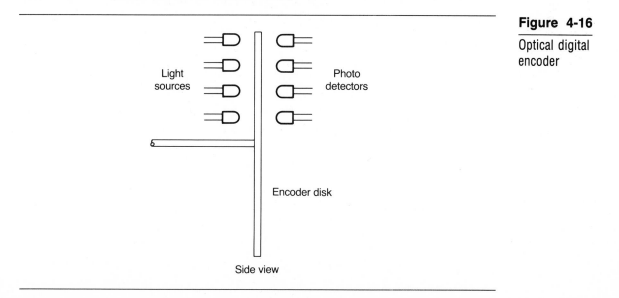

Figure 4-16

Optical digital encoder

The other type of encoder employs photoelectric principles to generate a binary code and is called an optical digital encoder. Two different techniques are employed to generate the binary word that identifies shaft location. One uses an opaque disk with transparent slots appearing in the tracks, the other a metal disk with holes are drilled in the tracks at predetermined intervals. Except for the type of disk used, the method of generating binary words is identical for both types of optical encoders. Each track has its own photodetector. A binary word representing the disk position is generated as the disk rotates past a light source and detectors, as seen in Figure 4-16. A binary one is detected whenever light shines through the disk and is received by the detector.

Angular digital encoders are available that generate a variety of different binary codes. Although the 8421 binary code is utilized in Figure 4-15, the Gray-code encoder is popular since only a single bit change occurs for adjacent binary numbers. The resolution and accuracy of the encoder depends upon the number of tracks and corresponding binary-word size. Disks with 8 to 16 tracks, generating 8- to 16-bit binary words, are common. A 12-track disk has a resolution of 1 in 4096 (2^{12}) and an accuracy of 0.0879^0/bit (360^0/4096).

4-7.4 Synchros

A synchro is a rotary transducer that changes angular displacement into an AC voltage or an AC voltage into angular displacement. As such, a synchro can be used to measure angular displacement, or it can be used to position an angular mechanical load from a remote position.

A synchro consists of a pair of devices similar in appearance to small motors connected as shown in Figure 4-17. One device is called a transmitter, the other a receiver. The transmitter and receiver are connected by a three-conductor bus. Both units are connected to an AC line source. Each unit has three stator windings, displaced 120 degrees, and a rotor. External connections to the stator windings are called S_1, S_2, and S_3, while the rotor terminals are identified as R_1 and R_2.

The transmitter operates as a variable transformer. The rotor serves as the primary winding and is connected to the AC line voltage. AC line current flowing through the rotor induces a voltage across the stator windings, with an amplitude that depends upon the angular position of the rotor. Maximum voltage is induced across a stator winding whenever the rotor and a stator winding are aligned. No

Figure 4-17

Synchro
connections

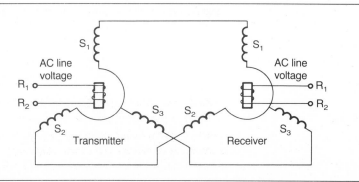

voltage is developed whenever the rotor is perpendicular to a stator. The voltages induced across the stator windings of the transmitter cause currents to flow through the corresponding stator windings in the receiver and establish magnetic fields about these stator windings. At the same time a magnetic field is established about the rotor in the receiver because of the line current flowing through it. The interaction of these two fields creates a torque, which causes the rotor to turn. The receiver rotor follows the transmitter rotor as long as the transmitter rotor is turned. When the transmitter rotor is stopped, the receiver rotor stops in the same position.

The action just described allows the synchro to be used to measure angular displacement. The shaft whose angular measurement is to be measured is coupled to the rotor of the transmitter. A calibrated dial is coupled to the rotor of the receiver. Whenever the transmitter rotor turns, the receiver rotor follows in synchronization even though there is no mechanical linkage between the rotors.

As previously mentioned, a synchro may also be used to position angular loads from a remote position. Although mechanical coupling through a shaft can be utilized in some applications, the distance between the control point and controlled device and the difficulty of transmitting mechanical energy around corners prohibits the use of mechanical coupling in many situations.

4-8 SHAFT-SPEED MEASUREMENT

4-8.1 Introduction

In many manufacturing and other industrial applications it is necessary to measure the speed of a rotating shaft. The tachometer is the sensor used to quantify this variable. A tachometer is a small generator mechanically coupled to the shaft whose speed is to be measured. Two different types of generators are employed. The amplitude of the tachometer voltage represents the rotational speed in one, while the frequency of tachometer voltage represents the speed in the other.

4-8.2 The magnitude tachometer

This type of tachometer is essentially a constant-field generator. The voltage developed by the tachometer is proportional to the speed of the rotating shaft. It is essential that the field strength remain constant. Variations in field intensity create voltage variations that are independent of shaft speed. Either a permanent magnet or saturated electromagnet may be used to provide the field. A voltmeter, calibrated in revolutions per minute (rpm), can be connected across the tachometer output terminals when a visual indication is required.

4-8.3 The frequency tachometer

Two different methods are used when the frequency of the tachometer voltage is used to quantify shaft speed. One method uses a conventional synchronous alternator. Permanent magnets mounted on the rotor ensure constant field intensity. The other method uses an alternator that has a permanent magnet stator with a pickup coil wound around it and a ferromagnetic toothed rotor. A voltage pulse is induced across the pickup coil every time a tooth passes by the stator.

Frequency tachometers are more linear and accurate than the magnitude type. Moreover, they are not affected by such external factors as temperature variations, shaft vibration, and output loading. However, they are more expensive than amplitude tachometers and require a more complex indicating unit.

4-9 FLUID-PRESSURE MEASUREMENT

4-9.1 Introduction

Many machines used in manufacturing employ hydraulic- or pneumatic-powered devices. As these devices are powered by a pressurized liquid or compressed air, it is important that the pressure of these media be measured for monitoring and control purposes. Furthermore, the pressure of liquids and gases present in the manufacture of petroleum, chemical, pharmaceutical, food, and other products must be continually monitored and controlled. Pressures encountered in manufacturing range from a low of almost a pure vacuum to a high of several thousand pounds per square inch.

Fluid pressure is the force per unit area produced by a gas or liquid acting on the surface of an object. Depending upon the reference, three different types of pressure measurements may be obtained. They include absolute, gage, and differential. A vacuum is used as the reference for measuring absolute pressure; atmospheric pressure ($14.696 \ \mathrm{lb/in^2}$ at sea level) is used to measure gage pressure; and any arbitrary pressure can be used in differential-pressure measurements. The pascal (Pa) is the S.I. unit of measurement for fluid pressure. One pascal is equal to one newton per square meter and represents a very small pressure unit. Pressure is usually recorded in kilopascal units. In the English system of measurement, fluid pressure is measured in pounds per square inch (psi). To differentiate among the various pressure measures, absolute pressure is often indicated as psia, gage pressure as psig, and differential pressure as psid.

Many techniques are used to measure fluid pressure. In most, pressure is converted into displacement and a sensor used to change displacement into an electrical signal. Displacement transducers used in fluid-pressure measurement include the bellows, the Bourdon tube, and the diaphragm. LVDTs, strain gages, potentiometers, capacitors, piezoresistive devices, and piezoelectric devices may all be used to convert displacement into an equivalent electrical signal.

4-9.2 The bellows

The bellows is a corrugated hollow cylindrical tube—made from brass, bronze, stainless steel, or an alloy of copper–zinc brass—that can expand or contract depending upon the pressure of the liquid or gas, as illustrated in Figure 4-18. The device converts pressure into shaft displacement. If the bellows is used as a me-

Figure 4-18

Bellows

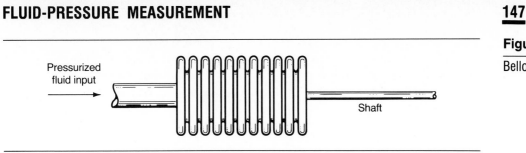

chanical pressure gage, a pointer is attached to the shaft, as seen in Figure 4-19(a). When an electrical output is required, a potentiometer and a voltage source can be utilized, as shown in Figure 4-19(b). The bellows shaft is mechanically coupled to the shaft of the potentiometer. The output voltage is proportional to the resistance of the potentiometer, which in turn is proportional to the bellows displacement and pressure. A more accurate and sensitive pressure measurement can be made if an LVDT is employed, as shown in Figure 4-19(c). The bellows shaft is coupled to the core of the LVDT. As the pressure changes, the core is displaced, causing a differential voltage to appear across the secondary windings.

The bellows may be used to measure pressures ranging from a vacuum to approximately 4000 pounds. It is capable of withstanding excessive pressure without damage and has excellent linearity. The major disadvantages of the bellows are that it has a poor resistance to vibration and a shorter lift span than the other types.

4-9.3 The Bourdon tube

This is a simple, popular method of measuring pressure. Several different forms of the device are available. All operate on the same principle, however. The Bourdon tube consists of a spiral flexible tube that is sealed at one end, as seen in Figure 4-20. The tube unwinds at the sealed end when the opposite end is connected to a pressurized fluid line. A mechanical pointer attached to the sealed end moves as the tube unwinds, allowing the device to be used as a pressure gage. When an electrical output is required, the sealed end can be coupled to the shaft of a potentiometer connected to form one leg of a bridge circuit.

4-9.4 Diaphragm pressure sensors

Pressure applied to the sensor causes a diaphragm to flex in these pressure sensors. A variety of different designs have been developed. The capacitance pressure sensor is one of the more popular of these. This device is composed of two metallic plates, which form a capacitor and a center diaphragm as illustrated in Figure 4-21. Pressure applied to the sensor causes the diaphragm to flex, changing the capacity of the capacitor. The capacitor is connected to, and forms one leg of, an AC bridge circuit. The bridge output voltage is proportional to the applied pressure.

With the recent emphasis in industrial automation a need for smaller, more accurate, and more sensitive pressure measurements has developed. Several solid-state pressure sensors have been developed to meet this demand. One of the more popular of these devices is the piezoresistor pressure sensor. A diaphragm and one

Figure 4-19

Bellows
pressure-indicating
methods

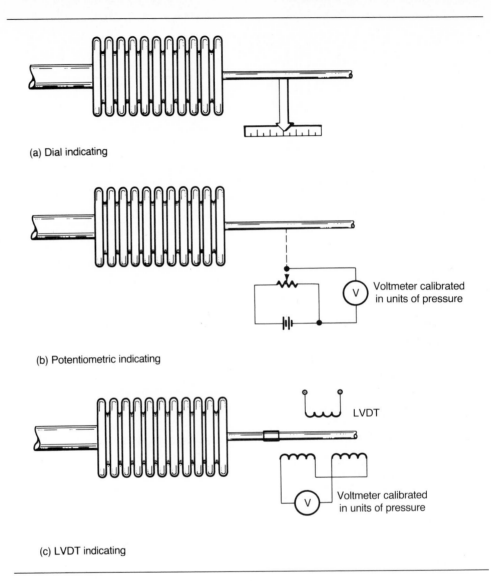

(a) Dial indicating

(b) Potentiometric indicating

(c) LVDT indicating

or more piezoresistor strain gages are all fabricated from a single silicon semi-conductor substrate. The strain gages form one or more legs of a Wheatstone bridge. The output voltage of the bridge is proportional to the pressure exerted on the diaphragm. Some models include the entire bridge, circuits for nonlinearity and temperature compensation, and an analog-to-digital (A/D) converter.

Solid-state pressure sensors are characterized by their small size, accuracy, ability to measure a wide range of pressures, and sensitivity as exhibited by a relatively high bridge output voltage. The major disadvantage of these sensors is their high cost. This is expected to decrease as the manufacturing techniques mature.

Figure 4-20

Bourdon tube

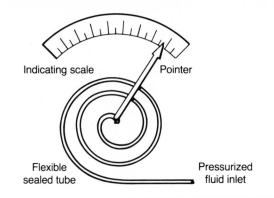

4-10 FLUID-FLOW MEASUREMENT

4-10.1 Introduction

In many manufacturing industries, the measurement of the flow of fluids such as chemicals, petroleum products, steam, water, and gas is exceeded only by temperature in importance. Fluid flow is the velocity at which a fluid flows through a transmission line such as a pipe, tube, hose, or duct. It can be measured either as a volume velocity whose units of measurement are cubic centimeters per second (cm^3/s), cubic meters per second (m^3/s), and gallons per minute (gal/min) or as a

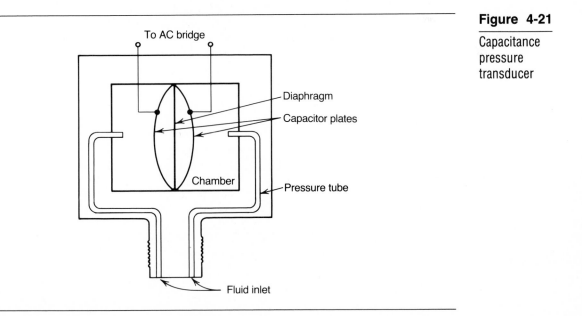

Figure 4-21

Capacitance pressure transducer

linear velocity, in which case the units of measurement include centimeters per second (cm/s), meters per second (m/s), inches per second (in/s), and feet per second (ft/s). Fluid flow is measured with a flowmeter.

Numerous flowmeters are available for measuring fluid flow through a transmission medium. The type used for a particular application depends upon such variables as fluid type, pressure, viscosity, density, cleanliness, and corrosiveness. Some of the more commonly used devices are discussed in the following sections.

4-10.2 The Venturi tube

The Venturi tube is a differential flowmeter. Differential flowmeters operate on the principle that a pressure differential is developed across the ends of a transmission line when a fluid is forced through the line. The velocity of the liquid depends upon the magnitude of the resulting differential pressure. The diagram of a Venturi-tube flowmeter appears in Figure 4-22. The meter is fabricated from a section of transmission line whose diameter is reduced in the center to restrict the flow of the fluid. Shunts are provided to allow for the measurement of pressure at both the full and restricted cross-sectional areas. Pressure sensors mounted on the shunts

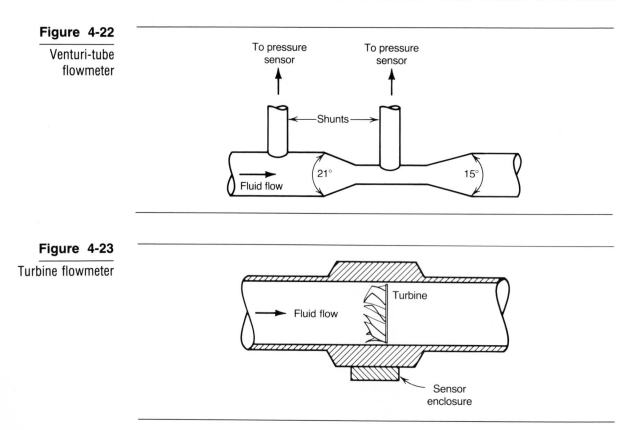

Figure 4-22

Venturi-tube flowmeter

Figure 4-23

Turbine flowmeter

measure pressure at both locations. Fluid velocity is represented by the pressure difference between the shunts. The greater the differential pressure, the less the flow rate.

4-10.3 The turbine flowmeter

This flowmeter consists of a multiblade rotor mounted in a section of transmission line, as seen in Figure 4-23. The rotor revolves as the fluid passes through the transmission line section. The speed of the rotor is proportional to the velocity of the fluid flowing through the line. Velocity can be determined electrically by counting the pulses produced by a magnetic pickup coil or by a photoelectric cell, or mechanically through gears coupled to a tachometer.

4-10.4 The thermal flowmeter

This is one of several mass flowmeters. Mass flowmeters measure mass rate of flow rather than volumetric flow. Thermal flowmeters operate independently of fluid pressure, viscosity, and density. They employ a heating element that is isolated from the fluid's flow path. The moving fluid conducts heat away from the heating element. The loss of heat is proportional to the flow rate. An RTD may be used as a leg in a Wheatstone bridge to convert the heat measurement into an electrical signal.

Each of the flowmeters just discussed has its own measurement characteristics. The Venturi tube is a very accurate fluid-flow sensor that can be used to measure the flow of both clean and dirty liquids as well as gas. Its insertion into the line does not affect the normal flow of the fluid, as approximately 95 percent of the pressure drops are returned to the transmission line. The turbine flowmeter can be used to measure the flow of clean liquids and gas and is highly accurate. Thermal flowmeters can be used to measure the flow rate of most liquids and gas. Thermal flowmeters are noted for their simplicity, sensitivity, and accuracy.

4-11 LIQUID-LEVEL MEASUREMENT

4-11.1 Introduction

The level or volume of a liquid must often be known. This might be the level of fuel oil in a tank, the amount of water in a reservoir, or the quantity of a chemical present in a container used in a particular manufacturing application. Several different methods are available to accomplish this measurement. Four commonly used methods include the float-operated rheostat, the differential-pressure transducer, the capacitive liquid-level sensor, and the sonic liquid-level sensor.

4-11.2 The float-operated rheostat liquid-level indicator

This is a simple and common method of monitoring liquid level and is the technique used to measure the quantity of fuel in the gasoline tank of a car. As shown in Figure 4-24, a ball float is attached to a rheostat. As the float rises or falls with a change in liquid level, the rheostat resistances decrease or increase accordingly, causing the current to increase or decrease. A milliammeter calibrated in tank-level or volume units is used as an indicator. Although simple, this type of liquid level indicator is not highly accurate and is vulnerable to breakdowns due to the moving parts forming the mechanism.

4-11.3 The differential-pressure transducer liquid-level indicator

A pressure transducer may be used to determine liquid level, as seen in Figure 4-25. This measurement technique is based on the principle that the weight of a liquid is proportional to the pressure exerted on the sensor. Any one of several pressure sensors may be used. The type of indicating unit employed is determined by the type of sensor utilized.

A similar but more accurate measuring technique is the hydrostatic-tank gaging (HTG) system. Instead of using a single pressure sensor, pressure sensors are mounted near the bottom, middle, and top of the tank.

4-11.4 The capacitive liquid-level indicator

This is the most common method of measuring liquid level. Depending upon the type of liquid and type of tank the liquid is stored in, a number of different techniques are utilized in capacitive liquid-level measurements. If the liquid is nonconductive and the tank is metallic, an insulted electrode is mounted parallel to the metal wall, as depicted in Figure 4-26(a). A capacitor is formed with the electrode and tank wall forming the plates and the liquid and insulation about the electrode forming the dielectric. The electrical capacity of the capacitor is proportional to the

Figure 4-24

Float-operated rheostat liquid-level indicator

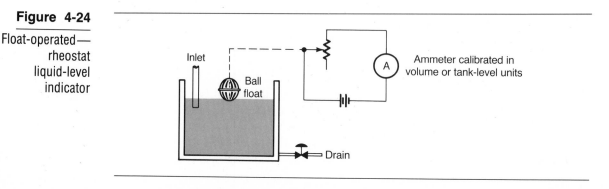

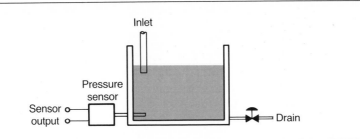

Figure 4-25

Differential-
pressure—
transducer
liquid-level
indicator

height of the liquid. The level is quantified by connecting an AC signal to the tank wall and electrode. The resulting reactance determines the intensity of the current as measured by an ammeter calibrated in tank-level or volume units.

Two insulted electrodes are used if a nonmetallic tank is used to hold a nonconductive liquid, as seen in Figure 4-26(b). In this application, the electrodes form the capacitor plates while the liquid and electrode insulation form the dielectric.

If the liquid is conductive, the same techniques are employed as before except that the electrodes and liquid form the capacitor plates and the electrode insulation serves as the dielectric.

Instead of being attached to the side of the tank, the electrodes may be in the form of probes, which are inserted into the tank. This type of probe typically has a capacitance of 10 pF when exposed to dry air and, depending upon the type of liquid, a capacitance of approximately 200 pF when fully immersed. An AC signal having a frequency in the range of 20 kHz to 1 MHz is used to excite the probe.

4-11.5 Sonic liquid-level measurement

Sometimes called ultrasonic liquid-level measurement, this measurement technique operates on the same principle as radar except that a lower frequency is employed. The sensing unit contains a transmitter and receiver. The transmitter consists of a pulse-forming network and an oscillator. The oscillator is turned on for a

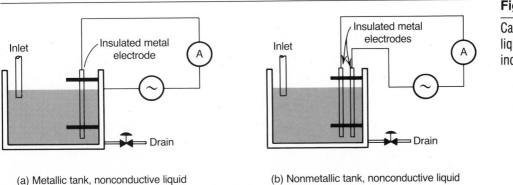

(a) Metallic tank, nonconductive liquid   (b) Nonmetallic tank, nonconductive liquid

Figure 4-26

Capacitive
liquid-level
indicator

short period of time by a pulse producing a short burst of energy that is directed downward in the reservoir to the liquid. The radiated energy strikes the surface of the liquid and is reflected back to the receiver, where it is detected. The level or volume of liquid in the reservoir is determined by measuring the time it takes the pulse to leave the transmitter and be reflected back to the receiver. The shorter the time, the more liquid in the tank. Oscillator frequencies used in sonic liquid-level measurement ranges from approximately 6 to 600 kHz.

4-12 PHOTO TRANSDUCERS

4-12.1 Introduction

This section deals with the principles of light and transducers that convert light energy into electrical energy and vice versa. Modern manufacturing utilizes these types of transducers extensively, and their applications are expected to increase as the techniques of automation continue to be refined.

4-12.2 Principles of light

Light is energy in the form of oscillating electromagnetic waves whose frequencies can be detected by the human eye. Light travels through space in the form of a transverse electromagnetic (TEM) wave in the same way as radio frequency (RF) waves. A TEM wave is composed of electrostatic (ϵ) and electromagnetic (H) fields that are transverse, or at right angles, to one another and to the direction of prop-

Figure 4-27

TEM wave

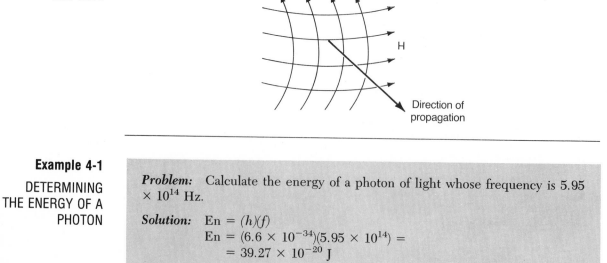

Direction of propagation

Example 4-1

DETERMINING THE ENERGY OF A PHOTON

Problem: Calculate the energy of a photon of light whose frequency is 5.95×10^{14} Hz.

Solution: $\text{En} = (h)(f)$
$\text{En} = (6.6 \times 10^{-34})(5.95 \times 10^{14}) =$
$= 39.27 \times 10^{-20}$ J

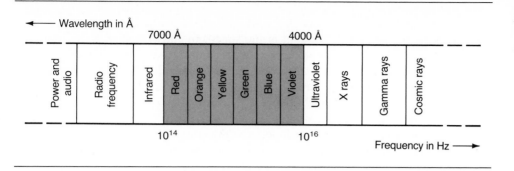

Figure 4-28

Electromagnetic
spectrum

agation, as seen in Figure 4-27. Light exits its source in a straight line and spreads uniformly outward in all directions.

Although light travels in the form of a TEM wave, wave motion does not fully describe all the properties of light. Early in the 20th century it was discovered that the emission and absorption of light occurs in finite "bundles" of energy called photons, or quanta. This has led physicists to conclude that light has a dual nature. It subscribes to laws which can be explained by transverse wave motion and by photons. Therefore, it is thought that light travels in transverse waves and sometimes acts as if it is composed of a series of small energy entities.

The energy of a photon is small and is a product of Planck's constant h and the frequency of the light wave, as expressed by Equation 4-1.

En = $(h)(f)$ in joules
Where
En represents the energy of the photon in joules
h is Planck's constant which is equal to 6.6×10^{-34} J/Hz
f is the frequency of the light wave in Hz

Eq. 4-1

The energy of a photon having a frequency of 5.95×10^{14} Hz is equal to 39.27×10^{-20} joules, as shown in Example 4-1. Although light travels through space in the same form as RF signals, its frequency is much higher, as shown by the electromag-

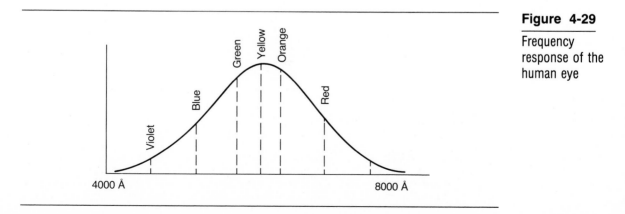

Figure 4-29

Frequency
response of the
human eye

netic spectrum appearing in Figure 4-28. The electromagnetic spectrum ranges from a few hertz (power frequencies) and extends outward through cosmic rays. Visible light forms a narrow segment of the spectrum and is located between infrared and ultraviolet waves. Visible light is made up of a series of colors whose frequencies lie between 10^{14} Hz for red and 10^{16} Hz for violet. Instead of expressing light as a unit of frequency, wavelength λ with units of angstroms (Å), is used. One angstrom equals one ten-billionth (1×10^{-10}) of a meter. The wavelength of visible light ranges from approimately 4000 Å for violet to about 7000 Å for red. The human eye recognizes different wavelengths within the visible-light segment of the electromagnetic spectrum as discrete colors. The eye operates as a bandpass filter whose maximum frequency response lies between the colors green and orange, as seen in Figure 4-29.

Light travels through space at approximately 3×10^8 m/s or 1.86×10^5 mi/s and moves at a reduced speed through a more dense medium. Whenever light enters a medium having a different density at an angle, it bends, or refracts. It refracts toward the normal plane, or surface, whenever it passes from a more to a less dense medium, as shown in Figure 4-30(a), and refracts away from the surface as it travels from a less to a more dense medium, as seen in Figure 4-30(b). When light passes from one medium density to another, no refraction occurs, if the light is perpendicular to the normal plane as illustrated in Figure 4-30(c).

When light rays strike a solid surface they are either absorbed, reflected, or a combination of both. Black surfaces absorb light, whereas white surfaces reflect light. Depending upon the smoothness of the surface, reflected rays are governed by the law of reflection, in which the angle of the reflected rays is equal to the angle of the incident rays.

Light can be focused to concentrate its intensity into parallel rays. This is accomplished through the use of a parabolic mirror, as seen in Figure 4-31. This is how light is focused in flashlights and automobile headlights. A lamp is placed at the

Figure 4-30

Wave action as light passes between two mediums having different densities

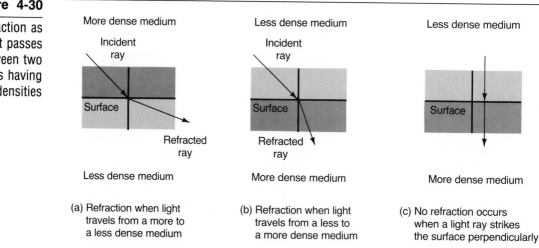

(a) Refraction when light travels from a more to a less dense medium

(b) Refraction when light travels from a less to a more dense medium

(c) No refraction occurs when a light ray strikes the surface perpendicularly

Figure 4-31

Parabolic mirror

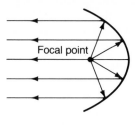

Focal point

focal point and the reflected rays leave the mirror as a parallel beam. A lens may be used to direct light. There are two types of lenses—converging and diverging. A converging lens is widest at the middle and becomes narrower at the edges, as seen in Figure 4-32(a). It is used to focus light into a very narrow beam. The diverging lens is illustrated in Figure 4-32(b). It is narrow in the center and becomes wider at the outside edges. This lens is used to spread light out.

Light is generated by the acceleration of electrons within an atom. Whenever an electron is given sufficient energy to elevate it into a higher energy level, the atom is said to be excited. It remains in this state for only a small fraction of a second before falling back into its original band. The energy released by the falling electron is in the form of electromagnetic radiation. The radiation frequency is dependent upon the type of atom. If the frequency is in the visible section of the electromagnetic spectrum, visible light is emitted.

There are three ways of exciting atoms to generate light—thermal excitation, electron collision, and photon collision. Thermal excitation occurs whenever a high-velocity atom collides with one that is moving slower. Kinetic energy is transferred to the slower atom, raising an electron to a higher energy level. This is how an incandescent light bulb produces light. Current flowing through a resistive filament causes it to become heated. This excites some of the filament atoms, which collide with others. The frequency of the resulting electromagnetic radiation lies in the light segment of the electromagnetic spectrum.

The electron method of generating light is based on the principle that the energy of an accelerating electron allowed to collide with an atom may be great enough to cause an electron in the atom to ascend into a higher energy level. Electromagnetic radiation is emitted when the electron falls back into its original orbit. This is how

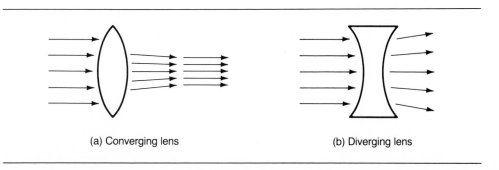

(a) Converging lens   (b) Diverging lens

Figure 4-32

Converging and diverging lenses

light is produced by fluorescent lamps and neon signs. A high DC voltage is applied across the end terminals of the tube to accelerate the electrons. As the electrons travel from the negative toward the positive terminal of the tube they collide with gas atoms, producing light.

The photon-collision method of generating light is based on the principle that whenever a photon collides with an atom, the kinetic energy of the photon is absorbed by the atom and an electron is raised to a higher energy level. When the electron drops to its original energy level, a photon is emitted from the atom. This is how light emitting diodes (LEDs) and lasers produce light.

Light sources produce either incoherent or coherent light. Incoherent, or chromatic, light is electromagnetic radiation having frequencies that extend across the entire visible light spectrum, and is white. Coherent, or monochromatic, light is produced by electromagnetic radiation having a single frequency and displays a particular color.

Photoelectronic, or optoelectronic, devices perform numerous control and sensing functions in manufacturing. They have no mechanical parts to wear out; they exert no mechanical force on objects; they do not generate electromagnetic fields; sensing can be accomplished from almost any distance; and they can be designed to respond to particular colors. Photoelectronic transducers may be grouped into three categories—electroluminescent sources, photovoltaic cells, and photoconductive transducers. These devices are discussed in the following sections.

4-12.3 Photoluminescent sources

These transducers convert electrical energy into light and include such devices as the incandescent light bulb, the fluorescent lamp, the light-emitting diode (LED), and the laser. The former two devices are primarily used for environmental lighting and will not be discussed.

The light-emitting diode: The light-emitting diode (LED) is a PN-junction diode, fabricated from either gallium arsenide (GaAs) or gallium arsenide phosphide (GaAsP), that operates on the principle of photon collision. The diode is operated in the forward-biased mode. In a conventional silicon or germanium PN-junction di-

Figure 4-33

Electron-hole
recombination in
an LED

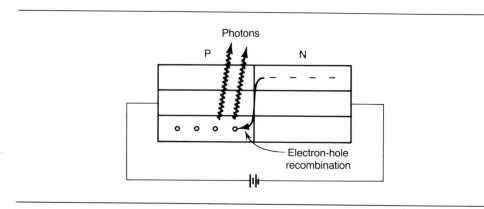

ode, heat is released when electrons in the conduction band of the N section cross the junction and recombine with holes in the valence band of the P section. In LEDs, energy released by electrons crossing the junction is in the form of heat and photons, as shown in Figure 4-33. The wavelength of the light emitted by the photons depends upon the type of semiconductor material utilized in the diode. The wavelength of GaAs is 9000 Å, which is outside the vision range of the human eye. This is the wavelength for infrared light. GaAsP LEDs emit radiant energy having a wavelength of 5400 Å. This is the wavelength of the color green and is near the center of the spectra response of the eye. The schematic symbol for the LED appears in Figure 4-34.

Light-emitting diodes have many industrial applications. The GaAsP device makes an excellent status indicator and is used to indicate the on/off status of machines, circuits, relays, and switches. Either the GaAs or GaAsP light-emitting diodes can be used as light sources for optical digital encoders. The GaAs device may be used whenever an invisible source of radiant energy is required.

The laser: The word laser is an acronym for *l*ight *a*mplification by *s*timulated *e*mission of *r*adiation. The laser produces monochromatic light through photon collision. Laser light can be created in some combinations of gases, in some crystals, and in some semiconductor materials.

When bombarded with very rapidly moving electrons, different types of gases create coherent light that can be used as a laser source. A popular type has been a gas combination of helium (He) and neon (Ne). This laser device is made up of a tube having terminals at either end. A fully silvered mirror is mounted on one end of the tube and a partially silvered mirror is mounted on the other end. A high DC potential is connected across the two terminals, causing electrons leaving the negative terminal to be accelerated through the tube at a very high velocity as they travel to the positive terminal. Some of these electrons collide with helium atoms, causing them to become excited. Some of these in turn collide with neon atoms, causing the excess energy contained by the helium atoms to be transferred to the neon atoms. The neon atoms disposes this excess energy by emitting photons of light. The mirrors located at the ends of the cylinder cause the newly generated light to be reflected back and forth along the tube. Additional atomic excitation is created with each reflection. The light eventually becomes intense enough to exit the partially silvered end of the tube focused into a very narrow, powerful beam of light.

Synthetic ruby is typically used as the crystal medium in crystal lasers. This laser is composed of a synthetic-ruby rod crystal, a fully silvered mirror attached at one end of the rod, and a partially silvered mirror attached at the opposite end, as seen

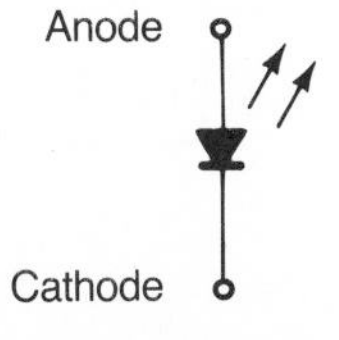

Anode

Cathode

Figure 4-34

Schematic symbol
for the LED

Figure 4-35

Synthetic ruby
laser

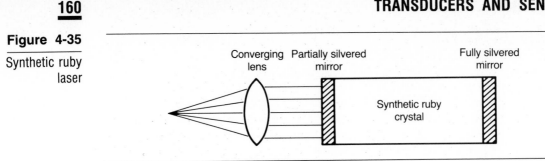

in Figure 4-35. Ruby atoms are initially excited by photons produced from an external source of light, such as a xenon flash tube. This is called pumping. Photons bombarding the ruby rod elevate many of the ruby atoms into the excited state. Photons having a frequency unique to ruby atoms are produced during the downward transition. These excite other ruby atoms, which in turn produce additional photons. This process continues, with light traveling back and forth as it is reflected from the mirrored ends. The light intensity continues to increase as more atoms are excited, until its strength is great enough to break through the partially silvered mirror and exit the rod. A converging lens mounted near the partially silvered end focuses the light into a highly intense, very narrow beam of red light.

A number of different semiconductor devices have been developed to produce light by laser. These are called laser diodes (LDs). One of the more common is the five-layer, four-junction, double-heterojunction, semiconductor laser diode illustrated in Figure 4-36. Layer 1 is an N-type gallium arsenide (GaAs) wafer; layer 2 is an N-type gallium aluminum arsenide (GaA1As) wafer; layer 3 is a P-type GaAs wafer; layer 4 is a P-type GaA1As wafer; while layer 5 is a P-type GaAs wafer. Laser generation occurs in layer 3. The device is forward biased by connecting a voltage source to terminals attached to layers 1 and 5. When the device is forward biased, electrons from layer 2 cross junction J_2 and recombine with holes in layer 3, initiating laser oscillation.

Figure 4-36

Semiconductor
laser

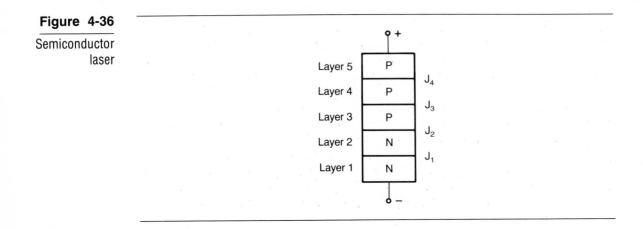

4-12.4 Photovoltaic cells

A photovoltaic cell converts light energy into voltage. Sometimes called solar cells or solar batteries, these are specially fabricated PN-junction semiconductor devices that generate voltage when stimulated by light.

Materials used in the fabrication of photovoltaic transducers include silicon, selenium, cadmium sulfide, and gallium arsenide. Silicon is the material most often used. A silicon photovoltaic cell operates in much the same fashion as a conventional PN-junction diode in the absence of light.

All photovoltaic cells operate on the same basic principle, although their method of fabrication differs. During the manufacture of any diode, a barrier potential is developed across the junction because of the normal diffusion of mobile carriers present in its vicinity. Whenever light of sufficient intensity strikes a photovoltaic cell, the energy released by the photons elevates valence electrons into the conduction band, creating additional minority carriers in the semiconductor material. Minority carriers (electrons) in the P section diffuse toward the N section and minority carriers (holes) in the N section diffuse toward the P section because of the polarity of the junction potential. This reduces the amplitude of the junction-barrier potential. As the diffusion process continues, the barrier potential is eventually overcome. The N section begins accumulating excess electrons while the P section attracts additional holes. This results in a difference of potential appearing across the anode and cathode terminals.

The voltage developed by a photovoltaic cell is proportional to the intensity of the light striking it, as shown in Figure 4-37(a). The symbol for the device appears in Figure 4-37(b).

These devices are more heavily doped than conventional diodes, which causes them to have a low internal resistance. This allows most of the generated voltage to appear across the output terminals when the cell is loaded. The no-load output voltage for a silicon photovoltaic cell is approximately 0.5 V. The cell is capable of supplying a maximum current of 100 mA to a load. Cells are often connected in series to provide a higher output voltage. They can be connected in parallel to increase

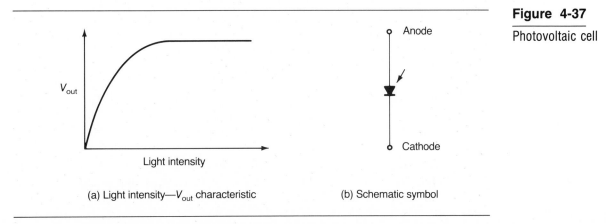

(a) Light intensity—V_{out} characteristic

(b) Schematic symbol

Figure 4-37

Photovoltaic cell

Figure 4-38

Light detector

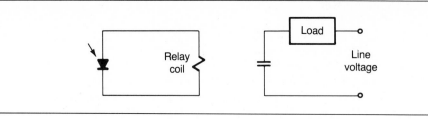

load-current capacity or they can be connected in series-parallel to increase both the voltage and current capacity.

The conversion efficiency of photovoltaic cells is low. Silicon cells have an efficiency of approximately 12 percent while the efficiency of gallium arsenide cells is about 17 percent.

Photovoltaic cells have a variety of applications. One of their more obvious uses is the conversion of sunlight into voltage, eliminating the need for a battery or generator to energize low-power circuits. They can be used as light detectors, as illustrated in Figure 4-38. A relay coil connected in the anode circuit is actuated whenever light strikes the diode. An amplifier can be connected between the diode and coil if additional power is required to actuate the relay.

4-12.5 Photoconductive transducers

These transducers convert changes in light intensity into changes in conductivity. The photodiode and phototransistor are the devices used to accomplish this change.

The photodiode: The fabrication of the photodiode is similar to that of a conventional silicon or germanium PN-junction diode, except that a lens is attached near the junction to focus light striking the junction. The photodiode is operated in the reverse-biased mode, as illustrated in Figure 4-39. In the absence of light a small temperature-dependent leakage current flows, just as in a conventional reverse-biased diode. In a photodiode this is called dark current. When light enters the junction through the lens, energy released by the photons is absorbed by the diode in the vicinity of the junction. This causes additional valence electrons to be elevated into the conduction band allowing the anode (reverse) current to increase. As seen by the volt–ampere characteristic curves appearing in Figure 4-39(b), the anode current is proportional to the intensity of the light as measured in units of lux. The schematic symbol for the device is shown in Figure 4-39(c).

The spectra response for silicon and germanium photodiodes appears in Figure 4-40. Both transducers are more sensitive to the longer light wavelengths. Since reverse current is utilized, the output of a photodiode is low and must usually be amplified.

The phototransitor: Phototransistors are available in both NPN and PNP form. As illustrated in Figure 4-41(a), they are similar in construction to a small-signal bipolar-junction transistor except that a lens has been attached to the base section. The device operates much like any other silicon or germanium transistor in the absence

Figure 4-39

Photodiode

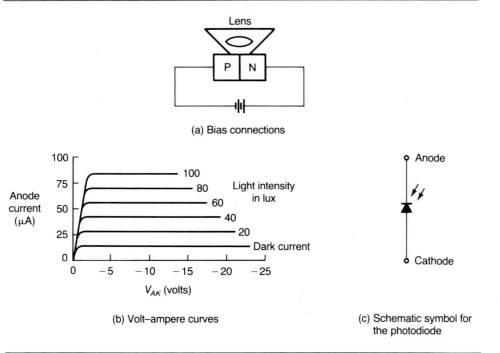

(a) Bias connections

(b) Volt–ampere curves

(c) Schematic symbol for the photodiode

of light. The emitter–base junction is forward biased and the base–collector junction is reverse biased. When light strikes the transistor, additional minority carriers are created by the photons absorbed by the semiconductor atoms in the base section, causing the base current to increase. This current is amplified by the h_{fe} value of the transistor and appears as output, or collector, current. The volt–ampere characteristic curves appear in Figure 4-41(b). Unlike a conventional BJT whose individual curves are formed by discrete base-current steps, the family of curves for a phototransistor is obtained from discrete values of light intensity. The symbol for the device appears in Figure 4-41(c).

Figure 4-40

Photodiode spectra responses

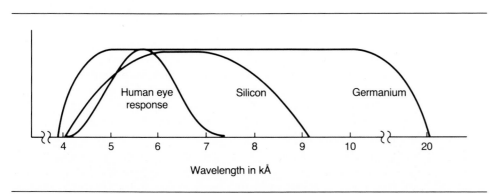

Figure 4-41

Phototransistor

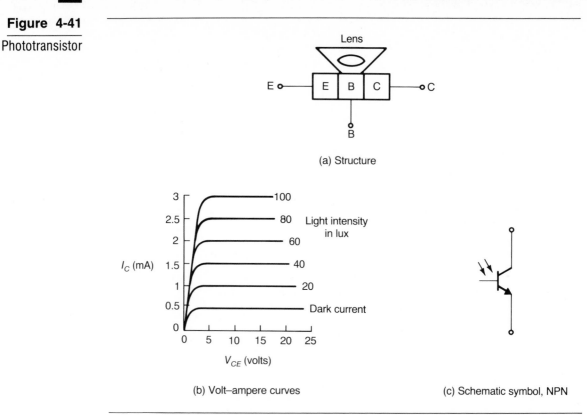

(a) Structure

(b) Volt–ampere curves

(c) Schematic symbol, NPN

Phototransistors are available in both two- and three-lead enclosures. Only the collector and emitter leads are utilized on a two-lead case, which is the more commonly used enclosure. Transistor control is affected only by the intensity of the light entering the base section. The third lead is used for making connection to the base in the three-lead case. This allows a Q-point to be established. Changes in light intensity causes the collector current to vary above and below the Q-point value.

A comparison of the family of volt–ampere characteristic curves for the photodiode in Figure 4-39(b) and the phototransistor in Figure 4-41(b) reveals that the characteristics for both devices are similar. This means that they can be used for many of the same applications. The phototransistor is much more sensitive, however. At a light intensity of 60 lux, the photodiode produces an output current of only 50 μA whereas the output current is 2 mA for the phototransistor. For the two devices shown, the phototransistor is 40 times more sensitive than the photodiode. Because of this, in many applications, the output current of a phototransistor does not have to be amplified prior to being used.

Photoconductive devices have a number of different applications in manufacturing and other industrial applications. They are small, rugged, and have long life spans. The spectra responses are such that they respond to a variety of light sources, including infrared. They may be employed in punched-tape readers on nu-

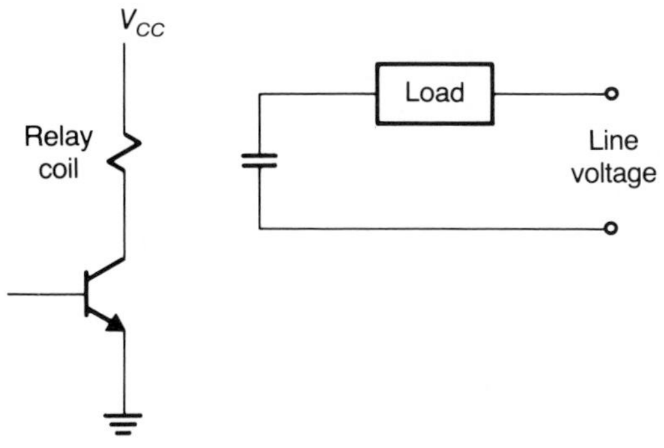

Figure 4-42

Phototransistor
operated as a light
actuated relay

merically controlled manufacturing machines, are often used as the detector in optical digital encoders, and may be utilized as light actuated relays, as seen in Figure 4-42. In this latter application, the relay coil is connected in the collector circuit of a phototransistor. In the absence of light the transistor is cut off and the relay is unenergized. When light of sufficient intensity strikes the base, the transistor begins conducting and the resulting collector current flowing through the relay coil actuates the relay. The NO contacts close allowing current to flow through the load. The load remains energized as long as light continues to strike the transistor.

4-13 PROXIMITY SENSORS

4-13.1 Introduction

Many manufacturing applications require that the presence, or absence, of an object be sensed without physical contact being made. Devices used for this are called proximity, or noncontact, sensors. Proximity sensors may be mounted on machine tools, materials-handling equipment, robots, inspection equipment, and packaging equipment. Depending upon the type, proximity sensors utilize principles of inductance, capacitance, ultrasonics, and photoelectricity.

4-13.2 Inductive proximity sensors

These sensors detect the presence of metallic objects only. Two different techniques are employed. In one, an RF oscillator is utilized. An RF signal is radiated from a coil located in the sensor head. Whenever the moving field comes in contact with a metallic object, the object absorbs part of the energy, attenuating the oscillator output signal. The amplitude decrease is sent to a waveshaping circuit, which develops a trigger pulse that can be used to acknowledge the presence of the object by counting it or actuate a switch which causes its movement to cease.

The other type of inductive proximity sensor utilizes the Hall-effect principle. This principle describes the method by which a voltage is produced by some semiconductor materials when the material is placed within a magnetic field while current is flowing through it. A semiconductor Hall-effect detector is attached to a stationary surface and a permanent magnet is mounted on the movable object. The field about the permanent magnet causes a voltage to be developed in the Hall-effect detector as the movable object approaches the stationary surface to which the detector is attached. The voltage is amplified and used to actuate a solenoid or relay causing the object to cease its movement.

4-13.3 Capacitive proximity sensors

Capacitive proximity sensors can detect the presence of any type of object or substance regardless of its composition. These sensors are especially sensitive to objects made of plastic or glass and low-density materials such as powders, paper, and dry cereal.

The capacitive proximity sensor utilizes an oscillator. The capacitor plates in the oscillator LC tank circuit are mounted to allow the moving object to pass between them, causing the oscillator frequency to change. The change in frequency is converted into a trigger that can be used to count the object, or to actuate a solenoid or relay to stop its movement.

4-13.4 Ultrasonic proximity sensors

These sensors employ an RF oscillator and detector, which are mounted side by side on a stationary surface. The oscillator signal is shaped and broadcast in the form of a narrow beam. Whenever a moving object passes through the beam, part of the radiated energy is reflected back to the detector. The distance between the moving object and the detector is determined by the time that it takes the signal to travel from the oscillator to the object and back to the detector.

Figure 4-43

Direct-scan system

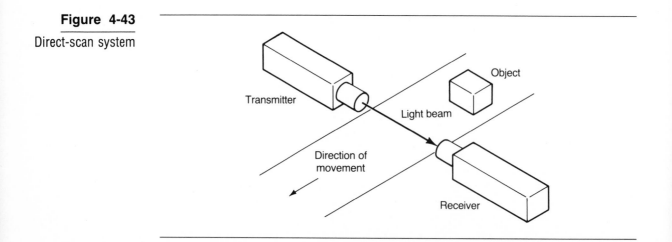

4-13.5 Photoelectric proximity sensors

These are the most commonly used proximity, or noncontact, sensors. They can detect all types of objects and substances regardless of size and composition of the materal. They have a wide sensing range from models capable of detecting objects as near as a fraction of an inch to those that can sense objects several hundred feet away. They are used for such diverse applications as parts counting, bin-level control, machine-feed control, inspection, quality control, labeling, registration control, and edge control.

Photoelectric sensors can be grouped into two categories—direct scan and reflective scan. All employ both a light source (transmitter) and a photodetector (receiver). Small incandescent bulbs, lasers, or LEDs may be used as a light source. Infrared LEDs are popular light sources, as these devices provide an invisible source of light. Since an infrared photodetector is required, ambient light does not affect the operation of the receiver.

In the direct-scan, sometimes called the thru-scan, system the transmitter and receiver are located opposite one another, as shown in Figure 4-43. The path for the moving object (sometimes referred to as the target) lies between the transmitter and receiver. A trigger pulse is created in the receiver whenever the object passes between the transmitter and receiver and interrupts the light beam.

The transmitter and receiver are placed side by side when reflective scan is employed. This sensing technique is used where space is limited and the receiver cannot be mounted across from the transmitter. There are three types of reflective-scan systems—retroreflective scan, specular scan, and diffuse scan.

A glass, plastic, or metal mirror reflects light beamed from the transmitter directly back to the receiver, as seen in Figure 4-44, when retroreflective scan is utilized. The receiver is triggered whenever a moving object passes through and interrupts the light beam.

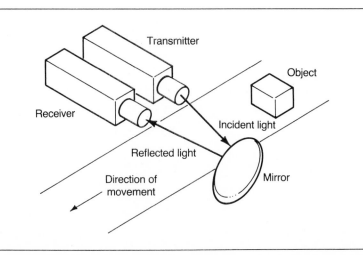

Figure 4-44

Retroreflective-scan system

Figure 4-45

Specular-scan
system

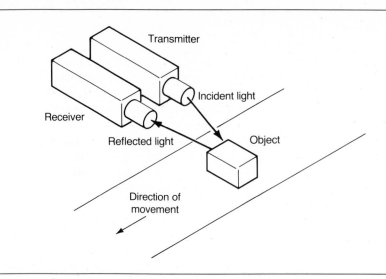

The moving object is used to reflect light when the specular-scan system is employed. Sometimes called the definite-reflective system, this type of photoelectric sensor may be used when the moving object has a surface that readily reflects light. Light from the transmitter strikes the moving object and is reflected back to the receiver, as illustrated in Figure 4-45.

The diffuse-scan system may be used to detect moving objects whose surface area is rough. Instead of reflecting light in one direction, the light is scattered in all directions when the moving object interrupts the light beam, as shown in Figure 4-46. The transmitter and receiver unit must be located close to the path of the

Figure 4-46

Diffuse-scan
system

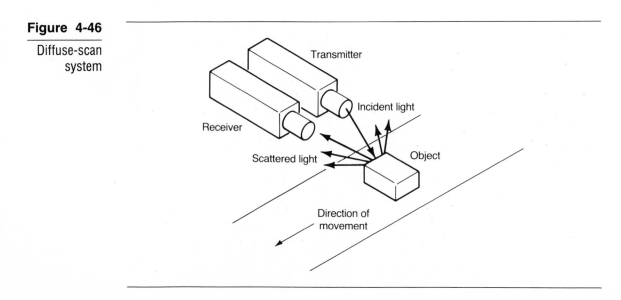

moving object, since the light is diffused and only a small amount of reflected light returns to the receiver.

Although photoelectric sensors have many advantages when compared to other proximity sensors, they can operate only in a clean atmospheric environment. Dust and moisture particles in the air cause the light beam to scatter, reducing the sensing range and increasing the risk of interference. Phototransistors are often used as the detector, as the output current of these devices is usually sufficient to actuate a relay, energize a solenoid, or drive a counter circuit.

Choosing a photoelectric sensor for a particular application depends upon such factors as the amount of space available for mounting the transmitter and receiver, the composition of the object to be detected, the scanning distance, the rate at which objects must be counted, and the ambient temperature.

Photoelectric sensors are usually mounted in sealed enclosures. NEMA switch-enclosure types 1, 3, 4, 12, and 13 may all be utilized. Most of these sensors are designed to be operated in a wide temperature range with −40 to 175° F being typical.

Summary

1. Transducers are devices that convert one form of energy into another. Sensors are transducers that convert physical phenomena into variables that can be quantified.

2. Temperature is a measure of the relative warmness or coldness of a material. Temperature is measured with a thermometer. The three scales used to measure temperature are Fahrenheit, Celsius, and Kelvin. Sensors used to measure temperature for industrial applications include the thermocouple, RTD, thermistor, and solid-state temperature sensor.

3. The thermocouple operates on the Seebeck effect. When heat is applied to the junction joining two dissimilar metals, a voltage is developed across the opposite open-ended terminals.

4. The RTD operates on the principle that the resistance of some metals is proportional to their temperature.

5. The thermistor is a temperature-sensitive resistor.

6. Relative humidity is the ratio of the actual amount of water vapor in the air at a particular temperature to the maximum amount the air can hold. Relative humidity is measured with a psychrometer and a hygrometer.

7. Stress is created by a force acting on a body. Strain is the resulting change in the shape of the body. Strain is measured with a strain gage. Three types of strain gages are available for strain measurement—the bonded-wire, semiconductor, and thin-film.

8. Acceleration is the rate at which the velocity of an object increases per unit of time. The accelerometer is used to measure this variable.

9. Linear displacement is the distance of an object from a reference point when both are located on the same plane. Linear displacement is measured with an LVDT.

10. Angular displacement refers to the angular position of an object with respect to a reference point. This variable can be measured with a potentiometer, an angular digital encoder, or a synchro.

11. The tachometer is used to measure shaft speed. This sensor is a small generator mechanically coupled to the shaft whose speed is to be measured.

12. Fluid pressure is the force exerted by a liquid or gas on a surface. Fluid pressure is measured by converting pressure into displacement and employing a sensor to change displacement into an electrical signal.

13. Fluid flow is the velocity at which a fluid flows through a transmission medium. This variable is measured with a flowmeter. Numerous types of flowmeters are available for measuring the rate of flow of different types of fluids. Three of the more commonly used types are the Venturi tube, the turbine flowmeter, and the thermal flowmeter.

14. Liquid level can be measured using a float-operated rheostat, a differential-pressure transducer, a capacitive liquid-level sensor, and sonic techniques.

15. Light is electromagnetic energy that travels through space as TEM waves.

16. Although light travels in the form of a TEM wave, it sometimes behaves as if it is composed of small entities of energy called photons.

17. Visible light is composed of a series of colors whose frequencies lie between 10^{14} and 10^{16} Hz and whose wavelengths range from 7000 to 4000 Å.

18. Light is generated by the acceleration of electrons within an atom. Methods of accelerating electrons to create light include thermal, electron collision, and photon collision.

19. Light sources produce either incoherent or coherent light. Incoherent light is white and includes colors whose frequencies extend across the entire light spectrum. Coherent light is monochromatic, or single-colored, and is produced by electromagnetic radiation having a single frequency.

20. Photoluminescent sources are transducers which convert electrical energy into light.

21. The LED is a GaAs or GaAsP PN diode used as a photoluminescent source.

22. The laser is a photoluminescent source that produces light from certain combinations of gases, from particlular crystals, or from specially prepared semiconductor materials.

23. Crystal lasers are usually fabricated from synthetic ruby crystals.

24. The five-layer, four-junction, double-heterojunction diode is a semiconductor diode used to develop laser light.

25. Photovoltaic cells convert light energy into voltage.

26. Photoconductive transducers convert changes in light intensity into changes in conductivity.

27. Proximity sensors are used to detect the presence or absence of an object.

28. Inductive proximity sensors detect the presence of metallic objects only. Two different techniques are employed. One technique utilizes an RF oscillator while the other uses a Hall-effect detector.

29. Photoelectric proximity sensors use both a light source and a light detector. Two different techniques are employed—direct scan and reflective scan. The light source and detector are placed opposite one another in the direct-scan system while the source and detector are placed side by side in the reflective-scan system. Light emitted by the source is reflected back to the detector by a mirror or the moving object.

Chapter Examination

1. Define the term *transducer*.

2. Define the term *sensor*.

3. Define the term *strain*.

4. List four ways of measuring liquid level.

5. List three ways of generating light.

6. Identify four ways of measuring temperature in industrial applications.

7. List three methods used to measure angular displacement.

8. What is a synchro?

9. Describe the operation of a synchro.

10. List three displacement transducers used in fluid-pressure measurements.

11. Which of the following is not a method used to measure angular displacement?
  a. LVDT
  b. synchro
  c. potentiometer
  d. digital encoder

12. Which of the following is not a type of strain gage?
  a. thin-film
  b. thin-wire
  c. bonded-wire
  d. semiconductor

13. Which of the following sensors is not used for measuring temperature?
  a. RTD
  b. VVC
  c. thermistor
  d. thermocouple

14. Which of the following is not a method of generating light?
  a. photon collision
  b. electron collision

   c. thermal excitation
   d. electron-photon recombination

15. Which of the following is not a photoluminescent source?
   a. LED
   b. laser
   c. photodiode
   d. incandescent bulb

16. Which of the following is not a type of laser?
   a. gas
   b. ruby crystal
   c. GaAsP PIN diode
   d. double-heterojunction semiconductor

17. Which of the following sensors is used to measure fluid pressure?
   a. turbine
   b. strain gage
   c. Bourdon tube
   d. psychrometer

18. The _____ is used to measure shaft speed.
   a. synchro
   b. tachometer
   c. hygrometer
   d. accelerometer

19. The _____ may be used to either measure shaft displacement or to position a shaft from a remote position.
   a. VVC
   b. LVDT
   c. synchro
   d. binary encoder

20. The _____ scan system is the type of photoelectric sensor used when a moving object ineterrupts light that is beamed from a transmitter and reflected back to the receiver by a mirror.
   a. direct
   b. diffuse
   c. specular
   d. retroreflective

21. The _____ scan system is the type of photoelectric sensor used when light that has been scattered by a moving object having a rough surface is received at the detector.
   a. direct-
   b. diffuse-
   c. specular-
   d. retroreflective-

22. Which of the following is not a transducer used to sense fluid pressure?
   a. bellows
   b. Bourdon tube

    c. Venturi tube

    d. solid-state pressure sensor

23. The LVDT is used to measure _____ .

    a. strain.

    b. liquid level.

    c. linear displacement.

    d. angular displacement.

24. The _____ is made up of two RTDs, a water reservoir, a fan, and a housing.

    a. hygrometer

    b. psychrometer

    c. accelerometer

    d. angular digital encoder

25. The _____ is a light source.

    a. laser

    b. photodiode

    c. phototransistor

    d. photovoltaic cell

26. The semiconductor strain gage operates on the principle of the _____ effect.

    a. Hall

    b. Seebeck

    c. piezoelectric

    d. piezoresistive

27. Which of the following is not a method of measuring angular displacement?

    a. LVDT

    b. synchro

    c. potentiometer

    d. digital encoder

28. _____ transducers convert changes in light intensity into changes in electrical conductivity.

    a. Optic

    b. Photovoltaic

    c. Photoconductive

    d. Photoluminescent

29. The _____ is an entity of light energy.

    a. lux

    b. Pascal

    c. photon

    d. coulomb

30. Which of the following is not a photoluminescent transducer?

    a. laser

    b. fluorescent lamp

    c. photovoltaic cell

    d. incandescent lamp

31. The thermocouple operates on the _____ effect.

    a. Hall

b. Seebeck

c. piezoelectric

d. piezoresistive

32. The ＿＿＿ is a temperature transducer which operates on the principle that the resistance of some metals is proportional to temperature.

a. RTD

b. VVC

c. thermistor

d. thermocouple

33. The wavelength of visible light ranges from

a. 50–800Å.

b. 200–3000Å.

c. 4000–7000Å.

d. 4500–9000Å.

34. A ＿＿＿ converts radiant energy into voltage.

a. photodiode

b. phototransistor

c. photovoltaic cell

d. photoluminescent cell

35. The Venturi tube is a type of ＿＿＿ transducer.

a. pressure

b. humidity

c. flow-rate

d. displacement

36. Which one of the following devices or techniques is *not* used for measuring liquid level?

a. LVDT

b. rheostat

c. capacitive

d. differential-pressure meter

37. Using Table 4-1, the relative humidity is ＿＿＿ percent if the ambient temperature is 80° F and the difference temperature is 5° F.

a. 24

b. 47

c. 66

d. 93

38. ＿＿＿ proximity sensors use an RF oscillator and a detector mounted side by side. An object is detected as it moves through the beam and part of the radiated energy is reflected back to the detector.

a. Inductive

b. Capacitive

c. Ultrasonic

d. Photoelectric

39. The quantity of moisture present in the air defines

a. humidity.

    b. dew point.

    c. saturation.

    d. relative humidity.

40. T F The tachometer is used to measure shaft speed. ___

41. T F Whenever a continuous relative-humidity measurement is re- ___
quired, a hygrometer is employed.

42. T F The accelerometer is used to measure acceleration and vibration. ___

43. T F Linear displacement is measured with a synchro. ___

44. T F The transmitter and receiver are located opposite one another ___
when a direct-scan photoelectric proximity sensor is used.

45. T F Inductive proximity sensors operate on the principle of the See- ___
beck effect.

46. T F Strain is created by a force acting on a body. Stress is the resulting ___
change in the shape of the body.

47. T F The Bourdon tube is a type of flowmeter. ___

48. T F Fluid pressure is measured by changing pressure into displace- ___
ment. A sensor is used to convert displacement into an electrical
signal.

49. T F The thermistor is more sensitive than the RTD and thermocouple ___
temperature sensors.

50. T F A 16-track binary encoder disc has an accuracy of 0.0109863°/bit. ___

51. T F Velocity is the rate at which the speed of an object increases per ___
unit of time.

52. T F Synchros may be used to position angular loads from a remote po- ___
sition.

53. T F Displacement transducers used in fluid-pressure measurement in- ___
clude LVDTs, RTDs, strain gages, and piezoelectric devices.

54. T F The photodiode is normally operated with reverse bias. ___

55. T F Helium and argon are gases commonly used in a gas laser. ___

CHAPTER
5

INTRODUCTION TO MICROPROCESSORS

OBJECTIVES

1. Define the term *microprocessor*.

2. Draw the block diagram of a basic microprocessor.

3. Identify the circuits forming a basic microprocessor and discuss the function of each.

4. List the buses used in a microprocessor and identify the types of signals carried by each.

5. Describe why the address word length is often greater than the data word length in a microprocessor.

6. List two methods used to identify peripherals connected to the MPU and discuss the principles of each.

7. Identify two methods employed to allow the microprocessor and input/output peripherals to communicate with one another and discuss the operating principles of each.

8. Describe the purpose of direct memory access (DMA).

9. Discuss the purpose of the stack, identify the type of information stored in the stack, and describe how the stack differs from RAM.

10. Identify the major 8-bit, 16-bit, and 32-bit Intel microprocessors and discuss the general characteristics of each.

11. Identify the major 8-bit, 16-bit, and 32-bit Motorola microprocessors and discuss the general characteristics of each.

5-1 INTRODUCTION

Microprocessors are a product of the work of scientists and engineers to develop smaller and more efficient active electronic devices. Active devices are capable of amplifying, generating, and switching electronic signals. The electron tube was the first active device. Patented in 1907, it provided the means for amplifying electrical signals. It took engineers but a short time to apply this device to the new and developing field of electronic communications.

The electron tube was an inefficient device. Scientists worked for decades developing better tubes while at the same time trying to find a more efficient substitute. A replacement was finally found in December of 1947 by three scientists working at Bell Laboratories. This device was the transistor. It took several years for the technology of producing transistors in large quantities to be developed before widespread applications were possible. By the early 1960s, however, transistors were replacing electron tubes as active devices in most electronic equipment.

Applications for electronics were primarily centered on communications through World War II. The development of the digital computer in the late 1940s, followed by the growth of the space program during the 1960s, created a need for smaller and more efficient types of active devices. This need was met by the introduction of the integrated circuit. Developed for military applications in 1962, integrated circuits were being used in space vehicles, computers, communications equipment, industrial electronics, and consumer products by the late 1960s. The growth and subsequent maturity of digital computers paralleled the development of the integrated circuit. It was only a matter of time before the two technologies merged to form a programmable integrated circuit—the microprocessor.

The first microprocessors were developed primarily for use in the defense industry. Applications in other areas soon followed. Today microprocessors are used in products as common as toys and as sophisticated as the most advanced telecommunications satelite. One of the largest users of microprocessors has been the manufacturing industry where they are used in product design, machine-tool and process controllers, automated materials-handling equipment, inspection, testing, and quality control.

5-2 MICROPROCESSOR ARCHITECTURE

5-2.1 Introduction

In its most elementary form, the microprocessor can be considered to be a versatile integrated circuit that can accept data from input devices and be programmed to manipulate the data to perform arithmetic, logic, data transfer, and decision-making operations on the data. The key to the mircoprocessor's versatility is its programmability. Any activity performed by a microprocessor can be duplicated by hard-wired logic circuits. With this method, the circuit has to be modified to accomodate any changes required by the activity being performed. To obtain the same changes when a microprocessor is utilized, all that is required is a change in the program. In addition, fewer circuits are needed since the microprocessor can be used to replace many different kinds of devices and circuits by being programmed to imitate their functions.

More technically, a microprocessor, sometimes referred to as an MPU, µp, or processor, is the central processor unit (CPU) of a computer fabricated on a single semiconductor substrate. As illustrated in Figure 5-1, a computer is made up of a CPU, memory, and input–output devices. The CPU forms the heart of the computer and contains the circuits that allow the computer to perform its operations. Memory is required to store the program or software. The program is made up of a list of data and commands that tell the CPU how to manipulate the data. The program is loaded into memory, via the CPU, through devices called input peripherals. These often include toggle, limit, push-button, and pressure switches; keypads; keyboards; and sensors. Output peripherals include those devices that are controlled by the computer or that provide a visual means of observing the results of the computer operations such video monitors and printers.

Like the CPU in the computer, the microprocessor is not a self-sufficient device. It requires supporting devices such as memory and input–output (I/O) peripherals. The microprocessor and its supporting devices are called a microprocessor-based system. Typical industrial and manufacturing microprocessor-based systems include process controllers, machine-tool controllers, robot controllers, and programmable controllers. In many manufacturing applications, microprocessor-based systems are connected to mini or mainframe computers. The microprocessor systems perform the slower, simpler, and more redundant operations required in a particular activity and transmit the results to the larger computer. This frees the large computer for faster and more complex operations.

Computers can be divided into three categories—mainframe computers, minicomputers, and microcomputers. The major differences between the categories are the length of the data words, the speed at which the data are processed, and the memory-addressing capability. Mainframe computers are the largest, fastest, and

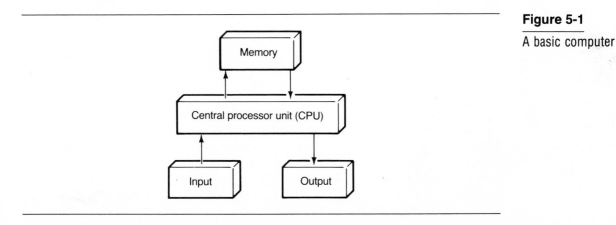

Figure 5-1

A basic computer

most expensive. Their cost makes their use prohibitive except in very large data-processing and high-level scientific applications. Minicomputers are scaled down mainframes developed for those applications where data word size, speed, and memory capacity are not primary requisites. A microcomputer, sometimes called a personal computer, is a microprocessor-based system designed for general computing applications. Central processor speed in mainframe computers is approximately 100 times faster than that for a typical microcomputer. However, with the advent of the 16- and 32-bit microprocessors and their subsequent use in microcomputers, it is becoming more difficult to differentiate between the categories. These microcomputers have the word size, processor speed, and memory capacity of many mainframe and minicomputers.

The circuits forming the internal structure of a microprocessor are called its architecture. The computing power and performance of a particular microprocessor are determined by factors such as the type of transistor technology employed in its fabrication, the number and types of internal circuits that it contains, data word length, address word length, the number of instructions developed for its use, and the method in which data are fetched from memory. Data word length refers to the number of bits which form the binary data words. Standard word lengths are 4, 8, 16, and 32 bits.

Although the architecture varies from one microprocessor to another, most contain the same general types of circuits. The diagram of a basic microprocessor is shown in Figure 5-2. The device is composed of several shift registers and combinational logic circuits, as discussed in the following sections.

5-2.2 The accumulator

This is one of the most important and the busiest shift register utilized in the MPU. It is used in all arithmetic and most other operations. In arithmetic operations, it stores one of the numbers to be operated on and, when the operation is complete, it stores the resultant. It can be loaded with data read from memory or from an input peripheral. In most processors, the accumulator has the same storage capacity as the word size for that processor. Some microprocessors do not have a register designated as an accumulator. In these devices a general-purpose register performs the accumulator's functions.

5-2.3 The arithmetic and logic unit (ALU)

The circuits constituting this unit perform arithmetic and logic operation on data. Addition is the most commonly performed arithmetic operation. Subtraction in most microprocessors is accomplished by using the two's complement and adding the two numbers together. Except for just a few microprocessors, multiplication and division operations are accomplished by performing repeated addition or subtraction operations. The most common logic operations performed are the OR, AND, and EXCLUSIVE-OR.

Figure 5-2

A basic
microprocessor

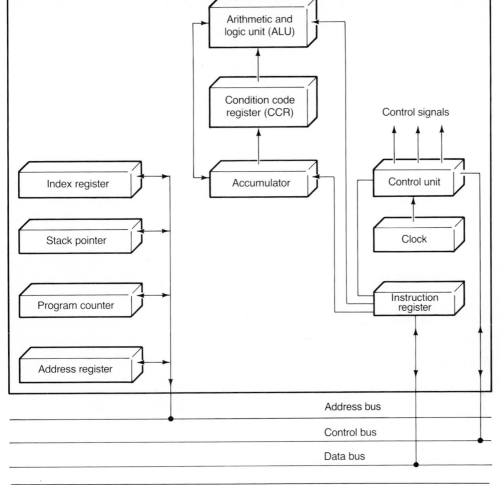

5-2.4 The condition code register (CCR)

This circuit is sometimes called the status or flag register. Each flip-flop forming the register is called a status flag and is assigned a particular monitoring function. Most flags monitor the contents of the accumulator. If interrupts are employed, one or more flags are used to indicate the interrupt status of the system. Upon executing an instruction in a typical microprocessor, one flag checks to see if the number in the accumulator is zero (zero flag), one checks to see if a carry has occurred from the most significant bit (carry flag), another checks to see if the number in the accumulator is negative (negative flag), and yet another flag is used to indicate that a peripheral requires service (interrupt flag). Depending upon the particular microprocessor, other flags may also be employed.

5-2.5 The program counter

The program counter is a register that stores the address of the next byte of data to be fetched from memory. When a program is to be executed, the starting address is placed in the program counter. Unless a jump or branch instruction is encountered, the program is executed in the sequence in which it was loaded into memory. Whenever data are to be fetched from memory, the program counter provides the memory address of the data and then increments to the address of the next memory segment. In this fashion the program counter keeps track of the data in the memory unit as the program is being executed.

5-2.6 The address register

This register stores the address of the memory segment or I/O peripheral being accessed while the address is being decoded. The address register can be loaded from the program counter, memory, or a register. In some microprocessors the address register is incorporated into the address bus and is called an address buffer.

5-2.7 The index register

This register is used to control loops in the program. It can access a subroutine stored in memory or a block of data with a single instruction.

5-2.8 The stack pointer

Information in the program counter, accumulator, and other key registers is temporarily placed in a special memory called the stack whenever a break in sequential program execution occurs. The stack pointer keeps track of the information stored in the stack during the duration of the break. This circuit performs the same function for the stack that the program counter performs for regular memory.

5-2.9 The instruction register

Sometimes called the data register, this circuit temporarily stores all data coming into or leaving the microprocessor. This register stores an instruction fetched from memory while it is being decoded by the control unit.

5-2.10 The control unit

Circuits forming the control unit decode the instruction fetched from memory that is stored in the instruction register and generate the control signals that cause the ALU to perform the desired arithmetic or logic operation. In some microprocessors, this circuit is called the control and timing unit.

5-2.11 The clock

The clock provides the timing pulses needed to synchronize the operation of the circuits within the microprocessor, memory, and I/O peripherals. In some microprocessors this is a separate integrated circuit; in others it is an "on board" circuit fabricated as part of the microprocessor. Some microprocessors require only a single train of clock pulses, whereas others require a dual set of timing pulses. To differentiate between the two, one pulse train is called phase 1 (ϕ1) and the other phase 2 (ϕ2).

5-3 BUSES

5-3.1 Introduction

The internal conductors, which connect the internal circuits together, and the external conductors, which connect the MPU with its memory devices and I/O peripherals, are called buses. Most microprocessors employ three bus systems—the address bus, the data bus, and the control bus.

5-3.2 The address bus

This bus carries the memory segment and I/O peripheral addresses from the program counter to the address register within the MPU. Externally, it connects the microprocessor, memory devices, and I/O peripherals. The address word is sent in parallel format. The number of conductors forming the address bus varies, depending upon the type of microprocessor. The size of the address word directly affects the amount of memory or I/O peripherals that can be accessed. In a few microprocessors the address word size is the same as the data word size. In an 8-bit processor this allows for accessing 256 (2^8) different memory segments or I/O peripherals. A microprocessor with an accessing capacity of this size has only limited applications. To increase addressing capability, an address word that is larger than the data word is utilized in most microprocessors. A two-byte address is commonly used in 8-bit microprocessors. This increases the addressing capability to 65,536 (2^{16}) bytes. Twenty or more address bits are utilized in 16- and 32-bit processors. This places their addressing capacity at one megabyte or more.

5-3.3 The data bus

This bus carries data or the information the microprocessor operates upon within the MPU and externally among the MPU, memory, and I/O peripherals. Data are transferred in parallel within the MPU and between the microprocessor and main memory. Each data bit requires a separate conductor. An 8-bit microprocessor has a data bus composed of eight individual conductors. Data that are moved between the MPU and the I/O peripherals may be sent in either parallel or serial fashion depending upon the type of peripheral employed. Only a single pair of conductors is required if the data are being transferred in serial fashion. In some microprocessors, the data bus is multiplexed with the address bus and carries address bits when it is not carrying data.

5-3.4 The control bus

The control bus is a collection of conductors which carry the signals required to implement control of the entire microprocessor-based system. Internally, the signals traveling on this bus control the ALU, causing the microprocessor to perform its operations. The external control bus carries control signals to and from the MPU, memory, and I/O peripherals causing data to be written into or read from these devices. The size of the control bus and the specific signals carried vary with the type of microprocessor.

5-3.5 Bus polarity

In addition to the types of signals carried, buses may also be classified as to polarity—the direction in which the signals travel. A bus in which the signal travels in one direction only is referred to as a unidirectional bus. If the signal may travel in either direction on the bus, at separate times, it is called a bidirectional bus. As seen in Figure 5-3, the address bus is unidirectional. The address always originates at the MPU and is sent to a memory IC or an I/O peripheral. The data bus is bidirectional. Data can both be sent to or received from the microprocessor at different times. The control bus is unidirectional. However, the signals traveling on this bus

Figure 5-3

Bus polarity

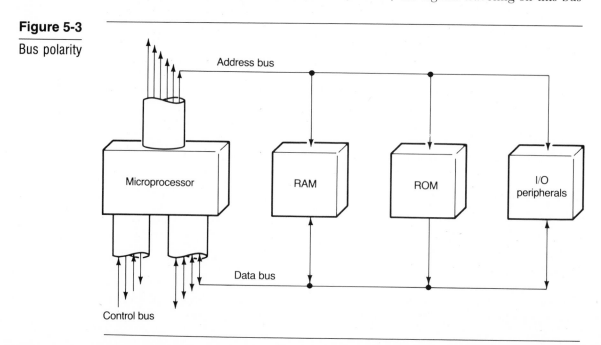

do not all flow in the same direction. Control signals exit the MPU on some of the conductors forming the control bus while they travel to the microprocessor on others.

5-4 COMMUNICATING WITH I/O PERIPHERALS

5-4.1 Introduction

There are hundreds of different kinds of microprocessor-based systems. Some are designed for general-purpose applications, such as microcomputers; others are employed in systems that have specific, or dedicated, applications—such as those used in machine tool controllers, robot controllers, and programmable controllers. Most of the latter systems utilize multiple I/O peripherals, as shown in Figure 5-4. This is possible because the microprocessor performs its internal operations many times more rapidly than its peripherals are capable of generating and using the data.

When multiple peripherals are used, some means must be provided for identifying the devices. Two techniques have been developed for this purpose. They include memory-mapped I/O and I/O-mapped I/O. When memory-mapped I/O is employed, the I/O peripherals are addressed and data are read from and written into them in the same fashion as memory segments. Each peripheral is assigned a unique address when I/O-mapped I/O is used. This is a device address and should not be confused with a memory address. Microprocessors are designed to utilize one type or the other. Memory-addressing capacity is reduced when memory-mapped I/O is used, since some addresses have to be reserved for peripheral ad-

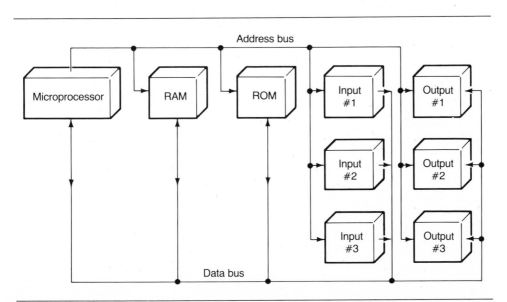

Figure 5-4

Microprocessor-based system utilizing multiple I/O peripherals

dressing. Control-bus pins have to be set aside for addressing peripherals when I/O-mapped I/O is employed. Memory-mapped I/O is utilized in most microprocessors. The resulting reduction in memory-addressing capacity is not significant in most microprocessor-based systems.

Although the microprocessor can be connected to many I/O peripherals, it can communicate with them only one at a time. This is analogous to the way in which humans talk to one another. When one is engaged in a conversation with several individuals within a group the mind focuses on the remarks made by individual members of the group. The microprocessor cannot do this automatically since it has no native intelligence. Two techniques have been developed to facilitate communication. One is called polling and the other is known as the interrupt. The method used depends on the architecture employed in the microprocessor. Both are discussed in the following sections.

5-4.2 Polling

Sometimes called programmed I/O, this is the simpler of the two methods. A polling loop is included as part of the program, as shown by the flow chart appearing in Figure 5-5. As the microprocessor proceeds through the program, it encounters the polling loops from time to time. Whenever these are encountered, the microprocessor stops whatever it is doing and checks the peripherals, one at a time, to determine if they require service. This is accomplished by checking for a set flag (flip-flop), which is usually connected to the peripheral. If no service is required of a particular peripheral, the microprocessor proceeds to the next peripheral, polls it, and continues the process until it reaches one that requires service. When this occurs, data are exchanged between that peripheral and the MPU and the microprocessor polls the next peripheral. Once all the peripherals have been polled, the microprocessor exits the loop.

Polling has the advantage of being simple and predictable. It requires no special control lines and data are exchanged between the MPU and peripheral only during the time the peripheral is being polled. The major disadvantage of polling is its inefficient use of software and time. Polling loops have to be written into the program and all the peripherals have to be checked whether they require service or not.

5-4.3 Interrupts

When this technique is employed, the peripheral needing service sends a signal, called an interrupt request, to the microprocessor which sets the interrupt request flag in the condition code register. At the end of each instruction, the microprocessor checks the state of this flag. If the flag is not set, the processor continues on to the next instruction. If the flag is set, the MPU services the peripheral requiring attention before going on to the next instruction. The program currently being executed is temporarily halted and the microprocessor branches to a peripheral subroutine in the program.

The interrupt method of communicating with the peripherals is a hardware tech-

Figure 5-5

Program flow chart
with polling loops

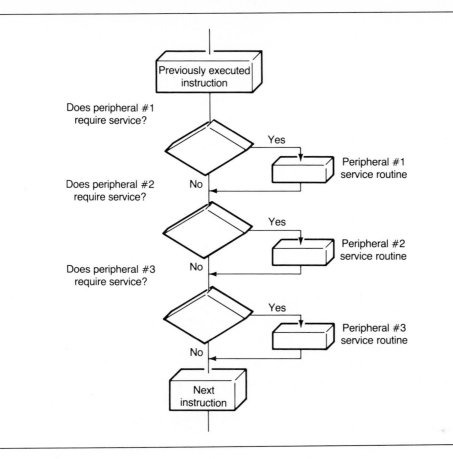

nique. It can be used only with those microprocessors that have interrupt-control capability. Most MPUs have this feature.

Two problems are associated with interrupts. One, a method must be provided to identify the peripheral requesting service; and two, some technique must be employed to establish a priority if two peripherals request attention simultaneously. Two methods are used to identify the peripheral requiring service. With one method, the microprocessor responds to an interrupt-request signal by sending an acknowledge signal to each peripheral one at a time. The peripheral requiring attention responds by sending a return signal to the microprocessor. When the microprocessor receives this signal it completes the instruction it is currently executing, places the information contained in the internal registers in the stack, and branches to the peripheral subroutine, as shown in Figure 5-6.

The other, and faster, technique is called vectored interrupt. Using this method, the peripheral needing service provides the microprocessor with both the interrupt-request signal and its address.

A number of microprocessors have provisions for two types of interrupts. These

Figure 5-6

Program flow chart
with interrupt loop

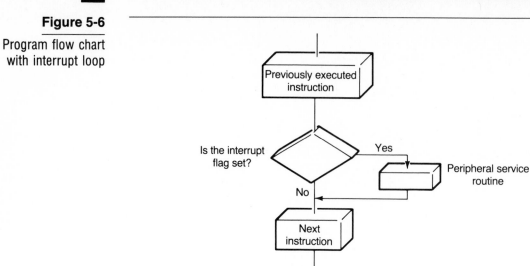

include a request for interrupt and a nonmaskable interrupt. When a request for interrupt is received, it may either be acted upon by the MPU or ignored. The nonmaskable interrupt cannot be ignored. When this interrupt is received, the microprocessor completes the instruction currently being executed and then clears its internal registers by transferring the information into the stack. The microprocessor branches to the subroutine in the program for that peripheral and the device is serviced. Information stored in the stack is written back into the original registers and the MPU begins normal program execution after the peripheral has been serviced.

As mentioned previously, a method of establishing peripheral priority must be provided in the event that two or more peripherals require service simultaneously. Since each peripheral can be addressed like memory, or has a unique device address, any number of techniques can be employed. In some systems, those peripherals sending the most important data to the microprocessor are assigned the highest priority. Priority is sometimes assigned by distance in applications where establishing by importance is not practical. In these cases, the peripherals located the greatest distance away from the microprocessors usually assigned the highest priority.

Manufacturers of many microprocessors produce a special priority-interrupt-controller (PIC) integrated circuit to be used with their processor for this purpose. This circuit provides the hardware necessary to implement the priority status of the peripherals.

5-4.4 Direct memory access

Direct memory access (DMA) lets a peripheral, such as floppy-disk mass memory, load data directly into main memory without going through the microprocessor. This allows for high-speed transfer of data from the peripheral to main memory, since the internal registers of the MPU do not have to be cleared by loading their contents into the stack. A special integrated circuit, called the direct memory access controller (DMAC), is used to accomplish this task.

5-5 THE STACK

The stack is a special memory used to store information contained in the internal registers of the MPU whenever a disruption in normal program execution occurs—a typical example being when an interrupt-request signal is received. Upon reception of an interrupt request, the microprocessor completes the execution of the current instruction and branches to a program subroutine that lets the MPU service the peripheral requiring attention. Information in the major registers—such as the program counter, instruction register, index register, accumulator, and CCR—is loaded into the stack. This frees the internal registers and allows the MPU to execute the instructions contained in the subroutine. Upon completion of the subroutine, the contents of the stack are loaded back into their original registers and the MPU branches back to the main program and resumes normal program execution.

In a few microprocessors, an internal register array is used as the stack. This is referred to as a hardware stack. In most microprocessors, part of the RAM is utilized as the stack. This is called a software stack. Usually only a few hundred bytes in size, any successive memory-byte segments can be established by the programmer for stack utilization. Even though RAM is often used as the stack, stack operates as a first-in, last-out (FILO) memory and is not random access. Special push instructions cause data to be loaded into the stack. Each time a push instruction is executed a new byte of data is loaded into the top of the stack. The bytes previously loaded are moved down one segment. After the MPU registers have been loaded

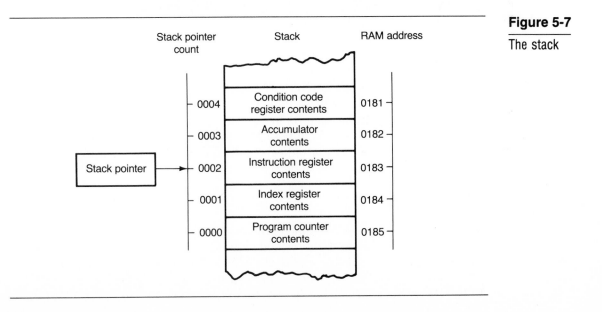

Figure 5-7

The stack

into the stack and the interrupt completed, special pop instructions cause the stack to be unloaded. The first byte to be unloaded from the stack is the last one placed there. As each succeeding pop instruction is executed the bytes are moved up one segment until the stack is cleared and all the data are loaded back into their original MPU registers.

The stack pointer is used to keep track of the information in the stack while the stack is being loaded and unloaded, as illustrated in Figure 5-7. The stack pointer decrements one count each time a new byte is loaded into the stack and increments one count every time a byte is unloaded.

5-6 THE 8080 FAMILY OF MICROPROCESSORS

5-6.1 Introduction

Since the first ones appeared in the early 1970s, scores of different microprocessors have been developed by a dozen or so different semiconductor manufacturers. Some of these were built for specific applications, while others were designed for general-purpose use. Although the basic operating principles for most micoprocessors are similar, each type has its own unique architecture and characteristics that differentiate it from the others. Microprocessor characteristics include items such as data word size, memory- and I/O-addressing capacity, case size, DC requirements, type of transistor technology employed, and maximum operating, or clock, speed.

Microprocessors initially used 4-bit data words. The first device to appear on the market was the 4004. Introduced by Intel Corporation in 1971, it was encapsulated in a 16-pin case and had 45 instructions. Intel soon followed this with an upgraded version called the 4040. This processor was housed in a 24-pin case and had 60 instructions. At about the same time Motorola, Texas Instruments, National Semiconductor, and Rockwell all began marketing their versions of the 4-bit microprocessor. Although few, if any, new applications are being developed for these processors today, many of these continue to be produced annually as replacements for those currently in use.

Eight-bit microprocessors soon followed the 4-bit MPUs and were the type primarily used from the mid 1970s through the early 1980s. These were followed by intermediate 8-bit, 16-bit, and 32-bit devices. Some of the more commonly used microprocessors are identified in the following sections to illustrate the growth and maturity of the microprocessor industry and to provide familiarity with the architecture and characteristics of these devices.

5-6.2 The 8080A

Intel's 8008 was the first 8-bit microprocessor to be marketed. Introduced in 1972, it was designed to be used as a cathode-ray-tube display controller. The company followed this a short time later with the 8080A. This was the first general purpose 8-bit microprocessor and became an immediate success.

The 8080A is an N-channel silicon-gate MOS (NMOS) microprocessor that uses a

Illustration 5-1

Intel 8080A
(Reprinted by
permission of Intel
Corporation.
Copyright 1979)

intel®

8080A/8080A-1/8080A-2
8-BIT N-CHANNEL MICROPROCESSOR

The 8080A is functionally and electrically compatible with the Intel® 8080.

- TTL Drive Capability
- 2 μs (−1:1.3 μs, −2:1.5 μs) Instruction Cycle
- Powerful Problem Solving Instruction Set
- 6 General Purpose Registers and an Accumulator
- 16-Bit Program Counter for Directly Addressing up to 64K Bytes of Memory

- 16-Bit Stack Pointer and Stack Manipulation Instructions for Rapid Switching of the Program Environment
- Decimal, Binary, and Double Precision Arithmetic
- Ability to Provide Priority Vectored Interrupts
- 512 Directly Addressed I/O Ports

The Intel® 8080A is a complete 8-bit parallel central processing unit (CPU). It is fabricated on a single LSI chip using Intel's n-channel silicon gate MOS process. This offers the user a high performance solution to control and processing applications.

The 8080A contains 6 8-bit general purpose working registers and an accumulator. The 6 general purpose registers may be addressed individually or in pairs providing both single and double precision operators. Arithmetic and logical instructions set or reset 4 testable flags. A fifth flag provides decimal arithmetic operation.

The 8080A has an external stack feature wherein any portion of memory may be used as a last in/first out stack to store/retrieve the contents of the accumulator, flags, program counter, and all of the 6 general purpose registers. The 16-bit stack pointer controls the addressing of this external stack. This stack gives the 8080A the ability to easily handle multiple level priority interrupts by rapidly storing and restoring processor status. It also provides almost unlimited subroutine nesting.

This microprocessor has been designed to simplify systems design. Separate 16-line address and 8-line bidirectional data busses are used to facilitate easy interface to memory and I/O. Signals to control the interface to memory and I/O are provided directly by the 8080A. Ultimate control of the address and data busses resides with the HOLD signal. It provides the ability to suspend processor operation and force the address and data busses into a high impedance state. This permits OR-tying these busses with other controlling devices for (DMA) direct memory access or multiprocessor operation.

8080A CPU FUNCTIONAL BLOCK DIAGRAM

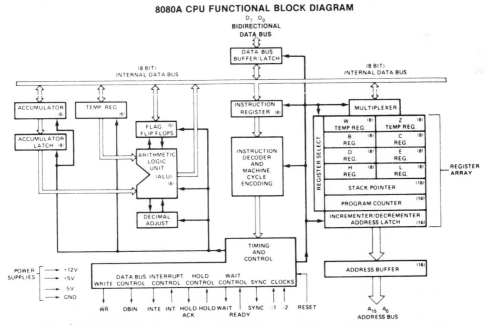

Figure 5-8

Pin diagram—
Intel 8080A

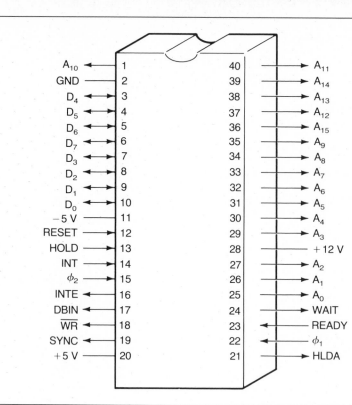

16-bit address capable of accessing 65,536 (65k) bytes of memory. As seen in Illustration 5-1 the architecture for this device includes an 8-bit accumulator, an ALU, a 5-bit condition code (flag) register, an 8-bit instruction register, an instruction decoder, and a timing and control unit. A seven-register array is included. The array includes a 16-bit stack pointer, a 16-bit program counter, six general-purpose registers (B–C, D–E, H–L), and two temporary registers (W–Z). The general purpose registers can be programmed to operate as three 16-bit registers to store addresses or six 8-bit registers to hold data. The W–Z registers are used for storing internal instructions and are not accessible to the programmer.

The pin, or case diagram appears in Figure 5-8. The 8080A is encapsulated in a 40-pin DIP case and is powered by three separate DC potentials. The case has 16 address pins, 8 data pins, and 12 pins that connect to the control bus. The four remaining pins are used to make the power-supply and ground connections. The 8080A requires two supporting integrated circuits—a clock generator and a system controller.

5-6.3 The 8085A

Although the 8080A was a popular microprocessor, the need for a separate system controller and three different DC potentials were obvious disadvantages. To over-

Illustration 5-2

Intel 8085A
(Reprinted by
permission of Intel
Corporation.
Copyright 1979)

intel®

8085A/8085A-2
SINGLE CHIP 8-BIT N-CHANNEL MICROPROCESSORS

- Single +5V Power Supply
- 100% Software Compatible with 8080A
- 1.3 μs Instruction Cycle (8085A); 0.8 μs (8085A-2)
- On-Chip Clock Generator (with External Crystal, LC or RC Network)
- On-Chip System Controller; Advanced Cycle Status Information Available for Large System Control

- Four Vectored Interrupt Inputs (One is non-Maskable) Plus an 8080A-compatible interrupt
- Serial In/Serial Out Port
- Decimal, Binary and Double Precision Arithmetic
- Direct Addressing Capability to 64k Bytes of Memory

The Intel® 8085A is a complete 8 bit parallel Central Processing Unit (CPU). Its instruction set is 100% software compatible with the 8080A microprocessor, and it is designed to improve the present 8080A's performance by higher system speed. Its high level of system integration allows a minimum system of three IC's [8085A (CPU), 8156 (RAM/IO) and 8355/8755A (ROM/PROM/IO)] while maintaining total system expandability. The 8085A-2 is a faster version of the 8085A.

The 8085A incorporates all of the features that the 8224 (clock generator) and 8228 (system controller) provided for the 8080A, thereby offering a high level of system integration.

The 8085A uses a multiplexed data bus. The address is split between the 8 bit address bus and the 8 bit data bus. The on-chip address latches of 8155/8156/8355/8755A memory products allow a direct interface with the 8085A.

MCS-80/85

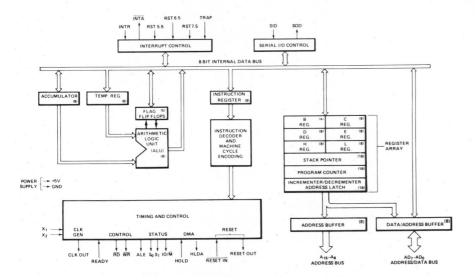

Figure 1. 8085A CPU Functional Block Diagram

come these and other less serious shortcomings, Intel developed the 8085A. Introduced in 1976, the device essentially consists of the 8080A MPU, a clock, and a system controller—all fabricated from one silicon substrate and powered by a single DC supply. The additional on-board circuitry created the need for extra pins. To facilitate its use in a standard 40-pin case, the 8085A is designed to time-share the address and data busses. A multiplexer allows the data bus to carry the lower eight address bits while the address bus carries the upper eight bits.

The data sheet for the device appears in Illustration 5-2. There is considerable similarity between the architectures for the 8085A and 8080A MPUs. The ALUs are identical in both processors. Since an on-board clock and system controller are included in the 8085A, the timing and control unit for this device is different. The register structure in each is similar except that the W—Z temporary register is eliminated in the 8085A. Software developed for the 8080A can be run on the 8085A.

The pin diagram is shown in Figure 5-9. The device has eight address pins, eight data pins, and 22 control pins. There is considerable difference between the signals present at the control pins of the 8085A and those of the 8080A. This is primarily due to the fact that an external system controller is not required for the 8085A.

Figure 5-9

Pin diagram—
Intel 8085A

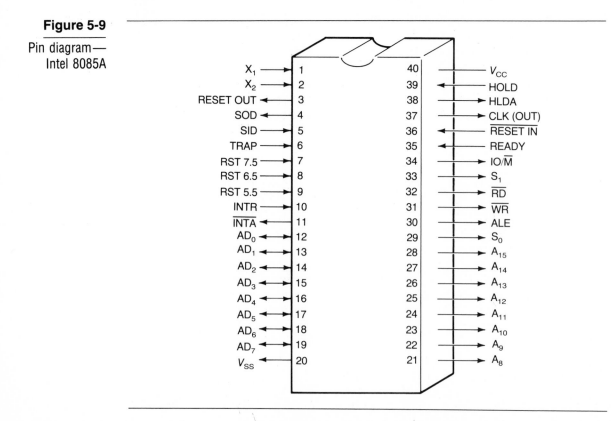

5-6.4 The 8088

The 8088, sometimes referred to as the iAPX 88/10, is the most powerful 8-bit microprocessor built by Intel. This MPU, along with a few similar types, such as Motorola's MC6809, is referred to as an intermediate, or in-between, 8-bit microprocessor. These devices have much of the processing power of a 16-bit MPU but utilize an 8-bit data word.

The data sheet for the device is seen in Illustration 5-3. This microprocessor employs a 20-bit address capable of accessing 1 Mbyte of memory. To facilitate its use in a 40-pin case, 12 of the address pins are shared with other bus pins. This is accomplished by multiplexing the least significant 8 bits of the address with the data bus and multiplexing the most significant 4 bits of the address with the control bus.

Architecture for the 8088 includes an ALU, a bus-interface unit, and an execution unit. The bus-interface unit contains five 16-bit registers, which include four segment registers (ES, CS, SS, DS) and an instruction pointer (IP). The segment registers are used when addressing particular memory segments. The instruction pointer performs the same function as the program counter in the previous microprocessors. Registers in the execution unit include four data registers (A,B,C,D) which can be operated as four 16-bit registers or eight 8-bit registers, two pointer registers (SP, BP) and two index registers (SI, DI).

External main memory is divided into 64 kbyte segments and is used for stack, for program storage, and for data storage. Each of the four segment registors points to one of the four memory segments. The extra segment (ES) register points to the memory segment used as a second storage area for data; the code segment (CS) register points to the memory segment where programs are stored; the stack segment (SS) register points to the memory segment used for stack memory; and the data segment (DS) register points to the memory segment that stores data.

5-6.5 The 8086

The 8086 is a first-generation 16-bit microprocessor. The 8080A and 8085A MPUs are LSI devices which employ NMOS transistor technology. The 8088 and 8086 are both VLSI devices which utilize HMOS technology. HMOS transistor technology permits the fabrication of approximately twice as many components on a semiconductor substrate of a given size when compared with conventional MOS. HMOS devices are also capable of operating at higher clock frequencies.

As shown in Illustration 5-4, the 8086 has a 16-bit ALU; a 16-bit flag register in which nine flags are used; a register file containing eight 8-bit registers that may be operated either as eight 8-bit registers or four 16-bit registers (which include an accumulator); a relocation and register file that includes a stack pointer, a base pointer, a source index, a destination index, and an instruction register, all 16 bits in size; and a control and timing unit. External memory is divided into 64 kbyte segments and is used for the same purposes and is addressed in the same fashion as the 8088.

The processor is housed in a 40-pin case and utilizes a 20-bit address capable of

Illustration 5-3

Intel 8088 (Reprinted by permission of Intel Corporation. Copyright 1981)

intel®

PRELIMINARY

iAPX 88/10 (8088) 8-BIT HMOS MICROPROCESSOR

- **8-Bit Data Bus Interface**
- **16-Bit Internal Architecture**
- **Direct Addressing Capability to 1 Mbyte of Memory**
- **Direct Software Compatibility with iAPX 86/10 (8086 CPU)**
- **14-Word by 16-Bit Register Set with Symmetrical Operations**
- **24 Operand Addressing Modes**
- **Byte, Word, and Block Operations**
- **8-Bit and 16-Bit Signed and Unsigned Arithmetic in Binary or Decimal, Including Multiply and Divide**
- **Compatible with 8155-2, 8755A-2 and 8185-2 Multiplexed Peripherals**

The Intel® iAPX 88/10 is a new generation, high performance microprocessor implemented in N-channel, depletion load, silicon gate technology (HMOS), and packaged in a 40-pin CerDIP package. The processor has attributes of both 8- and 16-bit microprocessors. It is directly compatible with iAPX 86/10 software and 8080/8085 hardware and peripherals.

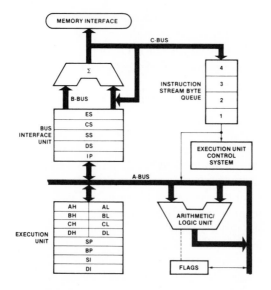

Figure 1. iAPX 88/10 CPU Functional Block Diagram

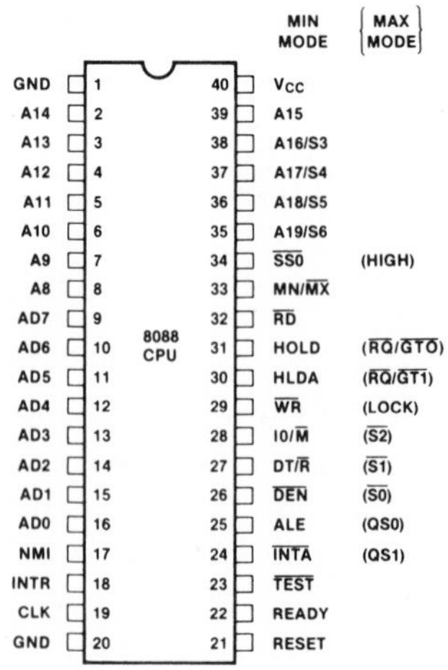

Figure 2. iAPX 88/10 Pin Configuration

Illustration 5-4

Intel 8086
(Reprinted by
permission of Intel
Corporation.
Copyright 1980)

intel®

8086/8086-2/8086-4
16-BIT HMOS MICROPROCESSOR

- Direct Addressing Capability to 1 MByte of Memory

- Assembly Language Compatible with 8080/8085

- 14 Word, By 16-Bit Register Set with Symmetrical Operations

- 24 Operand Addressing Modes

- Bit, Byte, Word, and Block Operations

- 8-and 16-Bit Signed and Unsigned Arithmetic in Binary or Decimal Including Multiply and Divide

- 5 MHz Clock Rate (8 MHz for 8086-2) (4 MHz for 8086-4)

- MULTIBUS™ System Compatible Interface

The Intel® 8086 is a new generation, high performance microprocessor implemented in N-channel, depletion load, silicon gate technology (HMOS), and packaged in a 40-pin CerDIP package. The processor has attributes of both 8- and 16-bit microprocessors. It addresses memory as a sequence of 8-bit bytes, but has a 16-bit wide physical path to memory for high performance.

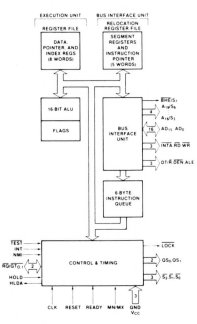

Figure 1. 8086 CPU Functional Block Diagram

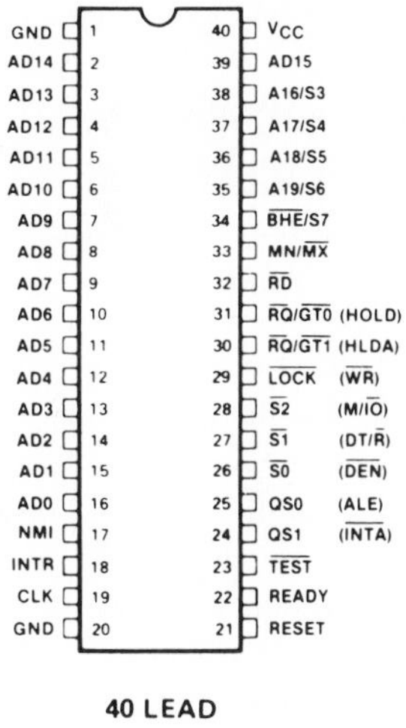

Figure 2. 8086 Pin Diagram

addressing 1 Mbyte of memory. To allow for its use in a 40-pin case, the lower two bytes of the address are multiplexed with the data bus while the upper four address bits are multiplexed with the control bus.

The 8086 is compatible with the software developed for the 8088 and the hardware and peripherals used with the 8080A and 8085A.

5-6.6 The 80286

Also known as the iAPX 286, this is a second-generation 16-bit microprocessor and is one of the most powerful 16-bit MPUs available. The 80286 contains approximately 140,000 transistors. System software and hardware developed for the 8086 can be used with this device.

5-6.7 The 80386

The 80386, or iAPX 386, is a first-generation 32-bit microporcessor. Introduced in 1985, the device contains almost 300,000 transistors and is fabricated from CHMOS II technology. CHMOS combines the high-density and high-frequency characteristics of HMOS with the low power consumption of CMOS. Both 12- and 16-MHz models are available, with instruction processing speeds of 3 to 4 million instructions per second (Mips). This high processing speed is made possible by the 80386's ability to fetch an instruction from memory while a second instruction is being decoded.

The 80386 is housed in a 132-lead ceramic pin-grid-array package. Software written for the 8086 and 80286 can be used with this microprocessor.

5-7 THE MC6800 FAMILY OF MICROPROCESSORS

5-7.1 The MC6800

The MC6800 was the first of several similar MPUs, produced by Motorola Incorporated, which have come to be known as the MC6800 family. The MC6800 was introduced at about the same time as the 8085. This 8-bit microprocessor is powered by a single DC source and utilizes a 16-bit address capable of accessing 65k bytes of memory. As seen by the data sheet appearing in Illustration 5-5, the MC6800 is housed in a 40-pin case and has 16 address pins, 8 data pins, and 11 control pins. Unlike the 8085, multiplexing is not used in the MC6800. The address, data, and control buses do not share signals. The device requires an external two-phase clock.

As shown by the simplified block diagram, or programming model, the MC6800 uses the same basic architectural scheme as the 8085, except for the number and types of registers used. This MPU has two 8-bit accumulators, a 16-bit stack pointer, a 16-bit index register, a 16-bit program counter, and an 8-bit condition code register that uses six flags. Although they are not shown, the device contains an ALU and a control unit. The MC6800 does not have the general-purpose regis-

Illustration 5-5

Motorola MC6800
(Courtesy of
Motorola, Inc.)

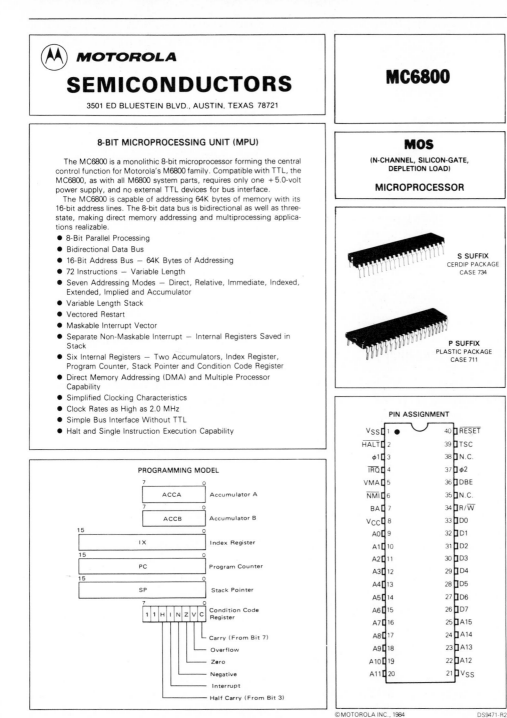

MOTOROLA

SEMICONDUCTORS

3501 ED BLUESTEIN BLVD., AUSTIN, TEXAS 78721

MC6800

8-BIT MICROPROCESSING UNIT (MPU)

The MC6800 is a monolithic 8-bit microprocessor forming the central control function for Motorola's M6800 family. Compatible with TTL, the MC6800, as with all M6800 system parts, requires only one +5.0-volt power supply, and no external TTL devices for bus interface.

The MC6800 is capable of addressing 64K bytes of memory with its 16-bit address lines. The 8-bit data bus is bidirectional as well as three-state, making direct memory addressing and multiprocessing applications realizable.

- 8-Bit Parallel Processing
- Bidirectional Data Bus
- 16-Bit Address Bus — 64K Bytes of Addressing
- 72 Instructions — Variable Length
- Seven Addressing Modes — Direct, Relative, Immediate, Indexed, Extended, Implied and Accumulator
- Variable Length Stack
- Vectored Restart
- Maskable Interrupt Vector
- Separate Non-Maskable Interrupt — Internal Registers Saved in Stack
- Six Internal Registers — Two Accumulators, Index Register, Program Counter, Stack Pointer and Condition Code Register
- Direct Memory Addressing (DMA) and Multiple Processor Capability
- Simplified Clocking Characteristics
- Clock Rates as High as 2.0 MHz
- Simple Bus Interface Without TTL
- Halt and Single Instruction Execution Capability

MOS

(N-CHANNEL, SILICON-GATE, DEPLETION LOAD)

MICROPROCESSOR

S SUFFIX
CERDIP PACKAGE
CASE 734

P SUFFIX
PLASTIC PACKAGE
CASE 711

PROGRAMMING MODEL

ACCA	Accumulator A
ACCB	Accumulator B
IX	Index Register
PC	Program Counter
SP	Stack Pointer
1 1 H I N Z V C	Condition Code Register

Carry (From Bit 7)
Overflow
Zero
Negative
Interrupt
Half Carry (From Bit 3)

PIN ASSIGNMENT

V_{SS}	1	40	RESET
HALT	2	39	TSC
$\phi 1$	3	38	N.C.
IRQ	4	37	$\phi 2$
VMA	5	36	DBE
NMI	6	35	N.C.
BA	7	34	R/W
V_{CC}	8	33	D0
A0	9	32	D1
A1	10	31	D2
A2	11	30	D3
A3	12	29	D4
A4	13	28	D5
A5	14	27	D6
A6	15	26	D7
A7	16	25	A15
A8	17	24	A14
A9	18	23	A13
A10	19	22	A12
A11	20	21	V_{SS}

DS9471-R2

ters that are used in the 8085. Instead, memory is used to hold the data that would normally be stored in these registers. This microprocessor is discussed in greater detail in Chapter 7.

5-7.2 The MC6809

This is an intermediate MPU and is the most poweful 8-bit microprocessor built by Motorola. As seen from the data sheet appearing in Illustration 5-6, the MC6809 is encapsulated in a 40-pin DIP case that has 16 address pins, 8 data pins, and 14 control pins. The device is normally capable of accessing 65k bytes of memory. This can be expanded to 2 Mbytes by using an ancillary integrated circuit called a memory management unit (MC6829).

The MC6809 has two 8-bit accumulators that may be operated as a single 16-bit accumulator or as two 8-bit accumulators. It has four 16-bit registers that can be used as index registers or stack pointers. An 8-bit direct-page (DP) register is used in addressing memory. The MC6809 has a 16-bit program counter and an 8-bit CCR that uses all eight flags. The device is compatible with the MC6800. Software and supporting integrated circuits developed for the MC6800 can be used with the MC6809.

5-7.3 The MC68000

Introduced in 1979, the MC68000 is a first-generation 16-bit VLSI HMOS microprocessor. As seen in Illustration 5-7, the device is housed in a 64-pin DIP case and has 24 address-bus pins, 16 data bus-pins, and 20 control-bus pins. None of the pins are multiplexed. Although the control bus is larger than those used in the MC6800 and MC6809 MPUs, it carries the same major signals. The architecture differs considerably from that used in the MC6800 and MC6809. The MC68000 utilizes eight 32-bit data registers, seven 32-bit address registers that can be programmed as fourteen 16-bit registers, two 32-bit stack pointers, a 32-bit program counter, and a 16-bit CCR. The eight data registers can be programmed to accept byte (8-bit), word (16-bit), or long-word (32-bit) data. Data are exchanged between the microprocessor, memory, and I/O peripherals 16 bits at a time. Data words are divided into two bytes, a lower and an upper, to allow the MC68000 to interface with 8-bit memory and I/O peripherals. This allows support ICs developed for the MC6800 and MC6809 to be used with this microprocessor. Four different versions of the MC68000 are available with maximum clock frequencies ranging from 4 to 10 MHz.

5-7.4 The MC68020

This is a first-generation 32-bit microprocessor. Introduced in 1984, the MC68020 is an HCMOS device that utilizes a 16.67 MHz clock and is capable of processing 2.5 million instructions per second. It can access over 4 Mbytes of memory. The device contains almost 200,000 transistors and is encapsulated in a square pin–grid–array package. Approximately 1 square inch in size, this type of package al-

Illustration 5-6

Motorola MC6809
(Courtesy of
Motorola, Inc.)

(Ⓜ) **MOTOROLA**

SEMICONDUCTORS

3501 ED BLUESTEIN BLVD., AUSTIN, TEXAS 78721

MC6809

8-BIT MICROPROCESSING UNIT

The MC6809 is a revolutionary high-performance 8-bit microprocessor which supports modern programming techniques such as position independence, reentrancy, and modular programming.

This third-generation addition to the M6800 Family has major architectural improvements which include additional registers, instructions, and addressing modes.

The basic instructions of any computer are greatly enhanced by the presence of powerful addressing modes. The MC6809 has the most complete set of addressing modes available on any 8-bit microprocessor today.

The MC6809 has hardware and software features which make it an ideal processor for higher level language execution or standard controller applications.

HMOS

(HIGH DENSITY N-CHANNEL, SILICON-GATE)

8-BIT MICROPROCESSING UNIT

P SUFFIX
PLASTIC PACKAGE
CASE 711

S SUFFIX
CERDIP PACKAGE
CASE 734

MC6800 COMPATIBLE
- Hardware — Interfaces with All M6800 Peripherals
- Software — Upward Source Code Compatible Instruction Set and Addressing Modes

ARCHITECTURAL FEATURES
- Two 16-Bit Index Registers
- Two 16-Bit Indexable Stack Pointers
- Two 8-Bit Accumulators can be Concatenated to Form One 16-Bit Accumulator
- Direct Page Register Allows Direct Addressing Throughout Memory

HARDWARE FEATURES
- On-Chip Oscillator (Crystal Frequency = 4 × E)
- $\overline{DMA}/\overline{BREQ}$ Allows DMA Operation on Memory Refresh
- Fast Interrupt Request Input Stacks Only Condition Code Register and Program Counter
- MRDY Input Extends Data Access Times for Use with Slow Memory
- Interrupt Acknowledge Output Allows Vectoring by Devices
- Sync Acknowledge Output Allows for Synchronization to External Event
- Single Bus-Cycle $\overline{RESET}$
- Single 5-Volt Supply Operation
- $\overline{NMI}$ Inhibited After $\overline{RESET}$ Until After First Load of Stack Pointer
- Early Address Valid Allows Use with Slower Memories
- Early Write Data for Dynamic Memories

SOFTWARE FEATURES
- 10 Addressing Modes
  - 6800 Upward Compatible Addressing Modes
  - Direct Addressing Anywhere in Memory Map
  - Long Relative Branches
  - Program Counter Relative
  - True Indirect Addressing
  - Expanded Indexed Addressing:
    - 0-, 5-, 8-, or 16-Bit Constant Offsets
    - 8- or 16-Bit Accumulator Offsets
    - Auto Increment/Decrement by 1 or 2
- Improved Stack Manipulation
- 1464 Instructions with Unique Addressing Modes
- 8 × 8 Unsigned Multiply
- 16-Bit Arithmetic
- Transfer/Exchange All Registers
- Push/Pull Any Registers or Any Set of Registers
- Load Effective Address

PIN ASSIGNMENT

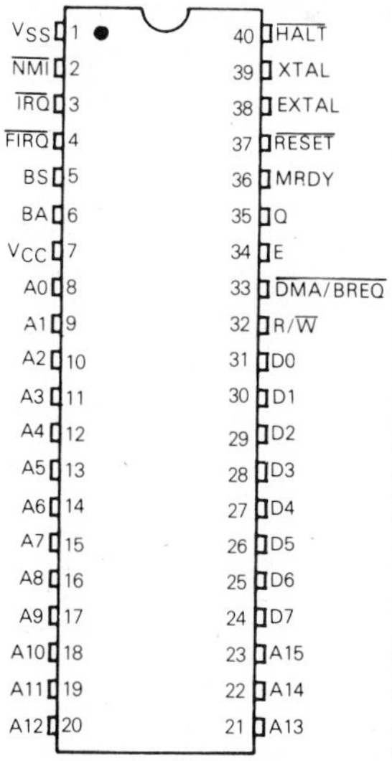

V$_{SS}$	1	40	HALT
$\overline{NMI}$	2	39	XTAL
$\overline{IRQ}$	3	38	EXTAL
$\overline{FIRQ}$	4	37	$\overline{RESET}$
BS	5	36	MRDY
BA	6	35	Q
V$_{CC}$	7	34	E
A0	8	33	$\overline{DMA}/\overline{BREQ}$
A1	9	32	R/$\overline{W}$
A2	10	31	D0
A3	11	30	D1
A4	12	29	D2
A5	13	28	D3
A6	14	27	D4
A7	15	26	D5
A8	16	25	D6
A9	17	24	D7
A10	18	23	A15
A11	19	22	A14
A12	20	21	A13

Illustration 5-7

Motorola MC68000
(Courtesy of
Motorola, Inc.)

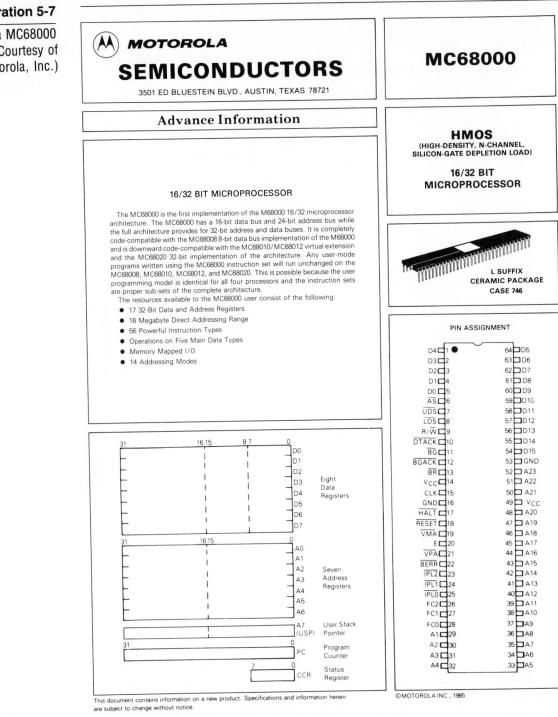

MOTOROLA

SEMICONDUCTORS

3501 ED BLUESTEIN BLVD., AUSTIN, TEXAS 78721

MC68000

Advance Information

HMOS
(HIGH-DENSITY, N-CHANNEL,
SILICON-GATE DEPLETION LOAD)

**16/32 BIT
MICROPROCESSOR**

16/32 BIT MICROPROCESSOR

The MC68000 is the first implementation of the M68000 16/32 microprocessor architecture. The MC68000 has a 16-bit data bus and 24-bit address bus while the full architecture provides for 32-bit address and data buses. It is completely code-compatible with the MC68008 8-bit data bus implementation of the M68000 and is downward code-compatible with the MC68010/MC68012 virtual extension and the MC68020 32-bit implementation of the architecture. Any user-mode programs written using the MC68000 instruction set will run unchanged on the MC68008, MC68010, MC68012, and MC68020. This is possible because the user programming model is identical for all four processors and the instruction sets are proper sub-sets of the complete architecture.

The resources available to the MC68000 user consist of the following:

- 17 32-Bit Data and Address Registers
- 16 Megabyte Direct Addressing Range
- 56 Powerful Instruction Types
- Operations on Five Main Data Types
- Memory Mapped I/O
- 14 Addressing Modes

**L SUFFIX
CERAMIC PACKAGE
CASE 746**

This document contains information on a new product. Specifications and information herein are subject to change without notice.

lows for the use of 120 pins. The MC68020 uses 114 of these pins, 32 connected to the address bus and 32 connected to the data bus.

Architecture forming the MC68020 includes a 32-bit program counter, a 32-bit vector base register, a 32-bit address register, a 32-bit control register, sixteen general-purpose 32-bit registers, and a 16-bit program counter.

5-7.5 The MC68030

The MC68030 is a second-generation 32-bit microprocessor whose performance is almost twice that of the MC68020. Introduced in 1987, the device has the same basic architecture as the MC68020. Processing speed has been increased and power consumption decreased by connecting many of the registers in parallel to allow multiple activities to occur simultaneously and by using CMOS technology. The device uses a clock frequency of approximately 30 MHz.

The MC68030 contains eight 32-bit data registers, seven 32-bit address registers. A 32-bit program counter, three 32-bit stack pointers, ten individual dedicated 32-bit registers, and three 16-bit registers including a CCR.

Summary

1. A microprocessor is a programmable integrated circuit that contains the CPU of a computer fabricated on a single semiconductor substrate.
2. A microprocessor-based system is made up of a microprocessor, memory, and I/O peripherals.
3. The internal circuits of a microprocessor are referred to as architecture.
4. The ALU contains the circuits that perform the arithmetic and logic operations.
5. The circuits forming the control unit decode the instructions fetched from memory and generate the control signals required to execute the instruction.
6. The CCR, sometimes called the flag or status register, monitors the contents of the accumulator.
7. The program counter keeps track of the instructions stored in memory as they are being executed.
8. The accumulator is the busiest register in the MPU. In arithmetic operations, it is used to store one of the numbers used in the operation. When the operation is completed, the resultant is stored in this register.
9. The instruction register temporarily stores all data entering and exiting the microprocessor.
10. The clock provides the synchronizing pulses required to control the timing of events within the microprocessor, memory, and I/O peripherals.
11. Buses are groups of conductors which connect the circuits within the MPU together and the MPU with its memory and I/O devices. Microprocessors utilize address, data, and control buses.

12. The microprocessor communicates with its I/O peripherals using either a polling or interrupt technique.

13. Direct memory access (DMA) is a technique that allows the microprocessor to be bypassed while data from mass memory are loaded into main memory.

14. The stack is a FILO memory used to store information currently contained in the internal registers of the MPU when normal program execution must be disrupted.

15. The 8080A was the first general-purpose 8-bit microprocessor. This MPU required three different DC potentials and two support chips.

16. The 8085A is an upgraded version of the 8080A that includes the 8080A architecture, and internal clock, and a system controller. It requires only a single DC power supply.

17. The 8088 is an 8-bit intermediate microprocessor. Intermediate MPUs have much of the processing power of a 16-bit microprocessor but use 8-bit data words.

18. The 8086 is a first-generation 16-bit microprocessor. Its architecture is similar to that of the 8088 except that a 16-bit data bus is used.

19. The 80286 is a second-generation 16-bit microprocessor.

20. The 80386 is a first-generation 32-bit microprocessor which utilizes CHMOS II transistor technology.

21. The MC6800 was the first of several microprocessors, built by Motorola Incorporated, known as the MC6800 family.

22. The MC6800 is an 8-bit microprocessor that employs two accumulators, an index register, a stack pointer, a program counter, a CCR, an instruction register, an ALU, and a control unit.

23. The MC6809 is an intermediate 8-bit microprocessor.

24. The MC68000 is a first-generation 16-bit microprocessor. This device is housed in a 64-pin DIP case.

25. The MC68020 is a first generation 32-bit microprocessor. This microprocessor uses HCMOS transistor technology.

26. The MC68030 is a second-generation 32-bit microprocessor which has almost twice the processing power of the MC68020.

Chapter Examination

1. What is a microprocessor?
2. What is the reason for the versatility of the microprocessor?
3. Draw a block diagram of a basic microprocessor.
4. List the three types of buses employed in microprocessors and identify the type of binary information carried by each.

5. What is the purpose of the clock in a microprocessor?

6. Which of the following registers temporarily stores the results of an arithmetic operation?
   a. accumulator
   b. stack pointer
   c. index register
   d. program counter

7. The _____ stores the address of the next byte of data to be fetched from memory.
   a. stack pointer
   b. index register
   c. program counter
   d. address register

8. This register monitors the contents in the accumulator and is made up of a series of flip-flops called flags.
   a. stack pointer
   b. index register
   c. program counter
   d. condition code register

9. This register stores an instruction while it is being decoded.
   a. stack pointer
   b. index register
   c. instruction register
   d. condition code register

10. The _____ generates the signals required to fetch and execute instructions stored in memory.
    a. control unit
    b. accumulator unit
    c. arithemtic and logic unit
    d. condition code register unit

11. The _____ is a memory used to store information contained in internal registers when a disruption in program execution occurs.
    a. stack
    b. accumulator
    c. index register
    d. instruction register

12. The _____ utilizes three DC voltages, an external clock, and an external system controller.
    a. 8080A
    b. 8085A
    c. MC6800
    d. MC6809

13. Which of the following is an intermediate MPU?
    a. 8086
    b. 8085A

    c. MC6800

    d. MC6809

14. _____ is a technique that allows data stored in mass memory to be loaded into main memory without going through the microprocessor
    a. DMA
    b. Polling
    c. Stacking
    d. Multiplexing

15. Which one of the following microprocessors uses a 16-bit ALU?
    a. 80386
    b. MC68000
    c. MC68020
    d. MC68030

16. An I/O peripheral that requires service sends a signal to the microprocessor when the _____ method is used to communicate between the MPU and its I/O peripherals.
    a. direct
    b. polling
    c. indirect
    d. interrupt

17. T F An 8-bit microprocessor must use an 8-bit address. ___

18. T F When interrupts are employed, the peripheral requiring attention sends an interrupt-request signal to the MPU. ___

19. T F The address register stores the address of the next byte of data to be fetched from memory. ___

20. T F The instruction register is used to store an instruction fetched from memory while it is being decoded. ___

21. T F The 8088 is capable of accessing a maximum of 65k bytes of memory. ___

22. T F Data are exchanged between the MC68000, memory, and I/O peripherals 16 bits at a time. ___

23. T F The data bus is bidirectional. Data can be transferred from the MPU to memory and I/O peripherals and vice versa. ___

24. T F The address bus is multiplexed with the data bus in the MC6800. ___

25. T F The address bus is bidirectional. Addresses can originate at the MPU, memory, or I/O peripherals. ___

26. T F Interrupts are a software technique used by microprocessors to communicate with I/O peripherals. ___

27. T F DMA is a technique, used in some microprocessors, that allows data stored in mass memory to be directly loaded into the accumulator of the microprocessor. ___

28. T F The nonmaskable interrupt can be either acted upon or ignored by the mircoprocessor. ___

29. T F The stack is a special control unit that lets the microprocessor ser- ___
vice multiple I/O peripherals simultaneously.

30. T F The MC6809 has a 16-bit accumulator that can be programmed to ___
operate either as two 8-bit accumulators or as a single 16-bit accumulator.

31. T F The 80386 uses a 32-bit program counter. ___

32. T F The MC68020 employs CMOS transistor technology. ___

CHAPTER
6

COMPONENTS AND TECHNIQUES USED TO FORM μP SYSTEMS

OBJECTIVES

1. Define the terms *opcode, operand,* and *instruction set.*

2. Describe the difference between machine- and assembly-language programming.

3. Identify the four general classes of microprocessor instructions.

4. Define the term *addressing mode* and identify seven modes commonly used with microprocessors.

5. List the two types of high-level programming languages and identify an example of each.

6. List six types of input peripherals.

7. List three types of output display peripherals.

8. Identify the two types of seven-segment LED displays and discuss the biasing requirements for each.

9. List three types of printers.

10. List the four standardized binary levels used in serial data transmission and identify the binary zero and one levels for each.

11. Define the term *baud rate.*

12. Describe the characteristics of asynchronous and synchronous serial transmission.

13. Define the term *parity,* identify two types of parity testing, and explain how each is achieved.

14. Identify the three modes of serial transmission and discuss the operating principles of each.

15. Discuss the purpose of a modem, identify the circuits forming the device, and describe its operation.

16. Identify the two categories of problems associated with interfacing peripherals with microprocessors.

17. Discuss the purpose of three-state buffers, identify their types, and describe how each is enabled.

18. Identify two ICs used to interface I/O peripherals to a microprocessor.

19. Identify two types of D/A converters and describe the operating principles of each.

20. identify two types of A/D converters and discuss the principles of operation for each.

21. Discuss the purpose of optical isolators as used in microprocessor interfacing.

22. Describe how relays, solenoids, DC stepping motors, fluid flow-control valves, and fluid direction-control valves are interfaced to a microprocessor.

23. Discuss the general techniques used to interface multiple I/O peripherals to a microprocessor.

6-1 INTRODUCTION

Regardless of its application and the type of system in which it is employed, only two things can be done to the microprocessor. It can be programmed and it can be interfaced. Programming involves listing the instructions needed to cause the microprocessor to execute a series of operations. Interfacing includes the circuits and techniques used to connect memory and I/O peripherals to the microprocessor.

6-2 ASSEMBLY-LANGUAGE PROGRAMMING TECHNIQUES

6-2.1 Introduction

Since microprocessors contain no native intelligence, they have to be programmed to perform their tasks. A program is a list of instructions that tells the microprocessor how to perform a particular operation. Instructions usually include both a command and data that are manipulated by the command. Since microprocessors respond to binary words only, the commands and data must be in binary format. Each different command is assigned a unique binary word called an operational code or opcode for short. Binary data manipulated by the microprocessor as instructed by the opcodes are called operands.

Each type of microprocessor has its own unique set of commands, or opcodes. Called an instruction set, these are the commands the microprocessor obeys. The number of commands appearing in the instruction set varies with the type of microprocessor. Some have fewer than 40 while others have more than 100.

Several techniques of programming microprocessors have been developed. The most elementary of these is called machine-language programming. With this method, the programming characters are in the same binary format as that used by the microprocessor. The program instructions, in the form of binary ones and zeros, are loaded into memory bit by bit.

Machine language is seldom used to program microprocessors today. Instead, one of several higher language forms is employed, one of which is assembly-language. In assembly-language programming, abbreviations called mnemonics are used to represent instructions. A hexadecimal number representing the opcode is assigned to each mnemonic. A code converter, called an assembler, changes the hexadecimal number into an equivalent binary number before it is placed in memory. Assembly-language programming allows the programmer to enter only the hexadecimal opcode, rather than the full binary word as in machine-language programming.

6-2.2 Classes of instructions

In general, microprocessors execute instructions by transferring data from memory or input peripherals to their internal registers, operating on the data, and sending the results back to memory or an output peripheral. Each different instruction causes a unique type of activity to occur within the microprocessor. Instructions that perform similar operations are grouped into categories called classes of instructions. Although instruction classes vary somewhat with the type of microprocessor, in general they include data movement, data manipulation, program manipulation, and program status.

Data-movement instructions cause data to be transferred between memory and the microprocessor's internal registers, between internal registers within the microprocessor, and between the microprocessor and its input/output (I/O) peripherals. Data-manipulation instructions cause data to be moved into the arithmetic and logic unit and include the arithmetic and logic instructions that cause data to be operated upon. Program-manipulation instructions cause the normal sequence of program execution to be altered. Instructions in this category include the jump and branch opcodes needed to handle program subroutines. Program status instructions are those that affect the operation of the complete microprocessor-based system.

6-2.3 Addressing modes

Program instructions consisting of opcodes and operands are loaded into memory prior to being executed. In certain circumstances addresses of operands may also be stored in memory. In an 8-bit microprocessor, the opcode occupies one byte of memory and the operand or operand address is located in a different memory segment byte. Often, the operand is placed in the memory segment immediately following the opcode. This has the advantage of requiring less memory and allows for faster instruction execution. Subroutines, information tables, and blocks of data required in the program may be more conveniently stored elsewhere in memory. In this situation, the address or partial address of the operand is located in the memory segment immediately following the opcode. The technique used to address the operand is referred to as the addressing mode. Most microprocessors use several different addressing modes. The more common modes are the inherent, immediate, direct, indirect, index, relative, and register. Those used with the Motorola MC6800 are discussed in Chapter 7.

6-3 HIGHER LEVEL PROGRAMMING LANGUAGES

Although assembly language has simplified the programming process, writing efficient programs still requires considerable time and skill. In addition, a new set of instructions has to be learned for each different type of microprocessor. Higher level languages, which utilize English words and decimal numbers to further simplify programming, have now been developed. They may be grouped into two categories—compiler and interpret. These are the languages used to program computers and many microprocessor-based systems. FORTRAN (*FOR*mula *TRAN*slation) and COBOL (*CO*mmon *B*usiness *O*riented *L*anguage) are examples of compiler languages. A compiler, usually in the form of a permanent internal program, converts the program into machine language. BASIC (*B*eginners *A*ll-purpose *S*ymbolic *I*nstruction *C*ode) is a popular interpret language. The interpreter, an internal permanent program, converts the interpret language statements into machine code.

The compiler and interpret languages just identified are used in applications in which the primary operations performed by the microprocessor are arithmetic and data manipulation. Other forms of these languages have been developed where the microprocessor is connected to a machine tool, a robot manipulator, or is used to control an industrial process. These are discussed in Chapter 8.

6-4 INPUT PERIPHERALS AND OUTPUT DISPLAY PERIPHERALS

6-4.1 Introduction

Peripherals include the input devices, which allow information to be loaded into memory, and output devices, which are controlled by the mircoprocessor or which provide a visual display of the results of operations performed on the input data by the microprocessor. Input peripherals normally include hexadecimal keypads, keyboards, and switches. Output devices, which physically implement an operation, include relays, solenoids, and motors. Output peripherals, which provide a display of the system status or results, include LEDs, video monitors, and printers.

6-4.2 Hexadecimal keypads

The hexadecimal keypad is commonly used to program and place data into a microprocessor when assembly language is employed. The device contains push-button

Figure 6-1

Hexadecimal
keypad

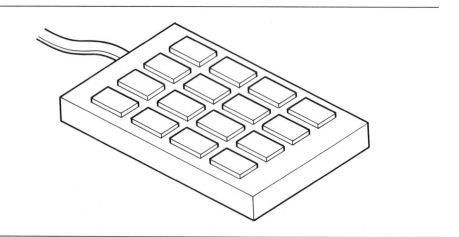

or touch switches for each hexadecimal alphanumeric character. A typical keyboard is depicted in Figure 6-1. The assembly-language program written in hexadecimal format is placed in memory by depressing the keys one at a time.

6-4.3 Keyboards

This is a popular input peripheral. With the large number of keys available for alphanumeric characters and symbols, a keyboard can be used to program a microprocessor with almost any type of language. A typical keyboard is illustrated in Figure 6-2. Encoders connected to the keys cause a binary word to be generated whenever a key is depressed.

6-4.4 Switches

Almost any type of switch can be used to place data into the microprocessor. Usually one side of the switch is connected to a DC source, as shown in Figure 6-3.

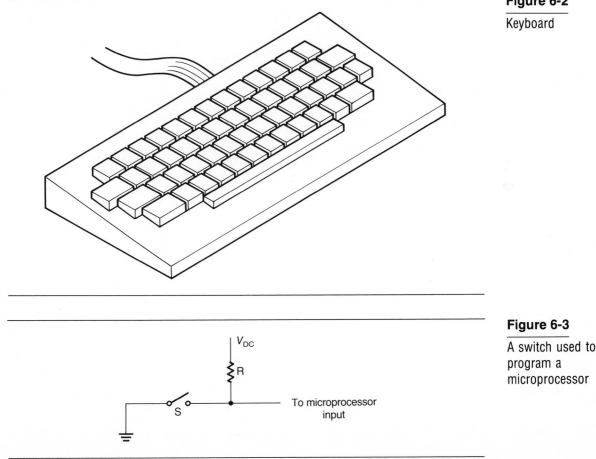

Figure 6-2

Keyboard

Figure 6-3

A switch used to program a microprocessor

Whenever the switch is closed a binary zero is produced. A binary one is represented when the switch is open. Toggle, push-button, limit, and pressure switches are commonly used as input devices in microprocessor systems used in manufacturing.

6-4.5 Light-emitting diodes (LEDs)

LEDs may be used to indicate the on–off status of a device or circuit or in an array to display an alphanumeric character.

The seven-segment LED is a common output display. This device is composed of seven (or eight if a decimal point is included) light-emitting diodes arranged in an array, as seen in Figure 6-4(a), and packaged as shown in Figure 6-4(b). The desired alphanumeric character is displayed whenever the proper diodes are forward biased. There are two types of displays—common anode and common cathode. In the common-anode type, the anodes are connected together, as illustrated in Figure 6-5. The desired diodes are turned on by applying a binary zero to their cathodes. The common-cathode display appears in Figure 6-6. The desired diodes are turned on by applying a binary-one potential to their anodes.

Figure 6-4

Seven-segment
LED array

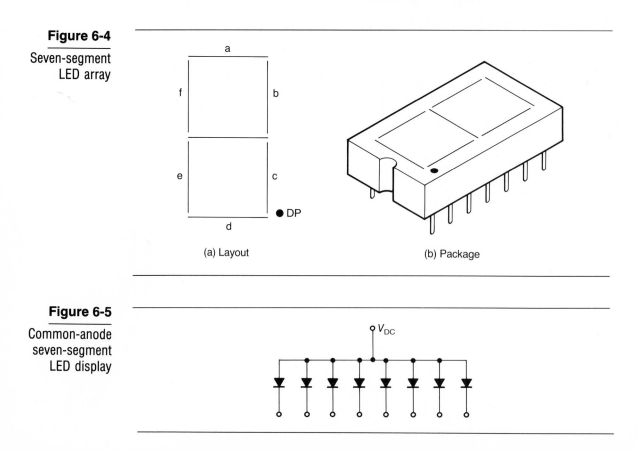

(a) Layout

(b) Package

Figure 6-5

Common-anode
seven-segment
LED display

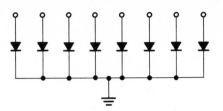

Figure 6-6

Common-cathode
seven-segment
LED display

6-4.6 Video monitors

This output peripheral is very popular and is used whenever a fast, inexpensive, temporary display is required. The video monitor operates much like a television receiver and includes many of the same circuits. Horizontal and vertical ramp waveforms generated internally create a series of horizontal lines on which alphanumeric characters and graphic symbols can be depicted, as shown in Figure 6-7. The characters and graphic symbols come from a permanent program stored in memory called a character generator. Although video monitors have the disadvantage of not providing permanent information, they do have the advantage of supplying an entire page of text or graphic material which can be updated and changed instantaneously. Color monitors are available which provide a color display.

6-4.7 Printers

Printers are output display peripherals which provide permanent, or hard copy, information, as seen in Figure 6-8. Many types of printers are available for computer and microprocessor systems. Some operate relatively slowly, producing copy in

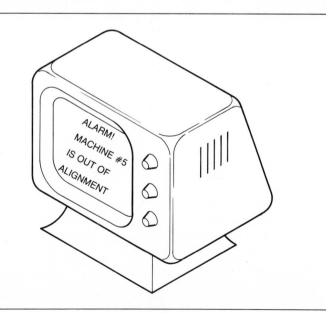

Figure 6-7

Video monitor

Figure 6-8

Printer

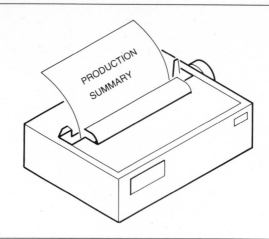

much the same fashion as the electric typewriter, while others operate very rapidly, printing an entire page almost instantaneously. Many printers can print graphic symbols as well as characters. Some printers produce monochrome copy while others generate copies having multiple colors.

Various methods are used to categorize the different types of printers. One method groups printers according to the technique used to form characters. These include impact overlay, dot-matrix, and page printers.

Impact overlay printers include the hammer, ball, and wheel types. These printers produce alphanumeric characters by the striking action of a character overlay against an inked ribbon and paper.

Several techniques of forming characters are employed in dot-matrix printers. One commonly used method employs pulsating electrodes, called print wires, mounted to a print head, as illustrated in Figure 6-9. When selected print wires are actuated, they strike an inked ribbon or specially coated paper, producing dots in the form of alphanumeric characters. The ink-jet printer is another popular type of dot-matrix printer. Ink ejected from orifices in the print head causes dots to be

Figure 6-9

Dot-matrix print head

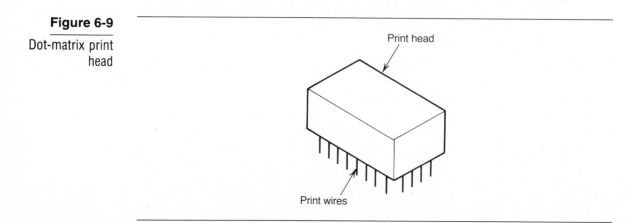

formed on the paper in patterns corresponding to alphanumeric characters. Dot-matrix printers are faster and more reliable than the impact overlay types. Printing speeds in the vicinity of 200 lines per minute are common.

A number of different types of page printers are available. These include electro-photographic laser printers, magnetographic printers, and ion-deposition printers. Printing speeds are very rapid—with some models capable of printing 90 pages per minute. Page printers are the fastest and most expensive. They are usually used with mainframe computers whose major activity is the very rapid production of information.

6-4.8 Terminals

A terminal is a printer and keyboard combination located at the same site, or work-station. A video monitor may also be included. Terminals perform input and output peripheral functions, as data can be both sent to and received from the microprocessor.

Many terminals contain their own microprocessor, which lets them perform many routine operations. Called smart, or intelligent, terminals, these free the host computer or microprocessor for more important activities.

6-5 PRINCIPLES OF SERIAL TRANSMISSION

6-5.1 Introduction

Binary data within the microprocessor always moves in a parallel format. This is done for the purpose of speed. Likewise, data transferred between the microprocessor and its main memory ICs are in parallel form. Data sent between the microprocessor and its I/O peripherals may either be in the form of parallel or serial binary bits. Since a separate conductor is required for each bit when parallel transmission is used, it is not practical to employ parallel transmission whenever the I/O peripherals are located far from the microprocessor. In addition, many peripherals generate or receive data only in serial format. These include switches, keypads, printers, video monitors, and many mass-memory mediums.

To provide a means of standardizing binary levels and transmission speeds, standards have been developed for the transmission of serial data. These are discussed in the following sections.

6-5.2 Binary-level standards for serial data

Four different amplitudes have been developed to define binary one and zero levels for serial transmission. These include the RS-232C, TTL, 20-mA current loop, and 60-mA current loop. The RS-232C standard defines a binary one as a voltage whose amplitude is -12 volts. A binary zero is represented by a voltage which has an amplitude of $+12$ volts. This is the most popular of the four amplitude standards. The TTL standard defines a binary one as +5 volts and a binary zero as 0 volts.

The 20- and 60-mA current loops are both current standards. A binary one is represented by the presence of 20 mA of current and a binary zero is represented by the absence of current when the 20-mA current loop is utilized. The presence or absence of 60 mA represents a binary one or zero respectively when the 60-mA current loop standard is employed.

Binary levels used in serial data transmission are sometimes referred to as marks and spaces. A mark indicates a binary one whereas a space represents the presence of a binary zero.

6-5.3 Baud rate

The rate at which binary bits are transmitted in serial format is referred to as baud rate. Baud rate requirements vary with the type of application. Standard baud rates are 50, 75, 110, 150, 300, 600, 1200, 2400, 4800, 9600, and 19,600 bits per second (bps).

6-5.4 Types of serial transmission

There are two types of serial transmission—asynchronous and synchronous. Asynchronous is the more common of the two. Any one of the four binary levels previously identified may be utilized with this type of transmission. Clock pulses are not used to control the timing sequence between the peripheral device and the microprocessor in asynchronous transmission. Since the data are transmitted in a train of bits, it becomes difficult to determine when one word stops and the next one begins unless clock pulses are present at the transmitter and receiver points, which may cause the receiver to lose its timing and scramble the incoming data. This problem is solved by using a series of start and stop bits to enclose the bits forming the word, as illustrated in Figure 6-10. The start bit is represented by a binary zero. This is followed by the 8-bit word with the least significant bit (D_0) appearing immediately after the start bit. Two stop bits, having binary one levels, follow the most significant bit (D_7). Therefore the total word is composed of 11 bits and includes 1 start bit, 8 data bits, and 2 stop bits.

Clock pulses are sent to both the microprocessor and peripheral when synchronous serial transmission is utilized. No start and stop bits are required. Binary data are sent very rapidly in the form of blocks. To ensure synchronization between the microprocessor and peripheral, synchronizing characters are transmitted at 100-

Figure 6-10

An asynchronous binary word

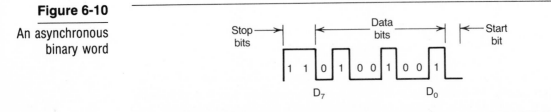

byte intervals. Since three fewer bits are transmitted per binary word, synchronous serial transmission is faster than the asynchronous type. Although asynchronous serial transmission is slower, it is nonetheless adequate in many applications. Most peripherals connected to a microprocessor operate much more slowly than the MPU. In those instances where high-speed peripherals are used, or where the microprocessor sends data in serial format to a larger computer, synchronous transmission is employed. Synchronous serial transmission has the disadvantage of requiring an extra pair of conductors to carry the clock pulses.

6-5.5 Parity

Parity is a technique used to check for errors in serial transmission. The chances that errors will occur increases as the distance between the microprocessor and peripheral device becomes greater. External phenomena such as magnetic fields may cause a voltage to be induced across the bus during the time that a binary zero is being sent. This can cause the zero to appear as a binary one, creating a bit error. Parity is accomplished by establishing either an even or odd number of binary one bits in the binary word being transmitted.

The most significant bit in the data word is reserved for the parity bit if parity is being utilized. Two techniques are employed—even and odd. If even parity is being used, the circuit which generates the parity bit (parity generator) checks the number of binary ones appearing in the word. If there are an odd number, the parity generator inserts a binary one into the most-significant-bit position to make the total number of binary ones even. If an even number of binary ones appears in the word, a binary zero is placed in the most significant bit position so that the total number of binary ones remains even. The same technique is used for odd parity except that the parity bit causes the total number of binary ones to be an odd number. If any one of the eight bits should change as the data are transmitted from the source to the destination, a parity detector, located at the destination, will recognize that the data are not valid.

6-5.6 Binary codes used in serial transmission

Although any binary code can be transmitted in serial format, most systems employ the American Standard for Information Interchange (ASCII). This is the standard code produced by most keyboards. Whenever a key is depressed, an encoder generates a seven-bit binary number associated with that particular key. A parity bit can be added to form an 8-bit data word.

6-5.7 Transmission modes

Three modes of transmission are possible whenever serial transmission is used—simplex, half-duplex, and full-duplex. The simplex mode is a technique used when one-way transmission is required. Data are transmitted from a peripheral device (transmitter) to the microprocessor (receiver) or from the microprocessor (trans-

Figure 6-11

Simplex operation

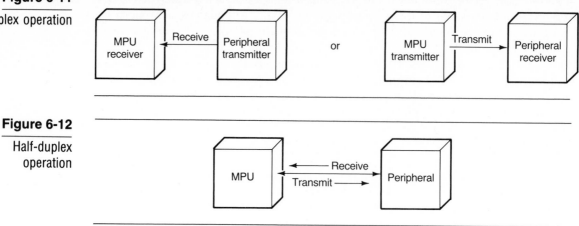

Figure 6-12

Half-duplex
operation

mitter) to a peripheral (receiver). The single-pair transmission line, or data bus, is either a transmit or receive line, but not both, as shown in Figure 6-11.

A single data-transmission-line pair is also employed in half-duplex operation. It, too, can be used as either a transmit or receive line, but not both at the same time. As illustrated in Figure 6-12, it serves as a receive line when data are sent from a peripheral to the microprocessor, and as a transmit line when data are transmitted from the microprocessor to a peripheral.

The full-duplex mode of operation requires two pairs of transmission lines, as seen in Figure 6-13. One pair serves as a transmit line allowing data to be sent from the mircoprocessor to a peripheral. The other is a receive line and carries data sent from a peripheral to the microprocessor. The microprocessor can both send and receive data simultaneously since two separate pairs of transmission lines are used.

6-5.8 Modems

Short for *mod*ulation–*dem*odulation, a modem is a device that allows the transmission of binary data over telephone lines. In applications where a peripheral or terminal is located a considerable distance from the microprocessor, it is not practical to use a conventional bus to connect the units together. Instead, telephone lines are used as the serial-data bus.

A modem consists of a modulator, demodulator, and filter, as shown in Figure 6-14. A modem is required at both the microprocessor and peripheral ends of the system. These are referred to as send and receive modems respectively. The mod-

Figure 6-13

Full-duplex
operation

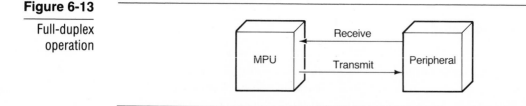

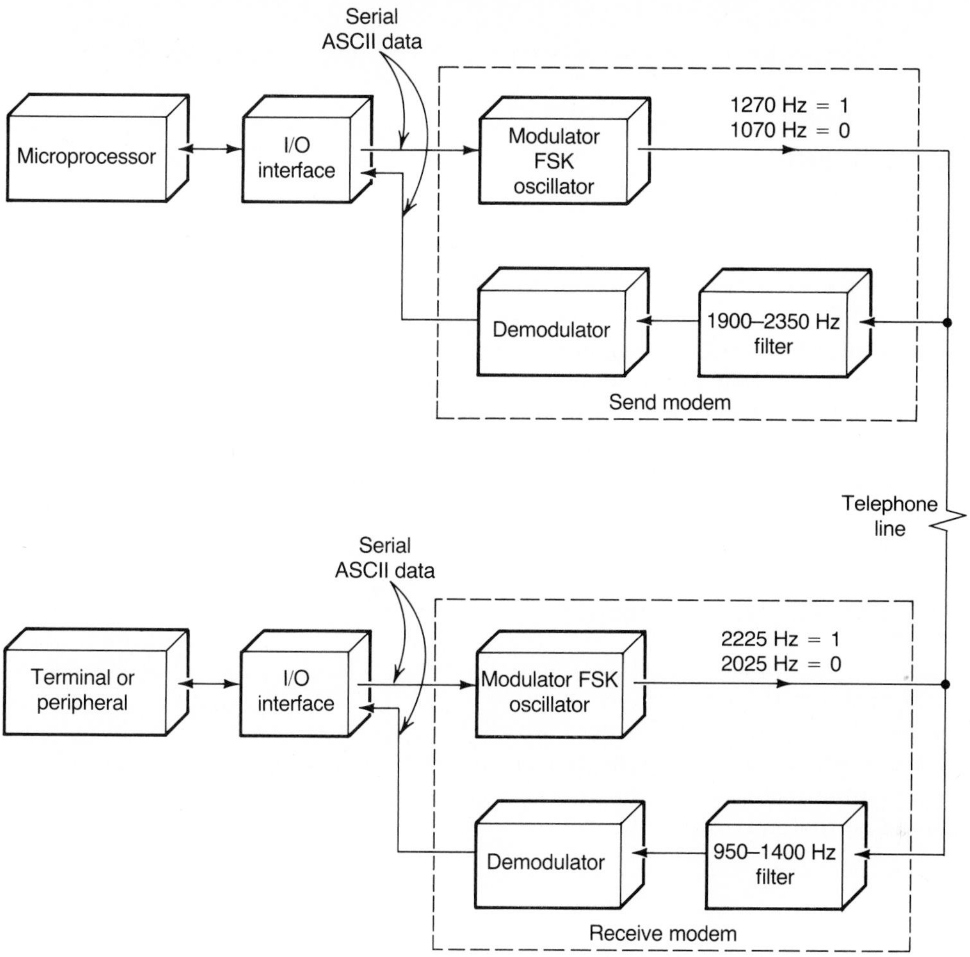

Figure 6-14

Microprocessor-based system employing modems

ulator in the send modem converts serial data coming from the microprocessor system, usually in the form of ASCII binary code, into audio tones, which can be transmitted through the telephone line. The modulator employs a technique called frequency-shift keying (FSK), in which binary ones are converted into audio tones, called marks, that have a frequency of 1270 Hz. Binary zeros are converted into 1070-Hz audio signals, called spaces. The FSK signal is transmitted through the telephone line to the receive modem. There, the FSK signal passes through a 950–1400-Hz bandpass filter and is reconverted into serial binary data by the demodulator and sent to the peripheral.

Whenever a peripheral sends data, the serial binary data, again usually is ASCII code form, are converted into mark and space signals having frequencies of 2225 and 2025 Hz, respectively, by the modulator in the receive modem. These audio signals are transmitted via the telephone line to a 1900–2350-Hz bandpass filter and demodulator in the send modem, where they are reconverted into ASCII code.

Data can be both sent to and received from either the microprocessor or peripheral at different times since two different mark and space frequencies are used.

6-6 MICROPROCESSOR INTERFACING

6-6.1 Introduction

A microprocessor-based system consists of a microprocessor connected to memory ICs and I/O peripherals via the address, data, and control buses. The design and implementation of such a system is seemingly simple and straightforward. It appears that one merely chooses a microprocessor, selects the required I/O peripherals, and adds sufficient memory for the processor to perform its activities. Unfortunately, this is usually not the case. There are inherent problems that must be solved before a system can become operational. These can be grouped into two broad categories—problems created by using a common data bus and those due to the incompatibility of the microprocessor and its I/O peripherals.

6-6.2 Common-data-bus problems

A simple microprocessor-based system appears in Figure 6-15. It includes a 4-bit microprocessor that utilizes an 8-bit address, RAM, ROM, an input peripheral, and an output peripheral. The devices are all connected to a common data bus. Some means must be provided to disconnect, or isolate, the memory and I/O peripheral devices from the data bus when they are not sending or receiving data. This allows data to be routed between the desired devices and prevents multiple binary words

Figure 6-15

A basic microprocessor-based system

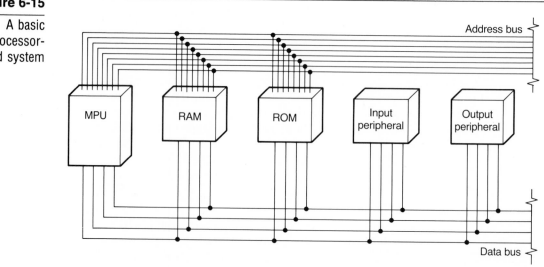

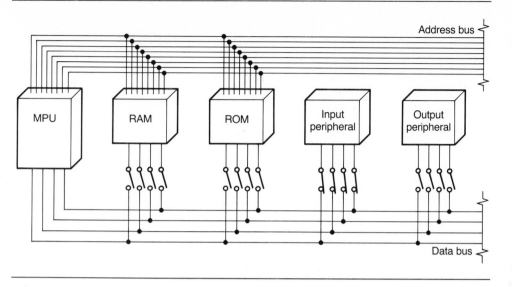

Figure 6-16

Toggle switches used to provide isolation

from appearing on the data bus simultaneously. Isolation could be accomplished by placing a toggle switch in each line leading to the data bus, as illustrated in Figure 6-16. When the switches are open the device is disconnected, or isolated, from the bus. Whenever that device must send or receive data, the switches are closed, thus enabling the device. Although this achieves the desired result, it is not a practical solution. Instead of manually operated switches, electronic switches called three-state buffers are employed.

A three-state buffer is an inverter with three terminals, shown in Figure 6-17. The circuit remains in the off or high-impedance state until a signal is applied to the enabling input causing the inverter to be turned on.

As illustrated in Figure 6-18, there are four types of three-state buffers. Two provide inversion between the input and output terminals, as seen in Figure 6-18(a) and (b), while two provide no inversion, as shown in Figure 6-18 (c) and (d). Two, (a) and (c), turn on whenever a binary one is applied to the enabling input, while a binary zero enables the buffers appearing in (b) and (d).

A three-state buffer array is connected between the RAM and data bus in Figure 6-19. The RAM is isolated from the bus until a logic level is applied to the enabling terminals of the buffers. This turns on the buffers connecting the RAM to the bus, allowing data to be written into or read from the memory device. In similar fashion, three-state buffers are connected between the data bus and each memory and pe-

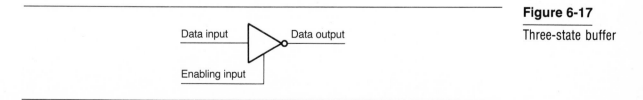

Figure 6-17

Three-state buffer

Figure 6-18

Types of
three-state buffers

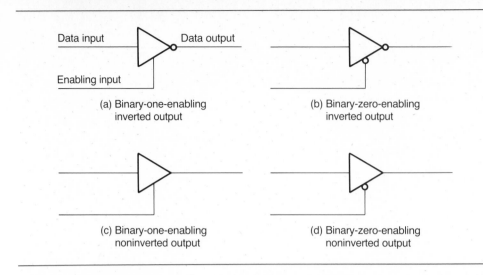

(a) Binary-one-enabling
inverted output

(b) Binary-zero-enabling
inverted output

(c) Binary-one-enabling
noninverted output

(d) Binary-zero-enabling
noninverted output

ripheral device forming a particular microprocessor system, as seen in Figure 6-20. An address decoder is connected between the address bus and the three-state–buffer-enabling input terminal for each memory and peripheral device. A unique address is assigned to each memory and peripheral device. This address should not be confused with the memory-segment address. Whenever data are to be sent to or recieved from a particular memory or I/O peripheral device, that device is first addressed. The address decoder associated with the device decodes the address and, depending upon the type of three-state buffer utilized, sends either a binary zero or binary one pulse to the enabling input of the three-state buffers connected to it.

The enabling pulse from the address decoder is often referred to as a chip-select (CS) or chip-enable (CE) signal. Clock pulses from the microprocessor are applied to the address decoders by way of the control bus to control the timing sequence

Figure 6-19

Three-state buffers
employed to
provide data-bus
isolation

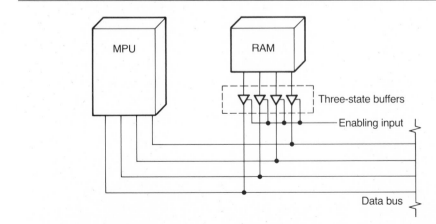

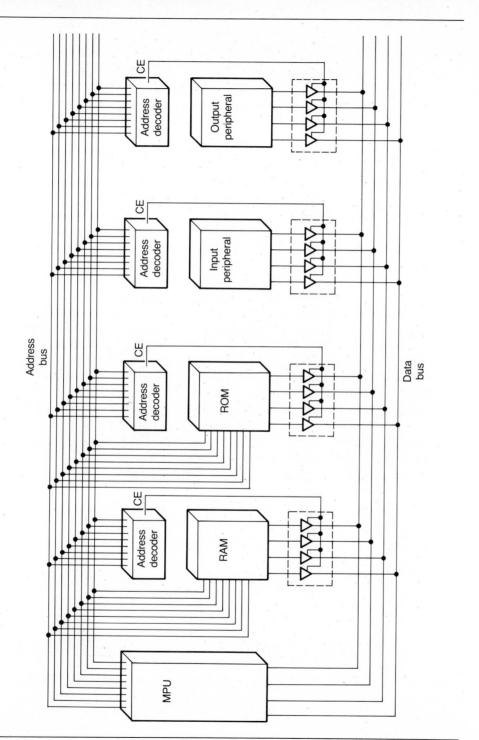

Figure 6-20

Microprocessor-based system with three-state buffers and address decoders

and the length of time that the enabling pulse is present. This ensures that the memory or peripheral device is enabled during the time that the data appears on the data bus.

The address decoder and three-state buffers may be built into the memory or peripheral device. This is commonly done with the RAM and ROM devices. External decoder and buffer ICs have to be included as part of the system if they are not part of the internal circuitry.

6-6.3 Incompatability

Often, the type of transistor technology used in memory and I/O peripherals differs from that used in the microprocessor. The microprocessor may be made from one type of semiconductor material while memory chips and ICs used in peripheral devices may be fabricated from another. Because of the electromechanical make up of many I/O peripherals, their operating speed may be considerably slower than that of the microprocessor. Variables such as operating speed, propagation delay, fan-in, fan-out, V_{CC} requirements, and binary voltage levels may also differ among the microprocessor, memory devices, and I/O peripherals. In addition, the data sent to the microprocessor from input peripherals such as switches, keypads, and keyboards are in serial format, whereas they must be in parallel form to be processed by the microprocessor. In similar fashion, some output peripherals require serial data while the data leaving the microprocessor are in parallel format. To complicate the interfacing problem even further, much of the input data are in the form of analog signals coming from sensors. These signals must be processed and converted into digital data. Likewise, many output peripherals require analog signals. The binary data coming from the microprocessor have to be changed into equivalent analog signals. Collectively, these problems present obstacles that could be difficult, if not impossible, to overcome. Fortunately, devices, circuits, buses, and techniques have been developed to overcome the effects of most obstacles. These are discussed in the following sections.

6-6.4 The peripheral-interface adapter

The peripheral-interface adapter (PIA) is a programmable integrated circuit designed to simplify the process of interfacing I/O peripherals to the microprocessor. Although any peripheral can be interfaced with a microprocessor using hard-wired logic circuits, the PIA is more flexible because it can be programmed to meet changing input/output circuit requirements. Most microprocessor manufacturers build PIA devices to be used with their major microprocessors. Depending upon the manufacturer, this circuit is known by a number of different names. It is sometimes called a programmable input/output (PIO), or a programmable peripheral interface (PPI). Whatever the name, all are programmable devices which are used to interface I/O peripherals to the microprocessor. Intel has developed the 8155 PPI for use with their 8-bit 8080 family processors whereas Motorola produces the MC6821 PIA for use with their MC6800 family microprocessors.

The PIA is connected between the microprocessor and I/O peripherals, as illustrated in Figure 6-21. It is connected to the microprocessor by way of the address, data, and control buses. It is connected to the I/O peripherals by peripheral data buses. Programmable internal registers within the PIA allow data to be transferred between the microprocessor and peripheral data buses. The peripheral data-bus terminals, sometimes referred to as ports, may be programmed to operate either as inputs or outputs. Data transferred between the PIA and peripherals may be either serial or parallel in format.

The PIA is addressed in the same fashion as main memory. The device and its internal registers are assigned addresses by the programmer. As the program is being executed, the PIA is accessed whenever data must be sent from an input peripheral to the microprocessor or whenever data must be transferred from the microprocessor to an output peripheral.

6-6.5 The RS-232 interface bus

Several active-interface buses have been developed to facilitate interfacing. These include logic circuits in addition to the individual bus conductors. Two of the more common of these are the RS-232 and the IEEE-488.

The RS-232 interface is an Electronics Industry Association (EIA) standard developed in 1969 that is widely used to interface serially operated peripherals. The standard has been revised periodically as more sophisticated equipment has been developed. The latest version is known as RS-232C.

The hardware used to implement the RS-232C consists of an interface board and a 25-conductor bus that carries control signals, data, ground, and the required DC potentials. It is frequently the interface used between the modem and microprocessor or peripheral. Because of its electrical limitations, the RS-232C bus length cannot exceed 50 feet. RS-232C interfacing circuitry is incorporated into most mainframe and minicomputers. Ports are provided in most microcomputers and other microprocessor-based equipment to allow for the use of this interface. The information carried on the individual bus lines is catalogued in Table 6-1.

The RS-232C has a maximum bit transmission rate of 20,000 bits per second. A newer standard called the RS-449 is gradually replacing the RS-232C. The RS-449

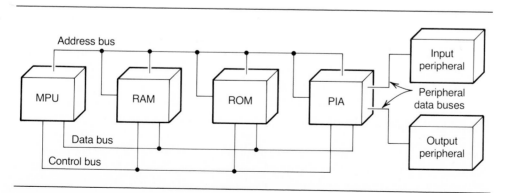

Figure 6-21

Interfacing I/O peripherals with a PIA

Table 6-1

RS-232C
INTERFACE-BUS
PIN ASSIGNMENT

PIN	FUNCTION
1	Protective ground
2	Transmitted data
3	Received data
4	Request to send (RTS)
5	Clear to send (CTS)
6	Data set ready (DSR)
7	Signal ground
8	Received-line signal detector
9	Used for data-set testing
10	Used for data-set testing
11	Not used
12	Secondary received-line signal detector
13	Secondary clear to send (SCTS)
14	Secondary transmitted data
15	Transmit signal timing
16	Secondary received data
17	Receiver signal timing
18	Not used
19	Secondary request to send (SRTS)
20	Data terminal ready (DTR)
21	Signal quality detector
22	Ring detector
23	Data signal-rate select
24	Transmit signal timing
25	Not used

permits bit-transmission rates up to 2 million bits per second and allows for greater distances between connected equipment.

6-6.6 The IEEE-488 interface bus

Sometimes called the general-purpose interface bus (GPIB), this parallel interface was originally developed by Hewlett-Packard Company and eventually adopted by the Institute of Electrical and Electronic Engineers (IEEE) as an interfacing stan-

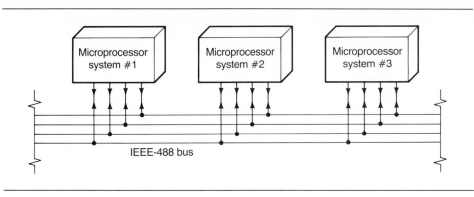

Figure 6-22

Interfacing
microprocessor-
based systems
together with an
IEEE-488 bus

dard. The IEEE-488 utilizes a 16-conductor bus, which may be used to carry parallel data between peripherals and the MPU or to connect multiple microprocessor-based systems together. A typical application appears in Figure 6-22. Most microprocessor manufacturers build an IC that interfaces their microprocessors with the IEEE-488 bus. Motorola produces the MC68488 for this purpose. This circuit makes their family of 8-bit MC6800 processors compatible with the IEEE-488 bus.

6-6.7 The universal asynchronous receiver transmitter

Since the microprocessor requires parallel data, the data generated by or sent to serially operated I/O peripherals must be converted to a parallel format or vice-versa. Serial in–parallel out or parallel in–serial out shift registers could be used to perform this conversion. However, additional hardware is required to insert parity bits, if used; start and stop bits for asynchronous transmission; and synchronizing bits for synchronous serial transmission. Control signals must be provided to insert these bits at the proper intervals.

Special ICs have been developed by most microprocessor manufacturers to simplify interfacing serial I/O peripherals to their microprocessors. Although marketed under various trade names they include the universal asynchronous receiver transmitter (UART), universal synchronous receiver transmitter (USRT), and the universal synchronous asynchronous receiver transmitter (USART).

The UART converts parallel binary data into asynchronous serial data and inserts the start, stop, and parity bits, or it changes asynchronous serial data into parallel data. The USRT is used for converting synchronous serial data into parallel data or vice versa while the USART can be used to convert either asynchronous or synchronous serial data into parallel data or parallel data into asynchronous or synchronous serial data.

The UART, or any one of the other devices, is connected directly to the microprocessor data bus, as seen in Figure 6-23. In this application, a microprocessor is connected to a remote terminal. Telephone lines are used as the serial data bus to link the microprocessor and the terminal. Parallel data from the microprocessor is converted into serial data by the UART. An RS-232C interface bus connects the

Figure 6-23

Parallel-to-serial
data conversion
using a UART

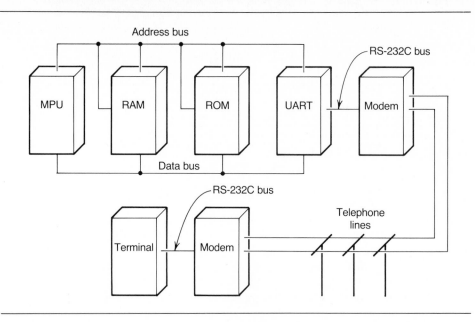

UART to a modem. A UART is not required at the terminal end of the system since a terminal receives and sends data in serial format.

6-6.8 D/A and A/D converters

Output data from microprocessors that operate valves, open and close heating and ventilation dampers, or energize relays and motors must be in analog form. Similarly, input data coming from sensors and linear electronic circuits is analog in format. Since the microprocessor operates only on digital data, the input signals must be converted into equivalent binary words and binary output data have to be changed into equivalent analog signals, if required.

Digital-to-analog converters: A digitial-to-analog (D/A) converter is used to convert binary words into equivalent analog voltages. Two different D/A converters are available—the binary-weighted and the R-2R. Both utilize a resistor current-divider ladder network to convert a weighted binary bit into a weighted branch current. The sum of the branch currents forms an output current proportional to the total weight of the binary word. An operational amplifier is used to convert the current into an equivalent voltage.

The diagram of a 4-bit binary-weighted D/A converter appears in Figure 6-24. Weighted resistors, connected in series with transistor switches, form the rungs of the ladder. Each resistor is twice the value of the next higher bit. This provides a means of converting a weighted binary bit into a weighted rung current. The rung currents are summed at the inverting input of the operational amplifier. The voltage across the output terminals of the amplifier is proportional to the weight of the binary word applied to the input terminals of the converter.

The R-2R D/A converter employs the same general ladder circuit as the binary-

Figure 6-24

Binary-weighted
D/A converter

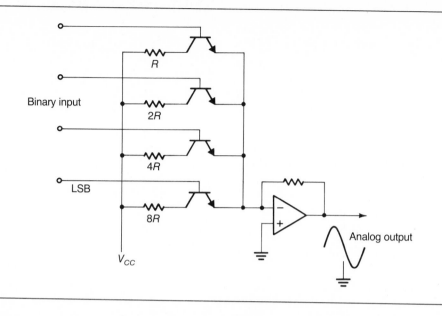

weighted converter except that the maximum difference between any two resistors is only a factor of two. This achieves the same binary weighting of branch currents but the current differential between the least and most significant bits is not nearly as great. In 8-, 16-, and 32-bit binary weighted D/A converters, the resistors in the lower bit rungs have tremendously high resistances. This results in extremely small weighted-rung currents. Noise may override these currents creating conversion errors.

Analog-to-digital converters: Circuits used to convert analog signals into equivalent binary words include the direct-comparison, parallel-conversion, successive-approximation, and dual-slope analog-to-digital (A/D) converters. The successive-approximation and dual-slope converters are by far the most commonly used.

   The successive-approximation A/D converter has the fastest conversion time. As illustrated in Figure 6-25, the circuit consists of a register, a D/A converter, and a comparator. A pulse from the clock is loaded into the most significant bit of the register. It is then shifted out of the register and into the D/A converter. The resulting analog signal is sent to the comparator, where it is compared with the input analog signal. If the amplitude of the input signal is less than that of the output of the D/A converter the bit in the register is set to zero. If the input-signal amplitude is greater than the D/A signal the binary one bit in the register is retained. A binary one bit is shifted into the second-most-significant bit of the register and is sent to the D/A converter. If the amplitude of the analog input signal is less than the output of the D/A converter, the bit in the register is changed to zero. The binary one bit is retained if the input signal is larger than the output of the D/A converter. Next, a binary one is shifted into the third-most-significant bit of the register and the process just described is repeated. This procedure continues until the least significant bit has been placed into the register and a comparison made between the

Figure 6-25

Successive-
approximation A/D
converter

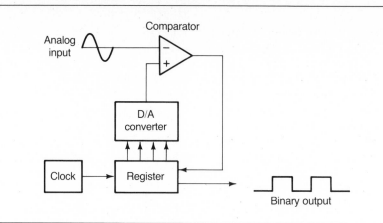

analog input signal and the output of the D/A converter. The binary zero and one bits stored in the register when the last comparison is completed form a binary word representing the analog input signal. The word is ready to be shifted out of the register and sent to the microprocessor.

The dual-slope A/D converter, sometimes called an analog-integration A/D converter, changes analog signals into digital data by causing the analog input voltage to charge a capacitor for a predetermined length of time. At the end of this period a reference voltage of opposite polarity is applied to the capacitor, causing it to discharge to zero. The discharge time is digitally counted. The binary number produced by the counter when the capacitor has fully discharged represents the binary equivalent of the input signal.

The diagram of a dual-slope A/D converter appears in Figure 6-26. An operational amplifier connected as an integrator forms the heart of the circuit. The analog input signal and a negative reference voltage are applied to the integrator through

Figure 6-26

Dual-slope A/D
converter

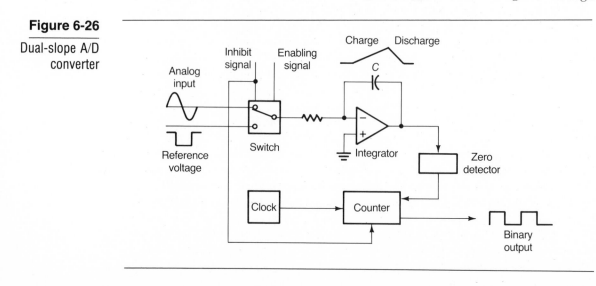

an electronic SPDT switch. A clock, counter, and zero detector form the remaining parts of the circuit. An enabling signal applied to the switch causes the analog input signal to be applied to the integrator for a predetermined length of time. The analog signal causes the feedback capacitor in the integrator circuit to charge. At the end of this period an inhibit signal applied to the switch disconnects the analog input signal and connects the negative reference voltage to the integrator input. At the same time, the counter is turned on. Application of the negative reference voltage causes the capacitor to discharge. When the capacitor voltage reaches zero volts it is detected by the zero detector and the counter is stopped. The resulting binary word in the counter represents the binary equivalent of the analog input signal.

Compared to the successive-approximation circuit, the dual-slope A/D converter is more accurate and is much less affected by noise. It is slower, however, because of the charging and discharging action of the capacitor. Conversion times of 10 to 50 milliseconds are typical.

Commercial D/A and A/D converters: As digital technology has moved into industrial-control applications, the use of D/A and A/D converters has increased substantially. These converters can be formed from circuits using discrete components; they can be manufactured as hybrid modules; and they can be fabricated in monolithic IC format. The latter accounts for the majority of those used today. Most of these are manufactured using BJT technology. It is simpler to fabricate operational amplifiers and other linear circuits using BJT technology rather than MOSFET, but BJT-type ICs have less component density and consume more power than do MOSFET ICs. A number of different CMOS D/A and A/D converters have been developed even though it is more difficult to produce the linear portions of the converters from this type of transistor technology, and considerable interest has been shown in BIMOS converters. This new transistor technology combines the best features of bipolar and field-effect technology.

Whatever the technology used, three major factors form the criteria used to select a converter for a particular application—accuracy, speed, and cost. The order of their importance depends upon the type of application.

The D/A and A/D converters previously discussed used 4-bit data words. This was done for the sake of simplicity. To increase conversion accuracy, longer data words are normally used in actual practice. Since the input signal of a D/A converter and the output of an A/D converter are serial in format, the binary word length of these devices does not have to match the data word of the microprocessor or computer. In many manufacturing applications, an 8-bit converter is adequate, as external factors such as temperature and noise may also affect conversion accuracy.

6-6.9 Optical isolation

In some applications the I/O peripherals are located in one building and the microprocessor is housed in another. The data bus connecting the units may be exposed to the elements. If lightning strikes the data bus, the interfacing circuitry and possibly the memory devices and microprocessor may be damaged. This can also occur if a telephone line is used as the data bus. In factory-floor applications, electromag-

Figure 6-27

Optical isolator

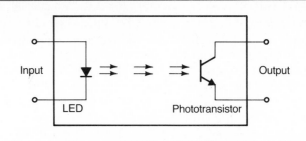

netic fields created by arcing generator and motor brushes, arc welders, and other sources of electromagnetic radiation may cause voltages to be induced across the data bus, which can damage circuits in the peripheral and microprocessor units. It is necessary to electrically isolate the data bus from the peripherals and microprocessor unit to protect the circuits. This is accomplished with an optical isolator, sometimes called an opto-isolator or optoisolator. An optical isolator is a device that converts electrical energy into radiant energy, which is transmitted a short distance and is then reconverted back into electrical data. A typical optical isolator appears in Figure 6-27. An LED used as a light source changes the electrical data into an equivalent radiant signal. This signal is transmitted to a photo-transistor that converts the radiant signal back into an electrical signal, which is sent on to its original destination.

6-7 INPUT/OUTPUT PERIPHERAL INTERFACING

6-7.1 Introduction

Input peripherals and circuits commonly used in microprocessor-based systems employed in manufacturing applications include keypads, keyboards, switches, A/D converters, and multiplexers. Keypads and keyboards are used to place the program and/or portions of the program variables into RAM. Other program variables such as displacement, pressure, flow rate, temperature, and strain are introduced to the system by way of switches, sensors, A/D converters, and multiplexers. Output peripherals include display units such as LEDs, video monitors, and printers; relays and solenoids to actuate loads; and motors. The PIA is used in most peripheral-interfacing applications. This device has internal registers and peripheral I/O data-bus ports that can be programmed to meet most interfacing requirements. To send data between a microprocessor and a peripheral connected to a PIA, the PIA is assigned an address by the programmer. Program subroutines that establish the internal PIA register and peripheral data-bus structures are developed for each peripheral, to allow data to pass between the peripheral and microprocessor. The general techniques used to interface some of the more commonly used peripherals are discussed in the following sections. Because of the proprietary characteristics of MPUs and PIAs, the particular manufacturer's data sheets for those devices must be consulted for specific interfacing solutions.

6-7.2 Interfacing relays to the microprocessor

Relays provide the means whereby high current loads, such as large motors, may be controlled by the microprocessor. The relay coil may be connected directly to an output port of the PIA when small relays are utilized. The relay is actuated whenever the PIA is addressed. The pickup current for large relays usually exceeds the output-port current capacity of the PIA. A buffer or current driver is used to provide the current to actuate the relay. The base terminal of the current driver is connected to an output port of the PIA. Depending upon the relay pickup current requirements, several different current-driver circuits may be employed. For light-to-moderate pickup currents, a large-signal transistor connected in the common-emitter configuration may be used as seen in Figure 6-28. A Darlington amplifier may be used when higher pickup currents are required. The PIA supplies the low current required to turn on the driver. The current needed to actuate the relay comes from the DC power supply. A reverse-biased diode connected to the coil protects the PIA and driver from the CEMF produced by the relay when the driver is turned off.

6-7.3 Interfacing solenoids to the microprocessor

Solenoids are interfaced to the microprocessor to provide programmable digital control for applications where the linear movement of a shaft actuates loads such as clutches and fluid-power flow-control and direction-control valves.

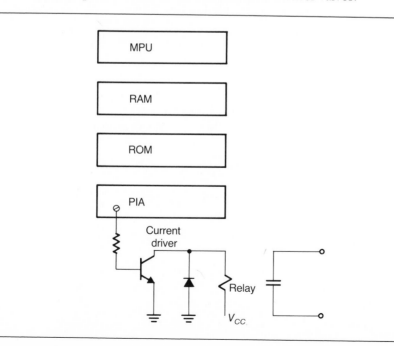

Figure 6-28

Interfacing a relay to a microprocessor

Solenoids are interfaced to the microprocessor in the same fashion as large relays. A current driver is connected to an output port of the PIA. The driver is turned on whenever the PIA is addressed. Current from the driver power supply actuates the solenoid.

6-7.4 Interfacing stepping motors to the microprocessor

The stepping-motor rotor moves in synchronization with the revolving stator field when binary pulses are applied in sequence to the four stator windings as discussed in Chapter 2. The number of steps to be rotated and the direction of rotation is determined from the stepping-motor subroutine developed as part of the operating program. For the stepping motor shown in Figure 6-29, clockwise rotation will occur when the binary word 0100 is applied to the stators in the sequence shown. Counterclockwise rotation occurs when the binary word 0001 is applied to stators in sequence. Whenever the binary word 0100 is applied to the stator windings four different times in the sequence shown, one complete rotor revolution occurs. All the stepping-motor-program subroutine has to do is to circulate the binary word 0100 the required number of times to provide the required rotor movement.

The stepping motor is interfaced to the microprocessor by connecting each stator winding to a PIA output port via a current driver, as illustrated in Figure 6-30. The voltage needed to drive the motor depends upon the type of motor and may range from 5 to 100 volts. Stator current is dependent upon the type of motor, the speed of rotation, and the amount of load coupled to the rotor.

6-7.5 Interfacing fluid-power systems to the microprocessor

As factory automation increases, the need for microprocessor control of fluid-power systems becomes more important. Microprocessor control of fluid-power systems involves the operation of flow- and direction-control valves using programmable binary signals. Flow-control valves control the operating speed of cylinders and mo-

---

Figure 6-29

Stepping-motor binary-word stator sequence

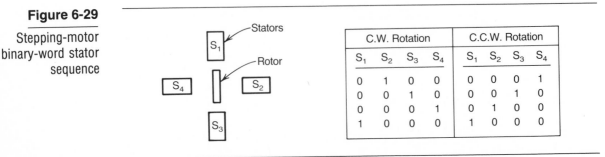

C.W. Rotation				C.C.W. Rotation			
S_1	S_2	S_3	S_4	S_1	S_2	S_3	S_4
0	1	0	0	0	0	0	1
0	0	1	0	0	0	1	0
0	0	0	1	0	1	0	0
1	0	0	0	1	0	0	0

Figure 6-30

Interfacing a
stepping motor to
a microprocessor

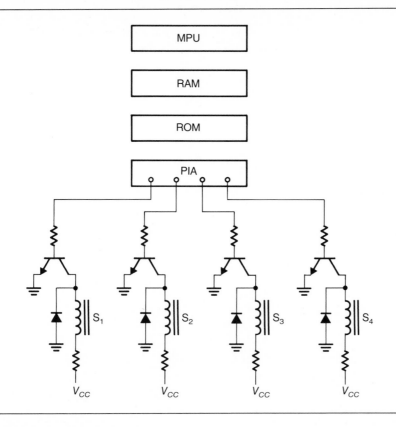

tors, while direction-control valves create the back-and-forth motion of cylinder pistons and determine the direction of rotation of fluid-power motors.

Interfacing flow-control valves: Fluid flow-control valves are usually operated by turning the valve stem radially. Microprocessor control may be accomplished by coupling either a stepping or torque motor to the valve stem. The stepping motor is interfaced to the microprocessor, as discussed in the last section.

The interfacing scheme used to interface a torque motor to the microprocessor appears in Figure 6-31. A D/A converter is connected to the output port of the PIA to convert the digital data into an equivalent analog signal, which is amplified by the current driver and applied to the stator winding of the motor.

The stepping motor is simpler and less expensive. Torque motors are used when the valve-actuating torque exceeds the torque produced by a stepping motor. In addition to requiring a D/A converter, the torque motor is considerably more expensive.

Interfacing direction-control valves: Direction-control valves are interfaced to the microprocessor by coupling a solenoid to the valve stem. A current driver connected to an output port of the PIA, as shown in Figure 6-32, provides the current required to actuate the solenoid.

Figure 6-31

Interfacing a
flow-control valve
to a
microprocessor

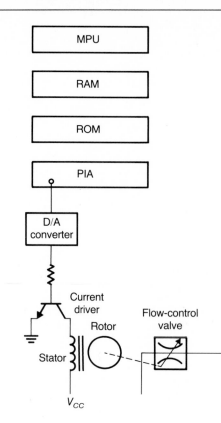

6-8 MULTICHANNEL MICROPROCESSOR-BASED SYSTEMS

6-8.1 Introduction

Microprocessor-based systems used in manufacturing applications almost always utilize multiple I/O peripherals. Most PIAs have two system-word-size peripheral data buses. These can be programmed to operate as a system-word-size input port and output port, two input ports, or two output ports. The PIA peripheral bus pins may also be programmed to allow each peripheral bus pin to operate as an independent entity. For an 8-bit microprocessor, this means that two parallel-operated 8-bit peripherals can be connected to the PIA. These may be an input and an output, two input peripherals, or two output peripherals. It also means that 16 different serially operated I/O peripherals can be connected to the PIA. Additional serially operated peripherals may be used if multiplexing is employed. Each conductor connected to the multiplexer or PIA from a serial peripheral may be called a channel. Some typical multichannel microprocessor-based systems appear in the following sections.

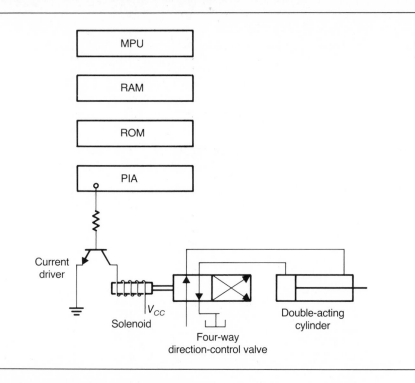

Figure 6-32

Interfacing a direction-control valve to a microprocessor

6-8.2 4-channel thermocouple input

The microprocessor system appearing in Figure 6-33 uses four thermocouples as inputs. This is a partial diagram, as no other inputs are shown and no outputs are indicated. Temperature, measured by four remote thermocouples is applied via transmission lines to amplifiers. The output signal from each amplifier is coupled

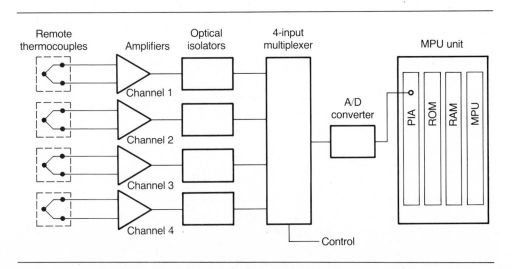

Figure 6-33

4-channel thermocouple input

Figure 6-34

15-channel I/O

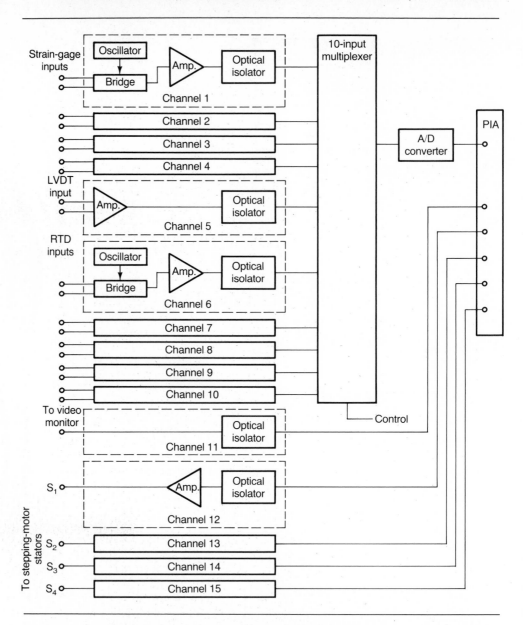

through an optical isolator to a 4-input multiplexer. The time-shared analog signals coming from the multiplexer are converted into binary data by a single A/D converter and connected to a single PIA peripheral data-bus pin. To increase the amplitude of the signal coming from the sensor before being sent to the A/D converter, amplifiers such as those shown in Figure 6-33 are often required as part of the interface. Filters may be included to remove noise which may have been picked up by the transmission line. Multiplexing input signals reduces data feedthrough speed but allows more peripherals to be connected to the PIA.

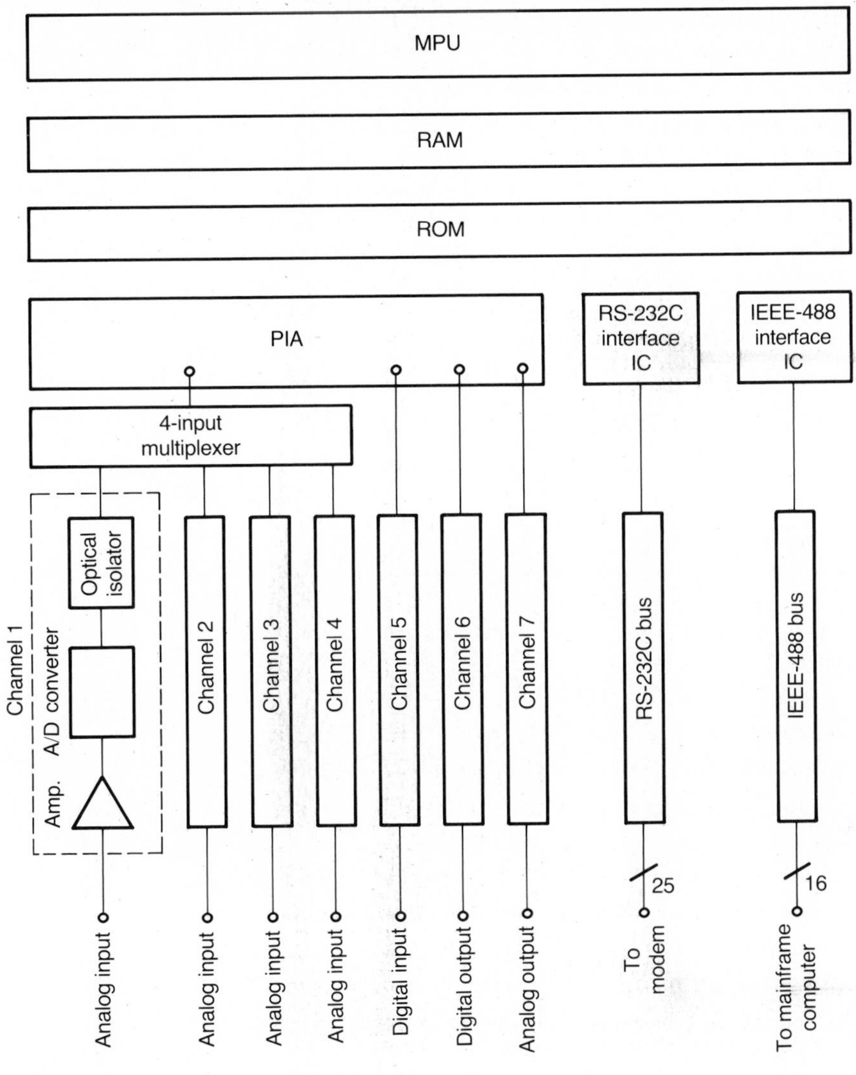

Figure 6-35

I/O interface with
RS-232C and
IEEE-488 bus
connections

6-8.3 15-channel I/O

Channels 1–4 in the system appearing in Figure 6-34 carry strain-gage signals. The analog signals from four remote strain gages are applied to strain-gage bridges. The output signal from each bridge is amplified, coupled through an optical isolator, and sent to a multiplexer. Linear-displacement data from an LVDT is sent through channel 5, while channels 6–10 include bridge circuits to measure analog temperature signals coming from remote RTDs. The analog signals from channels 1–10 are sent to a 10-input multiplexer. An A/D converter changes the time-shared multiplexer output signals into equivalent binary data, which are applied to the PIA.

Channel 11 carries binary data to a video monitor while channels 12–15 are connected to the stator windings of a stepping motor.

6-8.4 I/O interface with RS-232C and IEEE-488 bus connections

The system illustrated in Figure 6-35 utilizes multiple I/O channels carrying assorted analog and digital I/O signals, an RS-232C serial interface, and an IEEE-488 parallel interface.

Channels 1–4 carry analog input signals and are sent to a 4-input multiplexer. Unlike the systems appearing in Figures 6-33 and 34, where single A/D converters are connected to the output of the multiplexers, an A/D converter is included in each channel. An RS-232C bus is used to connect the microprocessor to a modem while an IEEE-488 bus connects the microprocessor to a mainframe computer.

Summary

1. A program is a list of instructions that tell the microprocessor how to perform a particular operation.
2. Instructions usually include both a command and data that are manipulated by the command.
3. An opcode is a binary word representing a command.
4. Operands are data that are manipulated by the microprocessor as instructed by the opcodes.
5. An instruction set is a list of all the commands the microprocessor responds to.
6. Mnemonics are abbreviations used to represent instructions in assembly-language programming.
7. An addressing mode is the technique used to address the operand.
8. I/O peripherals are input devices that allow information to be loaded into the main memory of the microprocessor system and output devices controlled by the microprocessor or that provide a visual indication of the results of operations performed by the microprocessor.
9. Input peripherals include hexadecimal keypads, keyboards, and switches.
10. Output peripherals that implement operations include relays, solenoids, and motors.
11. Output display peripherals include LEDs, video monitors, and printers.
12. Serial data may be transmitted in asynchronous or synchronous form. Start and stop bits enclose the binary word being transmitted when asynchronous transmission is employed. Synchronizing characters are transmitted periodically with the data when synchronous transmission is used.
13. Parity is a technique used to check for errors in serial transmission. Two types

of parity are used—even and odd. The most significant bit in the data word is reserved for the parity bit.

14. Three modes of transmission are possible when serial transmission is used—simplex, half-duplex, and full-duplex. The transmission line is either a transmit or receive line when simplex serial transmission is used. The transmission line is used for both transmission and reception in half-duplex operation but not at the same time. Two pairs of transmission lines are used in full-duplex transmission. One is used for transmission and the other is used for reception.

15. A modem is a device that allows binary data to be transmitted over telephone and voice-grade transmission lines. The modem is made up of a modulator, demodulator, and filter.

16. Problems associated with microprocessor interfacing include those arising from the use of a common data bus and from the incompatability of the microprocessor and its peripherals.

17. Three-state buffers connected between RAM, ROM, and I/O peripherals and the main data bus solve the common-data-bus problem. The buffers are enabled by addressing the device that is to send or receive data.

18. Interfacing ICs and buses solve many of the incompatability problems.

19. The PIA is a programmable integrated circuit used to interface I/O peripherals to the microprocessor.

20. The RS-232C is an active-interface bus that can be used to interface many serially operated peripherals to the microprocessor system.

21. The IEEE-488 interface bus may be used to connect parallel-operated peripherals to a microprocessor, connect microprocessors together, and connect an MPU to a mainframe computer.

22. The UART, USRT, and USART are interfacing ICs used to connect serial I/O peripherals to the microprocessor.

23. Digital-to-analog (D/A) converters change binary data into equivalent analog signals. There are two types of D/A converters—the binary weighted and the R-2R.

24. Analog-to-digital (A/D) converters change analog signals into binary data. The successive-approximation and dual-slope A/D converters are the two most commonly used types. In the successive-approximation A/D converter, binary bits are generated when the amplitude of the analog input signal is compared to the amplitude of the output signal of a D/A converter. In the dual-slope A/D converter, binary bits are generated by a counter that measures the discharge time of a capacitor.

25. Optical isolators are used to electrically isolate the I/O peripherals and microprocessor from the transmission line connecting the units together. An optical isolator converts electrical signals into radiant energy which is transmitted a short distance before being reconverted into electrical energy.

26. Program subroutines developed for each I/O peripheral connected to the PIA establish the PIA-register and peripheral data-bus structure to allow data to be sent between peripherals and the microprocessor.

27. Input circuits connected to the PIA in manufacturing microprocessor-based systems include sensors, bridges, amplifiers, optical isolators, A/D converters, and multiplexers.

28. Output devices connected to the PIA include relays, solenoids, and motors. Current drivers are often connected between the PIA and these devices. The PIA supplies the energy required to turn on the current driver while the driver's DC power supply provides the energy needed to actuate the relay, solenoid, or motor.

Chapter Examination

1. Discuss the major differences between asynchronous and synchronous serial transmission.

2. List an advantage and a disadvantage of using asynchronous serial transmission compared to the synchronous type.

3. Describe the process used to implement odd parity.

4. Draw a block diagram of a modem, label each block, and identify the mark and space frequencies utilized.

5. Describe how the three-state buffers used to isolate RAMs, ROMs, and I/O-peripheral-interfacing ICs are actuated to allow a particular IC to be connected to the common data bus.

6. The _____ interface bus may be used to interface serially operated I/O peripherals with the microprocessor.
   a. UART
   b. USRT
   c. RS-232C
   d. IEEE-488

7. A (an) _____ is the command portion of an assembly-language instruction.
   a. opcode
   b. operand
   c. mnemonic
   d. assembler

8. The _____ interface bus may be employed to interface parallel-operated I/O peripherals with a microprocessor.
   a. PIA
   b. UART
   c. RS-232C
   d. IEEE-488

9. A (an) _____ is the binary data that are manipulated by the microprocessor.
   a. opcode
   b. operand

    c. mnemonic

    d. assembler

10. A _____ is often used to program a microprocessor when assembly language is employed.

    a. sensor

    b. keypad

    c. switch

    d. keyboard

11. Which of the following is *not* a circuit utilized in a modem?

    a. modulator

    b. multiplexer

    c. demodulator

    d. bandpass filter

12. A (an) _____ is an abbreviation used in assembly-language programming to represent an instruction.

    a. opcode

    b. operand

    c. mnemonic

    d. assembler

13. Binary zero and one levels for the RS-232C standard are _____ and _____ respectively.

    a. 0; 5 V

    b. +12; −12 V

    c. 0; 20 mA

    d. 0; 60 mA

14. Which of the following is *not* a class of microprocessor instruction?

    a. data movement

    b. program status

    c. operand transfer

    d. program manipulation

15. The speed at which binary bits are transmitted in serial format is called

    a. baud rate.

    b. sending speed.

    c. formulation speed.

    d. transmission speed.

16. Which of the following is *not* an addressing mode?

    a. index

    b. stack

    c. direct

    d. register

17. Which of the following is *not* a mode of serial transmission?

    a. duplex

    b. simplex

    c. full-duplex

    d. half-duplex

18. Which of the following is *not* a type of printer?
    a. page
    b. impact
    c. holograph
    d. dot-matrix

19. _____ is a type of interpret language.
    a. BASIC
    b. COBOL
    c. FORTRAN
    d. Assembly

20. T F Current drivers are often required to provide the energy needed to actuate relays, solenoids, and motors. ___

21. T F The stepping motor is interfaced to a microprocessor by connecting each stator winding to a PIA peripheral data-bus connection via a current driver. ___

22. T F Data-movement instructions cause data to be transferred into the arithmetic and logic unit. ___

23. T F An addressing mode is the technique used to address the opcode of an instruction stored in memory. ___

24. T F Alphanumeric characters and graphic symbols used in video monitors are usually permanently stored in a PIA register. ___

25. T F A mark used in serial binary transmission represents a binary zero and a space represents a binary one. ___

26. T F The RS-232C, TTL, 20-mA, and 60-mA binary-level standards all may be used in asynchronous serial transmission. ___

27. T F Start and stop bits are not used in the transmission of synchronous serial data. ___

28. T F Parity is a technique used to check for errors in parallel data transmission. ___

29. T F Two pairs of transmission lines are required in half-duplex serial transmission. ___

30. T F A modem is a device that allows serial data to be transmitted over telephone lines. ___

31. T F A USRT is an IC used to convert parallel data into asynchronous serial data and vice versa. ___

32. T F Three-state buffers help solve the problem of interface incompatability between the microprocessor and the I/O peripherals. ___

33. T F Three-state buffers are enabled by an address-decoder signal. ___

34. T F The PIA is primarily used to convert parallel data into synchronous serial data. ___

35. T F The PIA and its internal registers are addressed in the same fashion as RAM and ROM. ___

36. T F An optical isolator is used to isolate the PIA from the microprocessor.   —

37. T F An operational amplifier connected as an integrator is used in a dual-slope A/D converter.   —

38. T F The RS-232C interface is used to convert synchronous serial data into equivalent asynchronous data.   —

39. T F Analog-to-digital conversion occurs in a successive-approximation A/D converter by a binary counter measuring the discharge time of a capacitor.   —

40. T F Synchronous binary data words are enclosed with start and stop bits.   —

CHAPTER 7

THE MC6800 MPU

OBJECTIVES

1. Identify the components forming the MC6800 architecture.

2. Identify the addressing modes used by the MC6800 and describe how the operand in each is accessed.

3. Identify the four instruction categories in which the MC6800 instructions are grouped and discuss the general functions of the instructions in each group.

4. List the control signals used in the MC6800 system and discuss the purpose of each.

5. Describe how programs are executed in the MC6800 system.

6. Define the term *MPU cycle* and discuss the relationship between MPU cycles and instruction execution time.

7. Describe the method by which main memory and I/O peripherals are addressed by the MC6800.

8. Discuss how data are transferred between the MC6800 and memory or I/O peripherals.

9. Identify the interrupt techniques used by the MC6800 and discuss the operating principles of each.

10. Discuss the function and basic operation of the MC6821 PIA.

11. Discuss the function and basic operation of the MC6850 ACIA.

12. Identify the purpose of the MC6843 FDC.

13. Identify the purpose of the MC6844 DMAC.

14. Identify the purpose of the MC6852 SSDA.

15. Identify the purpose of the MC6860 digital modem.

16. Identify the purpose of the MC68488 GPIA.

7-1 INTRODUCTION

Architecture, instruction sets, addressing modes, and control signals vary somewhat from one microprocessor to another. The more commonly used Intel and Motorola MPUs were introduced in Chapter 5. One of these devices, the MC6800 is discussed in greater detail in this chapter.

7-2 MC6800 ARCHITECTURE

The MC6800 is an 8-bit N-channel MOS (NMOS) monolithic microprocessor that contains a 16-bit program counter, a 16-bit stack pointer, a 16-bit index register, two 8-bit accumulators (A and B), an 8-bit instruction register, an 8-bit arithmetic and logic unit, an 8-bit condition code register, and an instruction decode and control unit. The MC6800 uses a 16-bit address, which provides 65,536 bytes of memory storage and addressing capability.

Figure 7-1

MC6800 block diagram. Courtesy of Motorola, Inc.

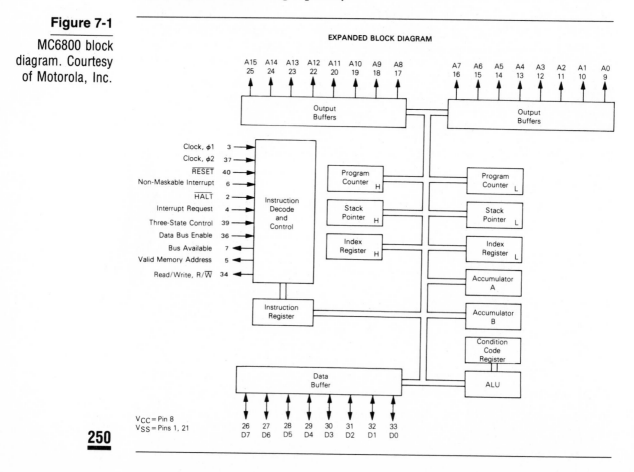

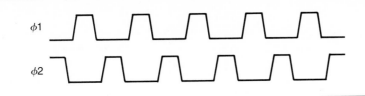

Figure 7-2

MC6800 two-phase nonoverlapping clock-pulse requirements

As seen in Figure 7-1, the 16-bit registers are divided into two single-byte storage mediums—high (H) and low (L). These registers are used to either store or help generate the 16-bit address. Six flags are utilized in the condition code register—the negative (N), zero (Z), overflow (V), carry (C), interrupt mask (I), and half-carry (H). These use bits 0–5. Bits 6 and 7 are not used and are maintained permanently at a binary one level.

The MC6800 requires an external 1-MHz clock that produces two nonoverlapping clock-pulse trains as shown in Figure 7-2. These synchronize the events that occur within the microprocessor, memory, and I/O peripherals. The MPU requires both $\phi1$ and $\phi2$ clock pulses while memory and I/O peripherals are synchronized with $\phi2$ clock pulses only.

7-3 MC6800 ADDRESSING MODES

7-3.1 Introduction

The MC6800 MPU has 72 instructions in its instruction set and utilizes seven addressing modes. Two or more addressing modes can be used with some instructions, making almost 200 different opcodes available to the programmer. Addressing modes include the implied, immediate, direct, extended, indexed, relative, and accumulator.

7-3.2 The implied-addressing mode

Sometimes called the inherent- or implicit-addressing mode, this instruction does not require an operand. The microprocessor executes the instruction without expecting an operand to follow. This is a one-byte instruction and requires only one byte of memory, as illustrated in Figure 7-3.

7-3.3 The immediate-addressing mode

This is a two-byte instruction in which the operand is located in the memory segment immediately following the opcode, as shown in Figure 7-4.

Figure 7-3

Implied-addressing
mode

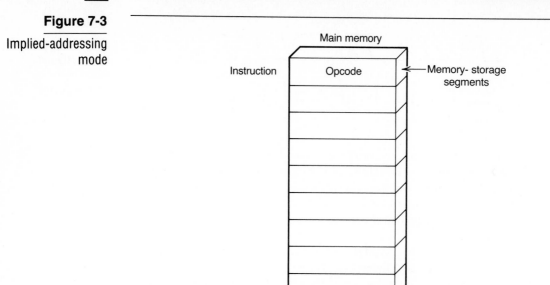

Figure 7-4

Immediate-
addressing mode

7-3.4 The direct-addressing mode

In the MC6800, the direct-addressing mode is a two-byte instruction where the address of the operand appears in the memory segment following the opcode, as seen in Figure 7-5. Since the address length is limited to a single byte, direct addressing can be used to address operands located only in the 255 memory segments following the opcode.

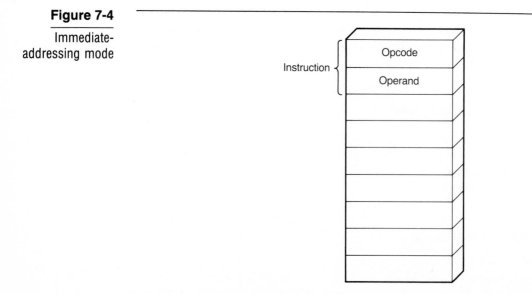

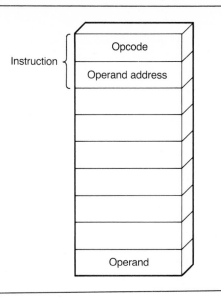

Figure 7-5

Direct-addressing
mode

7-3.5 The extended-addressing mode

This mode is identical to direct addressing except that it is a three-byte instruction. The first byte is the opcode and the second and third bytes are the address of the operand, as illustrated in Figure 7-6. Because a full 16-bit address is utilized, the operand can be stored anywhere in a 65-kbyte memory.

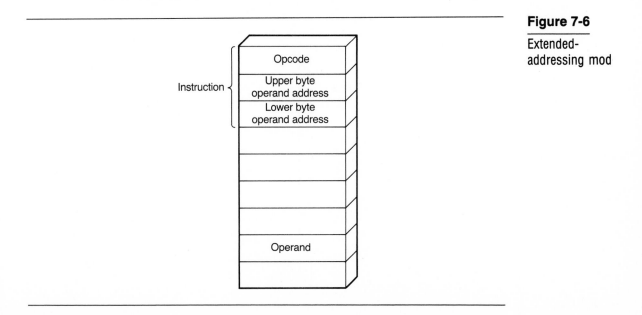

Figure 7-6

Extended-
addressing mod

7-3.6 The indexed-addressing mode

This is a two-byte instruction in which the first byte is the opcode and the second byte represents a number called an offset address. The offset address is added to the number appearing in the index register, as shown in Figure 7-7. The sum of these numbers represents the address where the operand is stored.

7-3.7 The relative-addressing mode

This mode is similar to the indexed mode except that the offset address is added to the contents of the program counter. The operand address is the sum of these two numbers, as illustrated in Figure 7-8.

7-3.8 The accumulator-addressing mode

This is a one-byte instruction in which the contents of one of the accumulators is the operand. An operand address is not required.

7-4 MC6800 INSTRUCTION CLASSIFICATION

7-4.1 Introduction

Motorola groups the 72 instructions for the MC6800 into four categories—accumulator and memory instructions, index-register and stack-manipulation instructions, jump-and-branch instructions, and condition-code-register instructions.

Figure 7-7

Indexed-addressing mode

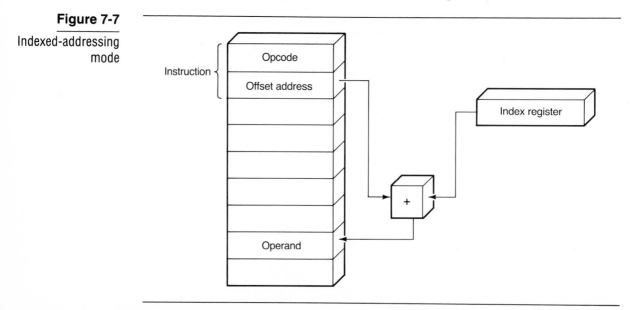

Figure 7-8

Relative-addressing mode

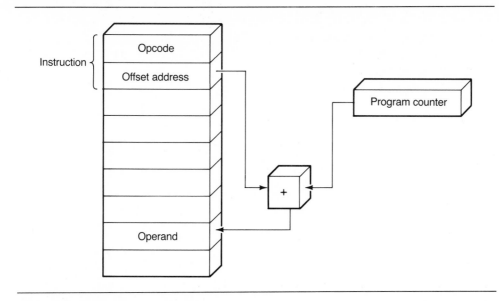

7-4.2 Accumulator and memory instructions

These instructions cause data to be moved between the two accumulators and memory or the I/O peripherals. They also include the arithmetic and logic instructions that let the microprocessor perform mathematical and logic operations. As shown in Table 7-1, 30 basic instructions are included in this group. Approximately 140 different opcodes are used to represent these instructions. The additional opcodes are created by the use of two accumulators and the large number of instructions which can be implemented using two or more addressing modes.

The load-accumulator (LDAA and LDAB) and store-accumulator (STAA and STAB) instructions are two of the most commonly used instructions in this category. The load-accumulator instructions cause a byte of data to be transferred from memory or an input peripheral into one of the accumulators. The immediate-, direct-, extended-, and indexed-addressing modes can be used with either of the two load-accumulator instructions.

The store-accumulator instructions (STAA and STAB) cause the contents of the accumulators to be loaded into an addressed memory segment or an output peripheral. The direct-, extended-, and indexed-addressing modes are used with these instructions.

7-4.3 Index-register and stack-manipulation instructions

Eleven instructions are included in this group, as listed in Table 7-2. The index-register and stack-manipulation instructions allow the index register and stack pointer to be loaded, incremented, decremented, and their contents stored.

ACCUMULATOR AND MEMORY OPERATIONS

OPERATIONS	MNEMONIC	IMMED OP	~	#	DIRECT OP	~	#	INDEX OP	~	#	EXTND OP	~	#	IMPLIED OP	~	#	BOOLEAN/ARITHMETIC OPERATION (All register labels refer to contents)	H	I	N	Z	V	C
Add	ADDA	3B	2	2	9B	3	2	AB	5	2	BB	4	3				A + M → A	‡	•	‡	‡	‡	‡
	ADDB	CB	2	2	DB	3	2	EB	5	2	FB	4	3				B + M → B	‡	•	‡	‡	‡	‡
Add Acmltrs	ABA													1B	2	1	A + B → A	‡	•	‡	‡	‡	‡
Add with Carry	ADCA	89	2	2	99	3	2	A9	5	2	B9	4	3				A + M + C → A	‡	•	‡	‡	‡	‡
	ADCB	C9	2	2	D9	3	2	E9	5	2	F9	4	3				B + M + C → B	‡	•	‡	‡	‡	‡
And	ANDA	84	2	2	94	3	2	A4	5	2	B4	4	3				A • M → A	•	•	‡	‡	R	•
	ANDB	C4	2	2	D4	3	2	E4	5	2	F4	4	3				B • M → B	•	•	‡	‡	R	•
Bit Test	BITA	85	2	2	95	3	2	A5	5	2	B5	4	3				A • M	•	•	‡	‡	R	•
	BITB	C5	2	2	D5	3	2	E5	5	2	F5	4	3				B • M	•	•	‡	‡	R	•
Clear	CLR							6F	7	2	7F	6	3				00 → M	•	•	R	S	R	R
	CLRA													4F	2	1	00 → A	•	•	R	S	R	R
	CLRB													5F	2	1	00 → B	•	•	R	S	R	R
Compare	CMPA	81	2	2	91	3	2	A1	5	2	B1	4	3				A - M	•	•	‡	‡	‡	‡
	CMPB	C1	2	2	D1	3	2	E1	5	2	F1	4	3				B - M	•	•	‡	‡	‡	‡
Compare Acmltrs	CBA													11	2	1	A - B	•	•	‡	‡	‡	‡
Complement, 1's	COM							63	7	2	73	6	3				M̄ → M	•	•	‡	‡	R	S
	COMA													43	2	1	Ā → A	•	•	‡	‡	R	S
	COMB													53	2	1	B̄ → B	•	•	‡	‡	R	S
Complement, 2's (Negate)	NEG							60	7	2	70	6	3				00 - M → M	•	•	‡	‡	①	②
	NEGA													40	2	1	00 - A → A	•	•	‡	‡	①	②
	NEGB													50	2	1	00 - B → B	•	•	‡	‡	①	②
Decimal Adjust, A	DAA													19	2	1	Converts Binary Add. of BCD Characters into BCD Format	•	•	‡	‡	‡	③
Decrement	DEC							6A	7	2	7A	6	3				M - 1 → M	•	•	‡	‡	④	•
	DECA													4A	2	1	A - 1 → A	•	•	‡	‡	④	•
	DECB													5A	2	1	B - 1 → B	•	•	‡	‡	④	•
Exclusive OR	EORA	88	2	2	98	3	2	A8	5	2	B8	4	3				A ⊕ M → A	•	•	‡	‡	R	•
	EORB	C8	2	2	D8	3	2	E8	5	2	F8	4	3				B ⊕ M → B	•	•	‡	‡	R	•
Increment	INC							6C	7	2	7C	6	3				M + 1 → M	•	•	‡	‡	⑤	•
	INCA													4C	2	1	A + 1 → A	•	•	‡	‡	⑤	•
	INCB													5C	2	1	B + 1 → B	•	•	‡	‡	⑤	•
Load Acmltr	LDAA	86	2	2	96	3	2	A6	5	2	B6	4	3				M → A	•	•	‡	‡	R	•
	LDAB	C6	2	2	D6	3	2	E6	5	2	F6	4	3				M → B	•	•	‡	‡	R	•
Or, Inclusive	ORAA	8A	2	2	9A	3	2	AA	5	2	BA	4	3				A + M → A	•	•	‡	‡	R	•
	ORAB	CA	2	2	DA	3	2	EA	5	2	FA	4	3				B + M → B	•	•	‡	‡	R	•
Push Data	PSHA													36	4	1	A → MSP, SP - 1 → SP	•	•	•	•	•	•
	PSHB													37	4	1	B → MSP, SP - 1 → SP	•	•	•	•	•	•
Pull Data	PULA													32	4	1	SP + 1 → SP, MSP → A	•	•	•	•	•	•
	PULB													33	4	1	SP + 1 → SP, MSP → B	•	•	•	•	•	•
Rotate Left	ROL							69	7	2	79	6	3				M	•	•	‡	‡	⑥	‡
	ROLA													49	2	1	A	•	•	‡	‡	⑥	‡
	ROLB													59	2	1	B	•	•	‡	‡	⑥	‡
Rotate Right	ROR							66	7	2	76	6	3				M	•	•	‡	‡	⑥	‡
	RORA													46	2	1	A	•	•	‡	‡	⑥	‡
	RORB													56	2	1	B	•	•	‡	‡	⑥	‡
Shift Left, Arithmetic	ASL							68	7	2	78	6	3				M	•	•	‡	‡	⑥	‡
	ASLA													48	2	1	A	•	•	‡	‡	⑥	‡
	ASLB													58	2	1	B	•	•	‡	‡	⑥	‡
Shift Right, Arithmetic	ASR							67	7	2	77	6	3				M	•	•	‡	‡	⑥	‡
	ASRA													47	2	1	A	•	•	‡	‡	⑥	‡
	ASRB													57	2	1	B	•	•	‡	‡	⑥	‡
Shift Right, Logic	LSR							64	7	2	74	6	3				M	•	•	R	‡	⑥	‡
	LSRA													44	2	1	A	•	•	R	‡	⑥	‡
	LSRB													54	2	1	B	•	•	R	‡	⑥	‡
Store Acmltr	STAA				97	4	2	A7	6	2	B7	5	3				A → M	•	•	‡	‡	R	•
	STAB				D7	4	2	E7	6	2	F7	5	3				B → M	•	•	‡	‡	R	•
Subtract	SUBA	80	2	2	90	3	2	A0	5	2	B0	4	3				A - M → A	•	•	‡	‡	‡	‡
	SUBB	C0	2	2	D0	3	2	E0	5	2	F0	4	3				B - M → B	•	•	‡	‡	‡	‡
Subtract Acmltrs	SBA													10	2	1	A - B → A	•	•	‡	‡	‡	‡
Subtr. with Carry	SBCA	82	2	2	92	3	2	A2	5	2	B2	4	3				A - M - C → A	•	•	‡	‡	‡	‡
	SBCB	C2	2	2	D2	3	2	E2	5	2	F2	4	3				B - M - C → B	•	•	‡	‡	‡	‡
Transfer Acmltrs	TAB													16	2	1	A → B	•	•	‡	‡	R	•
	TBA													17	2	1	B → A	•	•	‡	‡	R	•
Test, Zero or Minus	TST							6D	7	2	7D	6	3				M - 00	•	•	‡	‡	R	R
	TSTA													4D	2	1	A - 00	•	•	‡	‡	R	R
	TSTB													5D	2	1	B - 00	•	•	‡	‡	R	R

(Condition Code Register bits: 5 = H, 4 = I, 3 = N, 2 = Z, 1 = V, 0 = C)

LEGEND:

- OP — Operation Code (Hexadecimal);
- ~ — Number of MPU Cycles;
- # — Number of Program Bytes;
- + — Arithmetic Plus;
- − — Arithmetic Minus;
- • — Boolean AND;
- MSP — Contents of memory location pointed to be Stack Pointer;
- + — Boolean Inclusive OR;
- ⊙ — Boolean Exclusive OR;
- M̄ — Complement of M;
- → — Transfer Into;
- 0 — Bit = Zero;
- 00 — Byte = Zero;

CONDITION CODE SYMBOLS:

- H — Half-carry from bit 3;
- I — Interrupt mask
- N — Negative (sign bit)
- Z — Zero (byte)
- V — Overflow, 2's complement
- C — Carry from bit 7
- R — Reset Always
- S — Set Always
- ‡ — Test and set if true, cleared otherwise
- • — Not Affected

CONDITION CODE REGISTER NOTES:
(Bit set if test is true and cleared otherwise)

1	(Bit V)	Test: Result = 10000000?
2	(Bit C)	Test: Result = 00000000?
3	(Bit C)	Test: Decimal value of most significant BCD Character greater than nine? (Not cleared if previously set.)
4	(Bit V)	Test: Operand = 10000000 prior to execution?
5	(Bit V)	Test: Operand = 01111111 prior to execution?
6	(Bit V)	Test: Set equal to result of N⊕C after shift has occurred.

Note — Accumulator addressing mode instructions are included in the column for IMPLIED addressing

 MOTOROLA *Semiconductor Products Inc.*

7-4.4 Jump-and-branch instructions

There are 23 jump-and-branch instructions, as identified in Table 7-3. These instructions cause the microprocessor to branch, or jump ahead or behind to a subroutine in a program, when the appropriate flags in the condition code register are set. The jump-and-branch instructions perform the same basic operation. They allow for a change in program control from one memory location to another. The major difference between the two types of instructions is the method in which the subroutine is developed.

Table 7-1

MC6800
ACCUMULATOR
AND MEMORY
INSTRUCTIONS.
(COURTESY OF
MOTOROLA, INC.)
*SEE FACING
PAGE.*

Table 7-2

MC6800
INDEX-REGISTER
AND
STACK-
MANIPULATION
INSTRUCTIONS.
(COURTESY OF
MOTOROLA, INC.)

Table 7-3

MC6800
JUMP-
AND-BRANCH
INSTRUCTIONS.
(COURTESY OF
MOTOROLA, INC.)

INDEX REGISTER AND STACK POINTER INSTRUCTIONS

POINTER OPERATIONS	MNEMONIC	IMMED OP	~	=	DIRECT OP	~	=	INDEX OP	~	=	EXTND OP	~	=	IMPLIED OP	~	=	BOOLEAN/ARITHMETIC OPERATION	5 H	4 I	3 N	2 Z	1 V	0 C
Compare Index Reg	CPX	8C	3	3	9C	4	2	AC	6	2	8C	5	3				$X_H - M, X_L - (M + 1)$	●	●	①	⫶	②	●
Decrement Index Reg	DEX													09	4	1	$X - 1 \rightarrow X$	●	●	●	⫶	●	●
Decrement Stack Pntr	DES													34	4	1	$SP - 1 \rightarrow SP$	●	●	●	●	●	●
Increment Index Reg	INX													08	4	1	$X + 1 \rightarrow X$	●	●	●	⫶	●	●
Increment Stack Pntr	INS													31	4	1	$SP + 1 \rightarrow SP$	●	●	●	●	●	●
Load Index Reg	LDX	CE	3	3	DE	4	2	EE	6	2	FE	5	3				$M \rightarrow X_H, (M + 1) \rightarrow X_L$	●	●	③	⫶	R	●
Load Stack Pntr	LDS	8E	3	3	9E	4	2	AE	6	2	BE	5	3				$M \rightarrow SP_H, (M + 1) \rightarrow SP_L$	●	●	③	⫶	R	●
Store Index Reg	STX				DF	5	2	EF	7	2	FF	6	3				$X_H \rightarrow M, X_L \rightarrow (M + 1)$	●	●	③	⫶	R	●
Store Stack Pntr	STS				9F	5	2	AF	7	2	BF	6	3				$SP_H \rightarrow M, SP_L \rightarrow (M + 1)$	●	●	③	⫶	R	●
Indx Reg → Stack Pntr	TXS													35	4	1	$X - 1 \rightarrow SP$	●	●	●	●	●	●
Stack Pntr → Indx Reg	TSX													30	4	1	$SP + 1 \rightarrow X$	●	●	●	●	●	●

① (Bit N) Test: Sign bit of most significant (MS) byte of result = 1?
② (Bit V) Test: 2's complement overflow from subtraction of ms bytes?
③ (Bit N) Test: Result less than zero? (Bit 15 = 1)

JUMP AND BRANCH INSTRUCTIONS

OPERATIONS	MNEMONIC	RELATIVE OP	~	=	INDEX OP	~	=	EXTND OP	~	=	IMPLIED OP	~	=	BRANCH TEST	5 H	4 I	3 N	2 Z	1 V	0 C
Branch Always	BRA	20	4	2										None	●	●	●	●	●	●
Branch If Carry Clear	BCC	24	4	2										$C = 0$	●	●	●	●	●	●
Branch If Carry Set	BCS	25	4	2										$C = 1$	●	●	●	●	●	●
Branch If = Zero	BEQ	27	4	2										$Z = 1$	●	●	●	●	●	●
Branch If ≥ Zero	BGE	2C	4	2										$N \oplus V = 0$	●	●	●	●	●	●
Branch If > Zero	BGT	2E	4	2										$Z + (N \oplus V) = 0$	●	●	●	●	●	●
Branch If Higher	BHI	22	4	2										$C + Z = 0$	●	●	●	●	●	●
Branch If ≤ Zero	BLE	2F	4	2										$Z + (N \oplus V) = 1$	●	●	●	●	●	●
Branch If Lower Or Same	BLS	23	4	2										$C + Z = 1$	●	●	●	●	●	●
Branch If < Zero	BLT	2D	4	2										$N \oplus V = 1$	●	●	●	●	●	●
Branch If Minus	BMI	2B	4	2										$N = 1$	●	●	●	●	●	●
Branch If Not Equal Zero	BNE	26	4	2										$Z = 0$	●	●	●	●	●	●
Branch If Overflow Clear	BVC	28	4	2										$V = 0$	●	●	●	●	●	●
Branch If Overflow Set	BVS	29	4	2										$V = 1$	●	●	●	●	●	●
Branch If Plus	BPL	2A	4	2										$N = 0$	●	●	●	●	●	●
Branch To Subroutine	BSR	8D	8	2											●	●	●	●	●	●
Jump	JMP				6E	4	2	7E	3	3				See Special Operations	●	●	●	●	●	●
Jump To Subroutine	JSR				AD	8	2	BD	9	3					●	●	●	●	●	●
No Operation	NOP										01	2	1	Advances Prog. Cntr. Only	●	●	●	●	●	●
Return From Interrupt	RTI										3B	10	1					①		
Return From Subroutine	RTS										39	5	1		●	●	●	●	●	●
Software Interrupt	SWI										3F	12	1	See Special Operations	●	●	●	●	●	●
Wait for Interrupt *	WAI										3E	9	1		●	②	●	●	●	●

*WAI puts Address Bus, R/W, and Data Bus in the three-state mode while VMA is held low.

① (All) Load Condition Code Register from Stack. (See Special Operations)
② (Bit 1) Set when interrupt occurs. If previously set, a Non-Maskable Interrupt
    is required to exit the wait state.

Table 7-4

MC6800
CONDITION-
CODE-
REGISTER-
MANIPULATION
INSTRUCTIONS.
(COURTESY OF
MOTOROLA, INC.)

CONDITION CODE REGISTER INSTRUCTIONS

| OPERATIONS | MNEMONIC | IMPLIED | | | BOOLEAN OPERATION | COND. CODE REG. | | | | | |
		OP	~	=		5 H	4 I	3 N	2 Z	1 V	0 C
Clear Carry	CLC	0C	2	1	$0 \rightarrow C$	•	•	•	•	•	R
Clear Interrupt Mask	CLI	0E	2	1	$0 \rightarrow I$	•	R	•	•	•	•
Clear Overflow	CLV	0A	2	1	$0 \rightarrow V$	•	•	•	•	R	•
Set Carry	SEC	0D	2	1	$1 \rightarrow C$	•	•	•	•	•	S
Set Interrupt Mask	SEI	0F	2	1	$1 \rightarrow I$	•	S	•	•	•	•
Set Overflow	SEV	0B	2	1	$1 \rightarrow V$	•	•	•	•	S	•
Acmltr A → CCR	TAP	06	2	1	$A \rightarrow CCR$			①			
CCR → Acmltr A	TPA	07	2	1	$CCR \rightarrow A$	•	•	•	•	•	•

R = Reset
S = Set
• = Not affected

① (ALL) Set according to the contents of Accumulator A.

7-4.5 Condition-code-register instructions

Eight instructions are included in this group, as listed in Table 7-4. These instructions allow the programmer to control the condition code register while the program is being executed.

7-5 MC6800 CONTROL-BUS SIGNALS

The control bus is connected to 11 pins on the MC6800, as shown in Figure 7-9. These carry the signals—generated by the decoded instructions, the clock, and the

Figure 7-9

MC6800
control-bus
connections

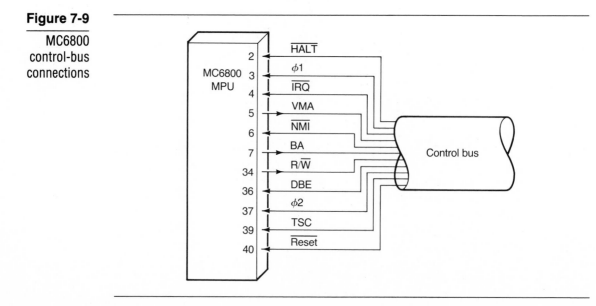

Table 7-5

MC6800
CONTROL-BUS
SIGNALS

- Pin 2 ($\overline{\text{Halt}}$)   : An input control signal that provides an external method of halting MC6800 MPU operations.

- Pin 3 ($\phi1$)   : Phase one clock input

- Pin 4 ($\overline{\text{IRQ}}$)   : Interrupt request. This is the maskable interrupt signal generated by a peripheral requiring service.

- Pin 5 (VMA)   : Valid memory address. An output signal that indicates to the I/O peripherals that the address on the address bus is valid.

- Pin 6 ($\overline{\text{NMI}}$)   : Nonmaskable interrupt. An input signal that initiates the nonmaskable interrupt sequence.

- Pin 7 (BA)   : Bus available. An output signal that informs the peripherals when the MC6800 is sending or receiving data on the data bus.

- Pin 34 (R/$\overline{\text{W}}$)   : Read/Write. An output signal that controls the polarity of the data bus. The MPU is receiving data when the signal is high and sending data whenever the signal is low.

- Pin 36 (DBE)   : Data-bus-enable. An input signal that controls the three-state buffers on the microprocessor end of the data bus.

- Pin 37 ($\phi2$)   : Phase two clock input.

- Pin 39 (TSC)   : Three-state control. This is an input signal that allows an external device to take control of the address bus, as in the case of DMA.

- Pin 40 ($\overline{\text{Reset}}$)   : An input signal used to initiate program execution from an external source.

I/O peripherals—that cause a program to be executed. A brief description of each of the control signals appears in Table 7-5.

7-6 MC6800 PROGRAM EXECUTION

7-6.1 Introduction

A number of activities occur simultaneously in the microprocessor, memory, and I/O peripherals to cause program execution. Instructions fetched from memory and decoded by the instruction-decode-and-control unit generate control signals that cause the transfer of data between memory or the I/O peripherals and the microprocessor and within the microprocessor.

7-6.2 Running the program

An assembly-language program that causes 3_{16} and 6_{16} to be added together appears in Table 7-6. The program has previously been loaded into memory. The op-

Table 7-6

ADD PROGRAM
FOR THE MC6800

ADDRESS	MNEMONIC	OPCODE	OPERAND OR OPERAND ADDRESS
0001_{16}	LDAA	86_{16}	
0002_{16}			03_{16}
0003_{16}	ADDA	$8B_{16}$	
0004_{16}			06_{16}
0005_{16}	STAA	97_{16}	
0006_{16}			0008_{16}
0007_{16}	WAI	$3E_{16}$	
0008_{16}			

codes and operands appear in hexadecimal format for the sake of convenience. Of course, the actual information stored in memory is in binary form.

The address of the first instruction (0001_{16}) is loaded into the program counter and the LDAA instruction having an opcode of 86_{16} is fetched from memory and loaded into the instruction-decode-and-control unit where it is decoded, as shown in Figure 7-10. The resulting control signals cause operand 03_{16} to be transferred from memory to accumulator A, as seen in Figure 7-11. The ADDA instruction, whose opcode is $8B_{16}$, is now fetched from memory and loaded into the instruction-decode-and-control unit and decoded, as illustrated in Figure 7-12. This generates control signals that cause operand 06_{16} to be transferred from memory to the arithmetic and logic unit. At the same time, the number 03_{16} is transferred from accumulator A to the arithmetic and logic unit. The two numbers are added together and the sum is loaded back into the accumulator, as seen in Figure 7-13. Next, the STAA instruction, whose opcode is 97_{16}, is fetched from memory and decoded, as shown in Figure 7-14. The resulting control signals cause the sum to be stored in memory at address 0008_{16}, as depicted in Figure 7-15. Finally, the WAI instruction, whose opcode is $3E_{16}$, is fetched from memory and loaded into the instruction-decode-and-control unit. The resulting control signals cause the microprocessor to halt its operations.

Four instructions utilizing three different addressing modes were needed to perform the addition operation. The LDAA and ADDA instructions utilized the immediate addressing mode; direct addressing was employed with the STAA instruction; while the implied mode was used to execute the WAI instruction.

7-6.3 Program termination

The last instruction in most programs is the wait for interrupt (WAI) instruction. This is the halt instruction for the MC6800. When this instruction is executed, the contents of the internal registers are loaded into the stack, just as they are for inter-

Figure 7-10

Fetching the LDAA opcode

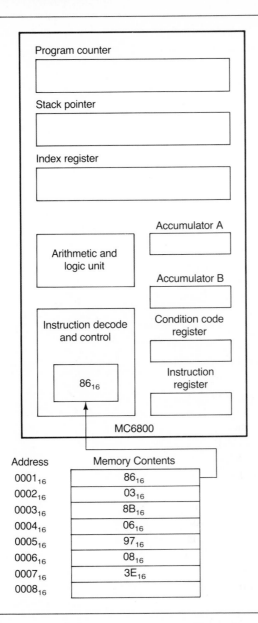

rupts. All operations cease and the microprocessor idles unless an interrupt occurs, in which case the MPU jumps to the interrupt subroutine and again commences operation.

7-6.4 MPU cycles

As previously mentioned, the MC6800 requires an external 1-MHz clock that produces two nonoverlapping clock-pulse trains to synchronize the activities occuring

Figure 7-11

Executing the
LDAA instruction

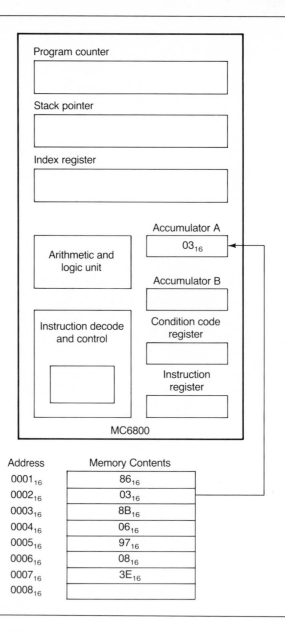

within the microprocessor, memory, and I/O peripherals. One clock pulse is referred to as one clock-pulse cycle. Each MPU cycle is 1 microsecond (μs) in duration.

The number of MPU cycles necessary to fetch an instruction from memory and perform its execution depends upon the addressing mode utilized. These are listed with the instructions shown in Tables 7-1 through 7-4 and appear under the ~ column. The ADDA instruction appearing in Table 7-1 requires two MPU cycles if

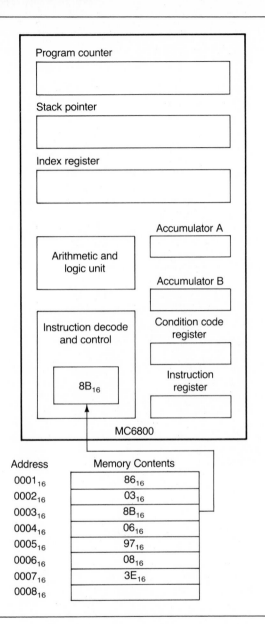

Figure 7-12

Fetching the ADDA
opcode

immediate addressing is utilized; for direct addressing, three cycles are required; indexed addressing requires five MPU cycles; while the extended addressing mode requires four MPU cycles. The execution time for instructions is proportional to the number of MPU cycles required to perform the operation. Although the indexed- and extended-addressing modes may provide more programming flexibility and greater programming efficiency than do some of the other addressing modes, they also require greater execution times.

Figure 7-13

Executing the
ADDA instruction

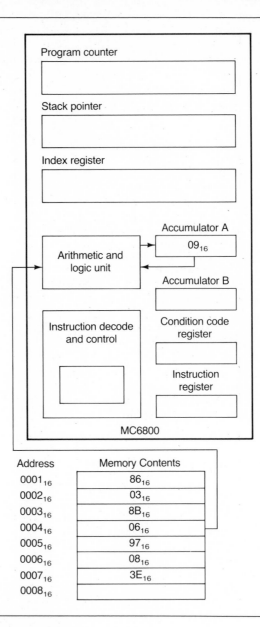

7-7 ADDRESSING MEMORY
AND I/O PERIPHERALS

7-7.1 Introduction

The MC6800 uses memory-mapped I/O to communicate with the main-memory ICs and I/O peripherals. This allows each memory IC and I/O peripheral to be ad-

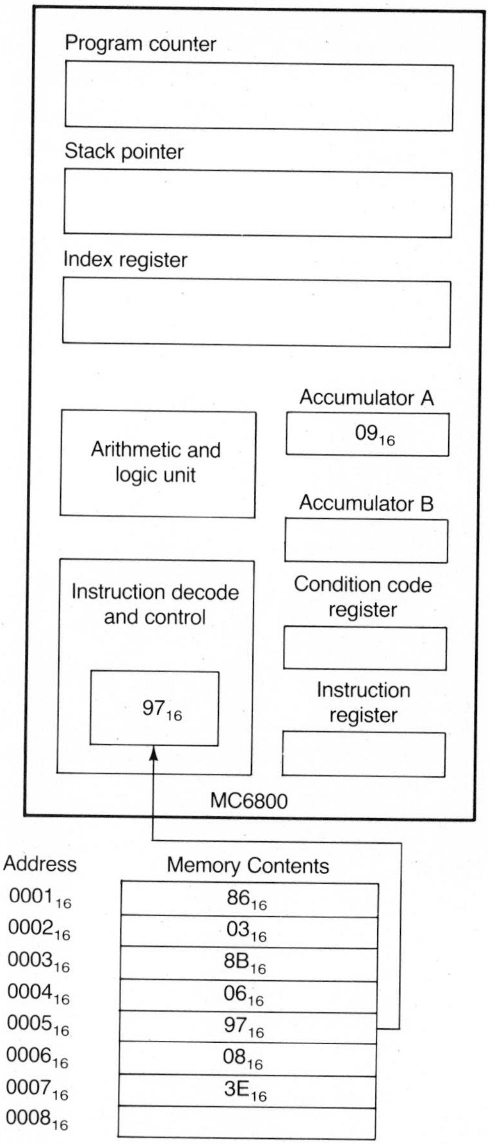

Figure 7-14

Fetching the STAA opcode

Program counter

Stack pointer

Index register

Accumulator A

09_{16}

Arithmetic and logic unit

Accumulator B

Instruction decode and control

Condition code register

97_{16}

Instruction register

MC6800

Address	Memory Contents
0001_{16}	86_{16}
0002_{16}	03_{16}
0003_{16}	$8B_{16}$
0004_{16}	06_{16}
0005_{16}	97_{16}
0006_{16}	08_{16}
0007_{16}	$3E_{16}$
0008_{16}	

dressed in the same fashion as a memory segment. Theoretically, this permits the connection of 65,536 memory and I/O peripheral-interface ICs to the MC6800. Main-memory storage capacity and the number of I/O peripherals required for a particular microprocessor-based system depends upon the system application. Few I/O peripherals are required in general-purpose systems such as microcomputers. Microcomputers require considerable RAM storage capacity and a relatively small amount of ROM. A system designed for a dedicated application, such as an indus-

Figure 7-15

Executing the
STAA instruction

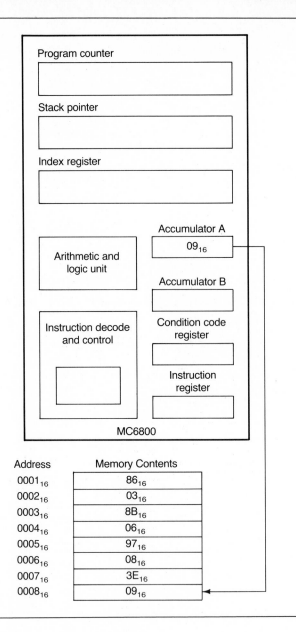

trial controller, often requires many I/O peripherals and a lesser amount of memory. In this type of system, main memory is often predominantly ROM with only a relatively small amount of RAM required. The major portion of the program is permanently stored in ROM. The only RAM required is that used to store the input—output variables.

7-7.2 Addressing the system components

Whatever the amount of RAM and ROM and number of I/O peripherals required in a particular system, RAM, ROM, and I/O interfacing ICs are all addressed in the same fashion. This is accomplished by connecting a decoder to the address bus for each memory and interface IC. Called a chip-address decoder, the output of this circuit is connected to the chip-enable (CE) input of each memory and peripheral interface chip.

As previously noted, a 16-bit address allows 65,536 different memory and peripheral interface ICs to be connected to the MC6800. Since few, if any, applications require that large a number of devices, the chip-address decoders do not have to be connected to all the address-bus lines. The chip-address decoders are connected to the upper eight address lines (A8–A15), while the lower eight lines (A0-A7) are connected to the memory chip, as shown in Figure 7-16. This allows

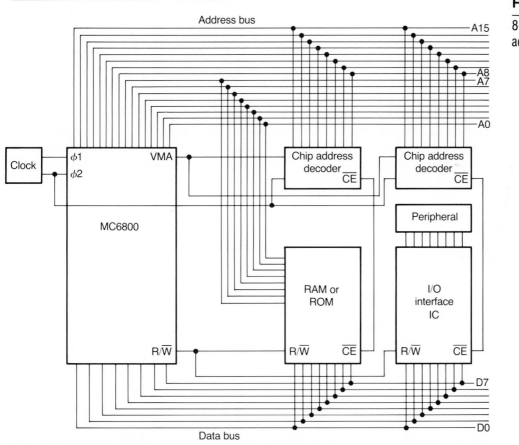

Figure 7-16

8-bit chip select address

256 different memory and I/O peripheral interface chips to be connected to the MPU and at the same time allows 256 bytes of memory to be accessed.

Each memory and I/O peripheral-interface IC is assigned an address that uses the upper eight address bits, while the storage segments within the memory device are assigned addresses from the lower eight address bits. The upper byte is used to address a particular memory or interface IC while the lower byte is used to access a particular storage segment within the memory chip. Whenever a memory-storage segment is addressed, the upper byte is detected by the memory chip-address decoder. At the same time, VMA and φ2 clock signals are sent to the decoder via the control bus. The address bits, VMA, and φ2 inputs force the signal from the output of the decoder to fall from a binary one to a binary zero. This signal is sent to the $\overline{CE}$ input of the memory chip, enabling the chip's three-state buffers and causing the data bus to be connected to the device. The lower address byte is decoded by an internal-memory address decoder, causing a particular memory segment to be selected. The R/$\overline{W}$ signal establishes the polarity of the data bus and determines whether data will be written into or read out of the addressed memory segment. Data are written into RAM if the R/$\overline{W}$ signal is binary zero, and read out of RAM or ROM whenever the signal is at a binary one level.

I/O peripherals are connected to the MC6800 data bus by interface chips. These are assigned an address that is decoded by the chip address decoder. Data are sent to or received from a peripheral-interface chip in the same fashion as a memory segment.

In applications where few memory ICs and I/O peripherals are required, fewer address lines need be dedicated to addressing memory and I/O devices. This increases the amount of memory which can be accessed. Only four address lines are connected to the chip-address decoder in Figure 7-17. This reduces the number of memory and interface ICs which can be connected to the microprocessor from 256 to 16 but increases the memory-accessing capability from 256 to 4096 bytes.

7-7.3 Transferring data between the MC6800 and the system components

Just as memory and I/O peripheral-interface ICs are addressed in the same fashion, data are transferred between the MC6800 and memory or I/O peripheral devices in the same way. As previously illustrated, the load accumulator (LDA) and store accumulator (STA) instructions are used to accomplish these transfers. Data are loaded into either accumulator by assigning an address to a memory segment or an input peripheral and using either the LDAA or LDAB instructions. This is illustrated in Figure 7-18. The input peripheral has previously been assigned the address 7000_{16} or 0111000000000000_2 and has generated data represented by the word 11011001_2. The peripheral will transfer the data into accumulator A when addressed by the instruction LDAA7000_{16}. The address appears on the address bus for a few microseconds and the upper four bits are detected by the peripheral chip-address decoder. The $\overline{CE}$ output signal falls to a binary-zero level, enabling the three-state buffers. This connects the input-peripheral interface IC to the main data

Figure 7-17

4-bit chip select
address

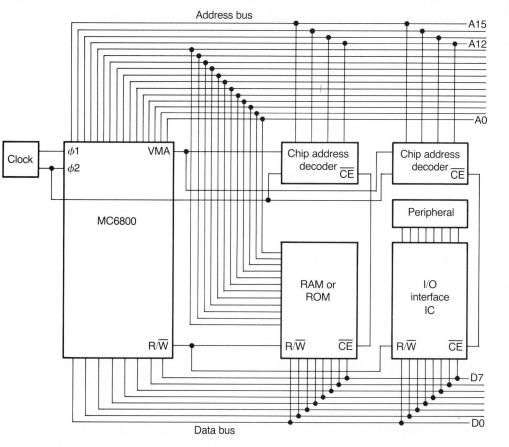

bus. At the same time, the decoded LDAA instruction forces the VMA and $R/\overline{W}$ signals to binary-one levels, causing the peripheral data to be loaded into accumulator A via the interface IC and the data bus, as seen in Figure 7-19.

Data are transferred from either accumulator to a memory segment or output peripheral by assigning an address to the memory segment or peripheral and using either the STAA or STAB instructions. Accumulator A in Figure 7-20 is storing 10110010_2 and the output peripheral has been assigned the address 4000_{16}, or 0100000000000000_2. The data in the accumulator will be transferred to the output peripheral when the STAA4000_{16} instruction is executed. Execution of this instruction connects the output peripheral-interface chip to the MPU data bus. The VMA signal rises to a binary-one level while the $R/\overline{W}$ signal falls to a binary zero. The contents are transferred to the peripheral, as shown in Figure 7-21.

7-7.4 MC6800 interrupts

The preceeding section described how data are transferred between the MC6800 and the system components. Consideration must be given as to when data will be

Figure 7-18

Addressing an
input peripheral

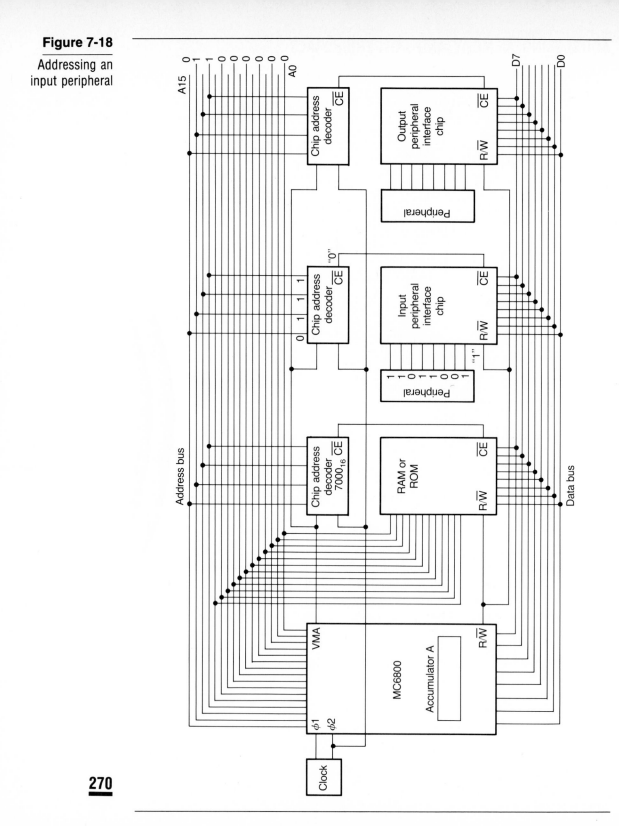

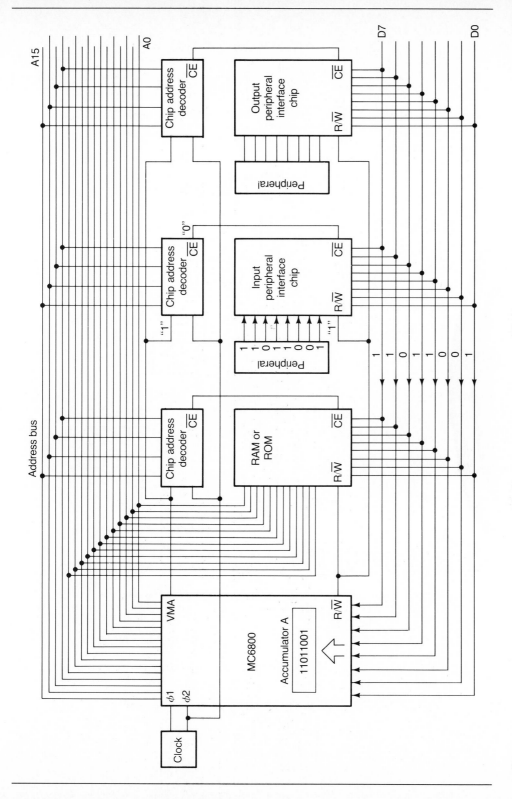

Figure 7-19

Transferring data between an input peripheral and accumulator

Figure 7-20

Addressing an
output peripheral

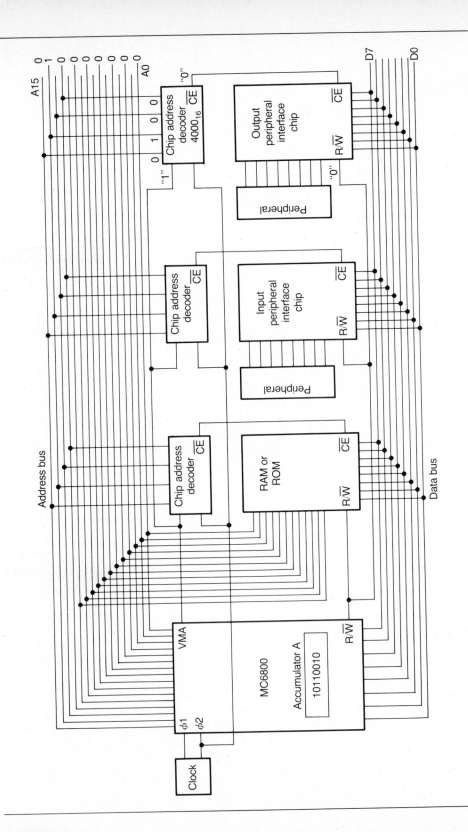

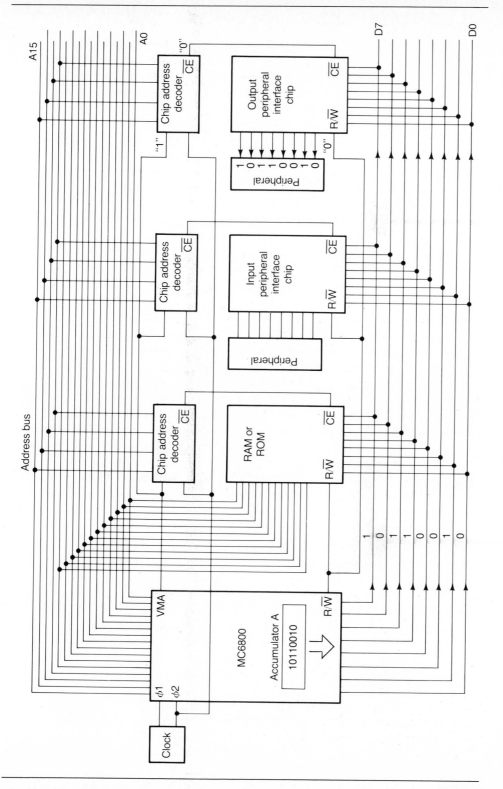

transferred. The LDA and STA instructions allow data to be moved into and out of the microprocessor at predetermined intervals. These instructions are part of the program and are stored in memory. Execution occurs during the normal sequence of running the program. In many industrial applications, data must be loaded into the microprocessor on an as-needed basis. For example, a microprocessor used in a machine-tool controller has several I/O peripherals. Input peripherals may include an LVDT to measure cutting distance, a tachometer to measure cutting-tool speed, and a flow-control sensor to measure coolant flow. If any of these variables deviate from their preset conditions, the microprocessor must be immediately notified so that it can generate the signals required for the output peripherals to take corrective action. Interrupts are used to perform this function. These were introduced in Chapter 6. Motorola has designed the MC6800 for both nonmaskable and maskable interrupt operation. Both types of interrupts require that a peripheral subroutine be stored somewhere in memory. Subroutine-program execution causes the microprocessor to service the input peripheral requiring attention. Two control lines are used for interrupt purposes—the nonmaskable interrupt ($\overline{\text{NMI}}$) and the interrupt request ($\overline{\text{IRQ}}$).

Nonmaskable interrupt: Whenever an input peripheral requires service, it causes the signal applied to the $\overline{\text{NMI}}$ control pin on the MPU to change from a binary-one to a binary-zero level. The microprocessor completes the execution of the current instruction and the contents of the condition code register, both accumulators, the index register, and the program counter are loaded into the stack. The interrupt (I) flag, or mask, in the condition code register is set. This causes the MPU to ignore other interrupt requests while the current request is being serviced. The microprocessor next jumps to the interrupt subroutine and initiates its execution.

The last instruction in the interrupt subroutine is the return from interrupt (RTI). This is one of the jump-and-branch instructions appearing in Table 7-3. The RTI instruction causes the MC6800 to return to that part of the program being executed before the interrupt occured. The stack contents are loaded back into their original registers, the interrupt mask is reset, and the microprocessor commences main-program execution.

Maskable interrupt: The MC6800 uses the interrupt-request ($\overline{\text{IRQ}}$) control line for maskable interrupts. An interrupt is received when this control signal falls from a binary-one to a binary-zero level. The microprocessor may or may not service the interrupt, depending upon the state of the I flag in the condition code register. The interrupt is ignored if the I flag is set and the microprocessor continues regular program execution. If the flag is in the reset state, the contents of the internal registers are loaded into the stack, the I flag is set to prevent the microprocessor from attempting to service additional interrupts, and the MPU jumps to the interrupt subroutine and begins subroutine-program execution. The last instruction in the subroutine is the return-from-interrupt instruction. When executed, this instruction causes the microprocessor to jump back to the main program. Data in the stack are loaded into their original registers, the interrupt flag is reset, and main-program execution resumes.

7-8 MC6800 PERIPHERAL-INTERFACE ICs

7-8.1 Introduction

Motorola has developed a series of supporting integrated circuits to facilitate interfacing the MC6800 with I/O peripherals. Some of these are programmable and are used for general-purpose interfacing applications, while others are designed for specific usage. The more common of these are discussed in the following sections.

7-8.2 The MC6821 Peripheral Interface Adapter

The Peripheral Interface Adapter (PIA) was introduced in Chapter 6. The MC6821 is a PIA that Motorola has developed for use with the MC6800 MPU. The MC6821 is an NMOS chip housed in a 40-pin DIP case, as seen in Illustration 7-1.

A block diagram of the PIA appears in Figure 7-22. The MC6821 is divided into two sides—an MPU side and a peripheral side. The peripheral side includes two identical sections, A and B. The device is connected to the MC6800 MPU through the address, data, and control buses. Five address-bus lines are connected to the PIA—A0, A1, A2, A14, and A15. They connect to the three chip-select pins (CS0, CS1, and $\overline{CS2}$) and the two register-select pins (RS0 and RS1). Control-bus-line connections include two interrupt requests ($\overline{IRQA}$ and $\overline{IRQB}$), read/write (R/$\overline{W}$), enable (E), and $\overline{Reset}$.

The CS0, CS1, and $\overline{CS2}$ lines are used to select the PIA. The device is selected by any address that causes lines A2 and A14 to be a binary one and A15 a binary zero. The interrupt-request outputs allow the PIA to initiate an interrupt. The enable input provides the system timing for the PIA and is connected to φ2 of the clock. The R/$\overline{W}$ line controls the polarity of the system data bus and determines whether the PIA will send data to or receive data from the MC6800 MPU.

Each section on the peripheral side of the PIA is connected to a peripheral by an 8-bit data bus (PA0–PA7 and PB0–PB7) and two control-interface lines (CA1, CA2 and CB1, CB2). The peripheral data buses can be programmed to allow the PIA to either send or receive data from the I/O peripherals. Sections A and B each contain three registers—a peripheral register (PRA and PRB), a data-direction register (DDRA and DDRB), and a control register (CRA and CRB). The peripheral registers are used to temporarily hold data being transferred between the microprocessor and I/O peripherals and act as buffers. The buffering action is important when interfacing I/O peripherals whose operating speeds are not the same as that of the MPU. Most peripherals operate much more slowly than the microprocessor.

The data-direction registers control the polarity of the peripheral data-bus lines. The flip-flops forming these registers configure the data-bus lines for data transmission when set. In the reset state, the peripheral data-bus lines are configured to receive data from the peripherals. For 8-bit parallel-operated I/O peripherals, the peripheral data-bus lines can be configured to allow one 8-bit input peripheral and one 8-bit output peripheral to be connected to the PIA, as shown in Figure 7-23(a);

Illustration 7-1

(Courtesy of
Motorola, Inc.)

MOTOROLA

SEMICONDUCTORS

3501 ED BLUESTEIN BLVD., AUSTIN, TEXAS 78721

MC6821

PERIPHERAL INTERFACE ADAPTER (PIA)

The MC6821 Peripheral Interface Adapter provides the universal means of interfacing peripheral equipment to the M6800 family of microprocessors. This device is capable of interfacing the MPU to peripherals through two 8-bit bidirectional peripheral data buses and four control lines. No external logic is required for interfacing to most peripheral devices.

The functional configuration of the PIA is programmed by the MPU during system initialization. Each of the peripheral data lines can be programmed to act as an input or output, and each of the four control/interrupt lines may be programmed for one of several control modes. This allows a high degree of flexibility in the overall operation of the interface.

- 8-Bit Bidirectional Data Bus for Communication with the MPU
- Two Bidirectional 8-Bit Buses for Interface to Peripherals
- Two Programmable Control Registers
- Two Programmable Data Direction Registers
- Four Individually-Controlled Interrupt Input Lines; Two Usable as Peripheral Control Outputs
- Handshake Control Logic for Input and Output Peripheral Operation
- High-Impedance Three-State and Direct Transistor Drive Peripheral Lines
- Program Controlled Interrupt and Interrupt Disable Capability
- CMOS Drive Capability on Side A Peripheral Lines
- Two TTL Drive Capability on All A and B Side Buffers
- TTL-Compatible
- Static Operation

NMOS
(N-CHANNEL, SILICON-GATE, DEPLETION LOAD)

PERIPHERAL INTERFACE ADAPTER

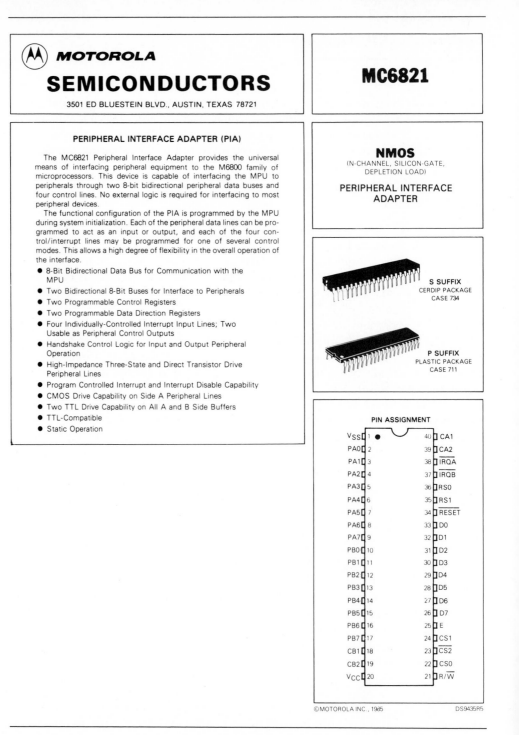

S SUFFIX
CERDIP PACKAGE
CASE 734

P SUFFIX
PLASTIC PACKAGE
CASE 711

PIN ASSIGNMENT

V$_{SS}$	1	40	CA1
PA0	2	39	CA2
PA1	3	38	$\overline{IRQA}$
PA2	4	37	$\overline{IRQB}$
PA3	5	36	RS0
PA4	6	35	RS1
PA5	7	34	$\overline{RESET}$
PA6	8	33	D0
PA7	9	32	D1
PB0	10	31	D2
PB1	11	30	D3
PB2	12	29	D4
PB3	13	28	D5
PB4	14	27	D6
PB5	15	26	D7
PB6	16	25	E
PB7	17	24	CS1
CB1	18	23	$\overline{CS2}$
CB2	19	22	CS0
V$_{CC}$	20	21	R/$\overline{W}$

DS9435R5

Figure 7-22
MC6821 PIA

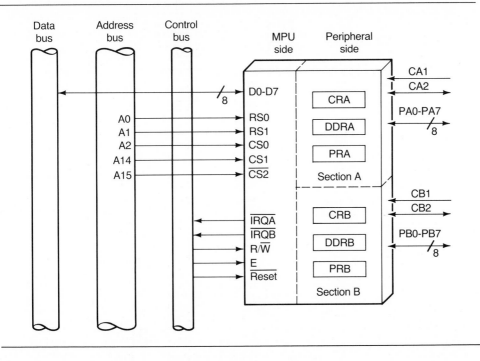

two 8-bit input peripherals, as seen in Figure 7-23(b); or two 8-bit output peripherals, as illustrated in Figure 7-23(c). The peripheral data buses can be configured to accomodate 16 different serially operated I/O peripherals for either input or output applications.

The control registers allow the microprocessor to control the functions of the four control-interface lines. They also allow the MPU to enable the two interrupt lines and monitor the status of the condition-code-register interrupt flag in the MPU.

7-8.3 The MC6850 Asynchronous Communications Interface Adapter

The MC6850 Asynchronous Communications Interface Adapter (ACIA) is a universal asynchronous receiver transmitter (UART) designed to be used with the MC6800 MPU. The UART was introduced in Chapter 6. The MC6850 ACIA is an NMOS device housed in a 24-pin DIP case. As seen from the data sheet in Illustration 7-2, the device is composed of data-bus registers, control circuits, and transmitter and receiver registers that facilitate the transfer of asynchronous data between the MC6800 and serially operated I/O peripherals.

The block diagram appears in Figure 7-24. The ACIA is divided into two sides—an MPU side and a peripheral side. The MPU side is connected to the system-address, data, and control buses. Four address-bus lines are connected to the RS, CS0, CS1, and $\overline{CS2}$ pins. The RS input is used to select the internal registers while the CS0, CS1, and $\overline{CS2}$ inputs are used as chip selects to address the ACIA. Ter-

Figure 7-23
_____
PIA 8-bit
peripheral
connections

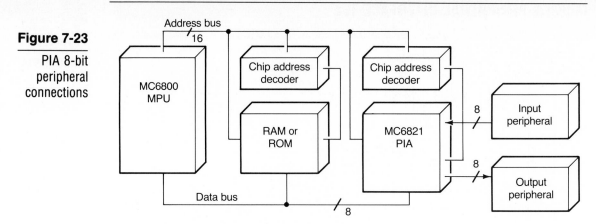

(a) PIA configured for 8-bit input–output operation

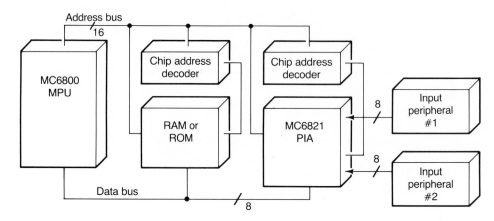

(b) PIA configured for two 8-bit input peripherals

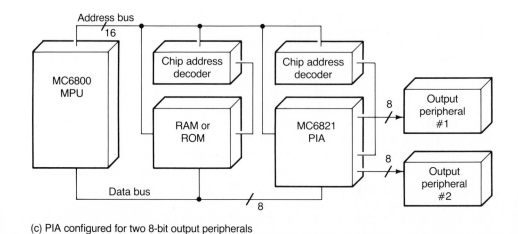

(c) PIA configured for two 8-bit output peripherals

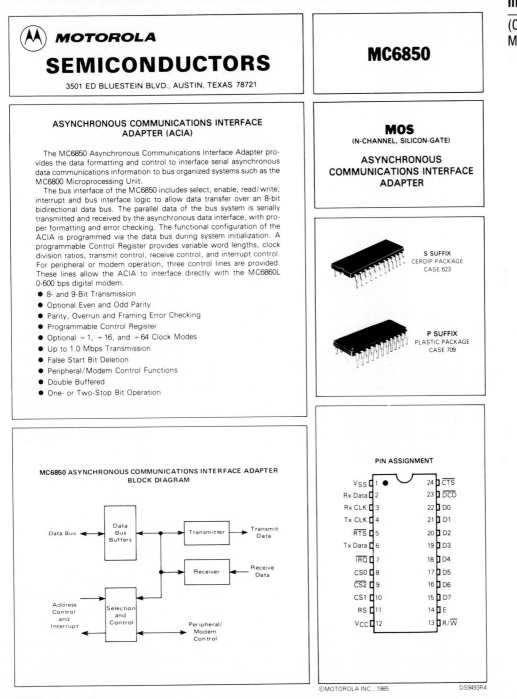

⊕ *MOTOROLA*

SEMICONDUCTORS

3501 ED BLUESTEIN BLVD., AUSTIN, TEXAS 78721

MC6850

ASYNCHRONOUS COMMUNICATIONS INTERFACE ADAPTER (ACIA)

The MC6850 Asynchronous Communications Interface Adapter provides the data formatting and control to interface serial asynchronous data communications information to bus organized systems such as the MC6800 Microprocessing Unit.

The bus interface of the MC6850 includes select, enable, read/write, interrupt and bus interface logic to allow data transfer over an 8-bit bidirectional data bus. The parallel data of the bus system is serially transmitted and received by the asynchronous data interface, with proper formatting and error checking. The functional configuration of the ACIA is programmed via the data bus during system initialization. A programmable Control Register provides variable word lengths, clock division ratios, transmit control, receive control, and interrupt control. For peripheral or modem operation, three control lines are provided. These lines allow the ACIA to interface directly with the MC6860L 0-600 bps digital modem.

- 8- and 9-Bit Transmission
- Optional Even and Odd Parity
- Parity, Overrun and Framing Error Checking
- Programmable Control Register
- Optional ÷1, ÷16, and ÷64 Clock Modes
- Up to 1.0 Mbps Transmission
- False Start Bit Deletion
- Peripheral/Modem Control Functions
- Double Buffered
- One- or Two-Stop Bit Operation

MOS
(N-CHANNEL, SILICON-GATE)

ASYNCHRONOUS COMMUNICATIONS INTERFACE ADAPTER

S SUFFIX
CERDIP PACKAGE
CASE 623

P SUFFIX
PLASTIC PACKAGE
CASE 709

MC6850 ASYNCHRONOUS COMMUNICATIONS INTERFACE ADAPTER
BLOCK DIAGRAM

Data Bus

Data Bus Buffers

Transmitter — Transmit Data

Receiver — Receive Data

Address Control and Interrupt

Selection and Control

Peripheral/Modem Control

PIN ASSIGNMENT

Vss	1	24	$\overline{CTS}$
Rx Data	2	23	$\overline{DCD}$
Rx CLK	3	22	D0
Tx CLK	4	21	D1
$\overline{RTS}$	5	20	D2
Tx Data	6	19	D3
$\overline{IRQ}$	7	18	D4
CS0	8	17	D5
$\overline{CS2}$	9	16	D6
CS1	10	15	D7
RS	11	14	E
VCC	12	13	$R/\overline{W}$

DS9493R4

Figure 7-24

MC6850 ACIA

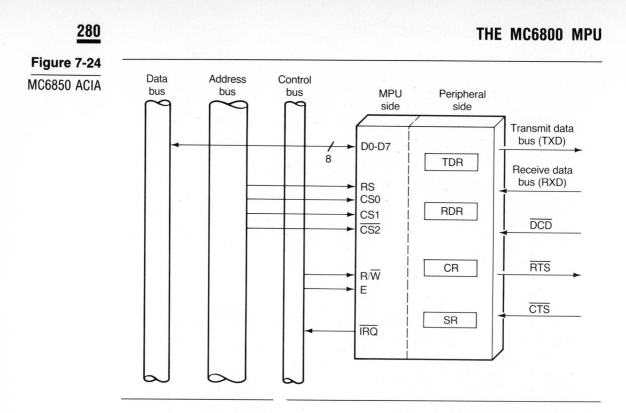

minals D0–D7 connect the device to the main data bus. Three control-bus lines are utilized. Control signals include an enable (E) input, which comes from the ϕ2 clock output, an interrupt-request output ($\overline{\text{IRQ}}$), and read/write (R/$\overline{\text{W}}$). The E control signal synchronizes the timing of the transmit and receive serial-data lines on the peripheral side of the ACIA and determines when data will be transferred between the ACIA and the MPU. The $\overline{\text{IRQ}}$ output allows the ACIA to initiate an interrupt. The R/$\overline{\text{W}}$ establishes the polarity of the system data bus.

The peripheral side of the ACIA contains four registers which are accessible to the programmer. These include the transmit-data register (TDR), the receive-data register (RDR), the control register (CR), and the status register (SR). The transmit- and receive-data registers are accessed when the RS input is at a binary-one level, while the control and status registers are addressed whenever the RS line is binary zero. Five signals affect the transfer of data between the ACIA and peripherals. They include request to send ($\overline{\text{RTS}}$), clear to send ($\overline{\text{CTS}}$), data-carrier detect ($\overline{\text{DCD}}$), transmit data (TXD), and receive data (RXD). The $\overline{\text{RTS}}$, $\overline{\text{CTS}}$, and $\overline{\text{DCD}}$ lines are used when the ACIA is interfaced with a modem. The TXD and RXD are serial-data lines on which data are either sent to or received from a peripheral.

The ACIA is connected to the MC6800 system as shown in Figure 7-25. The input peripheral sends asynchronous serial data to the ACIA via the receive-data line. The serial data are converted into parallel data within the ACIA and transferred to the MPU by way of the 8-bit system-data bus. When the MC6800 is sending data to an output peripheral, the parallel data enters the ACIA through the 8-bit system-

Figure 7-25

Connecting the
ACIA to the
MC6800

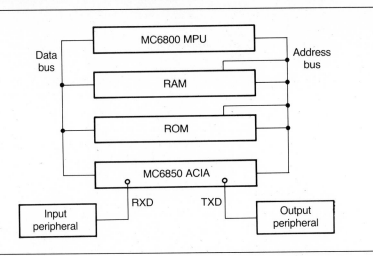

data bus. After being converted into asynchronous serial data, the data exits the ACIA via the transmit-data line and is sent to the output peripheral.

7-8.4 The MC6843 Floppy Disk Controller

The MC6843 Floppy Disk Controller (FDC) is an interface chip that allows the MC6800 MPU to be interfaced with a floppy-disk drive.

7-8.5 The MC6844 Direct Memory Access Controller

The MC6844 Direct Memory Access Controller (DMAC) provides direct memory access (DMA) between main and mass memory in an MC6800 microprocessor-based system.

7-8.6 The MC6852 Synchronous Serial Data Adapter

The MC6852 Synchronous Serial Data Adapter (SSDA) is a serial interface used to transfer synchronous serial data between the MC6800 MPU and serially operated I/O peripherals. This is Motorola's version of the universal synchronous receiver transmitter (USRT) identified in Chapter 6.

7-8.7 The MC6860 Digital Modem

This is a single-chip modem that converts serial data into audio mark and space signals which can be transmitted over telephone and other voice-grade transmission lines. The MC6860 is interfaced to the MC6800 MPU by connecting it to an

Figure 7-26

Connecting the
MC6860 digital
modem to the
MC6800

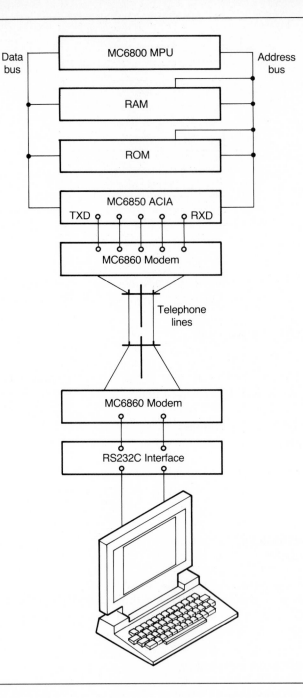

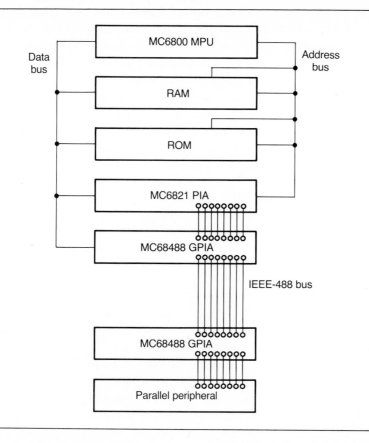

Figure 7-27

Connecting the
MC68488 GPIA to
the MC6800

MC6850 ACIA. A typical application appears in Figure 7-26. A remote terminal is connected to the MPU via telephone lines. An MC6860 is required at both ends of the telephone lines.

7-8.8 The MC68488 General Purpose Interface Adapter

The MC68488 General Purpose Interface Adapter (GPIA) allows the IEEE-488 General Purpose Interface Bus to be used with the MC6800. Interfacing is accomplished by connecting the MC68488 to an MC6821 PIA, as illustrated in Figure 7-27.

Summary

1 **1.** The MC6800 is an 8-bit microprocessor that utilizes a 16-bit address capable of accessing 65,536 bytes of memory.

2. MC6800 architecture includes a 16-bit program counter, a 16-bit index register, a 16-bit stack pointer, two 8-bit accumulators, an 8-bit instruction register, an 8-bit CCR, an ALU, and an instruction-decode-and-control unit.

3. The implied-, immediate-, direct-, extended-, indexed-, relative-, and accumulator-addressing modes are employed in the MC6800.

4. The implied-addressing mode is a one-byte instruction that requires no operand. The immediate-addressing mode is a two-byte instruction in which the operand is stored in the memory segment following the opcode.

5. The direct- and extended-addressing modes are similar. The address of the operand is located in the memory segment following the opcode for both. A single-byte address is used with direct addressing whereas a two-byte address is employed with extended addressing.

6. The indexed and relative addressing modes are both two-byte instructions. The first byte is the opcode while the second byte is an offset address. The offset address is added to the contents of the index register for indexed addressing while the offset address if added to the contents of the program counter when relative addressing is used.

7. The instruction set for the MC6800 is made up of 72 instructions, which can be grouped into four categories. These include accumulator and memory instructions, index-register and stack-manipulation instructions, jump-and-branch instructions, and condition-code-register instructions.

8. The load-accumulator (LDAA and LDAB) instructions cause data to be loaded into an accumulator from RAM, ROM, or an input peripheral.

9. The store-accumulator (STAA and STAB) instructions cause the accumulator contents to be loaded into RAM or an output peripheral.

10. The MC6800 uses an 11-line control bus. The R/$\overline{W}$ connects to pin 34 on the case and controls the polarity of the data bus. The BA signal connects to pin 7 and informs the I/O peripherals when the MPU is sending or receiving data.

11. The MC6800 requires an external clock that produces two nonoverlapping clock pulse trains having a frequency of 1 MHz.

12. Memory-mapped I/O is used by the MC6800 to communicate with its memory and I/O peripheral devices. Memory ICs and peripherals are addressed in the same fashion as memory-storage segments.

13. Nonmaskable and maskable interrupts are both employed in the MC6800. The I flag is set in the CCR while an interrupt is being serviced.

14. The MC6821 PIA is a general-purpose programmable interface IC. The PIA contains an MPU side and a peripheral side. The peripheral side is divided into two sections. Each section contains three registers. These include the peripheral register, data-direction register, and control register.

15. The peripheral registers temporarily store data being transferred between the MPU and the I/O peripherals. The data-direction registers control the polarity of the peripheral data bus lines while the control registers govern the four peripheral control lines.

16. The MC6850 ACIA is used to interface peripherals which utilize asynchronous serial data to the MC6800. The ACIA is divided into an MPU side and a peripheral side. The peripheral side contains a transmit-data register, a receive-data register, a control register, and a status register.

17. The MC6840 PTM is a programmable timer used to generate continuous or one-shot pulses for external timing and for delay-, interval-, and cycle-timer applications.

18. The MC6843 FDC is used to interface the MC6800 MPU with a floppy-disk drive.

19. The MC6852 SSDA is a serial-interface chip used to transfer synchronous data between serially operated I/O peripherals and the MC6800.

20. The MC6860 is a modem that converts serial data into audio tones, which can be transmitted through telephone and other voice-grade transmission lines.

21. The MC68488 interfaces the MC6800 MPU with the IEEE-488 GPIB.

Chapter Examination

1. Which of the following control signals establishes the polarity of the MC6800 data bus?
   a. DBE
   b. R/$\overline{\text{W}}$
   c. VMA
   d. $\overline{\text{IRQ}}$

2. Which of the following instructions causes the MC6800 to return from an interrupt subroutine to the main program?
   a. INX
   b. WAI
   c. RTI
   d. INC

3. The _____ instruction causes data to be transferred from memory or an input peripheral to the accumulator.
   a. LDA
   b. STA
   c. ADA
   d. ACC

4. How many opcodes are available for the LDAA instruction?
   a. one
   b. two
   c. four
   d. three

5. The _____ is a control signal which indicates to the I/O peripherals that the address appearing on the address bus is valid.
   a. TSC
   b. $\overline{\text{NMI}}$
   c. VMA
   d. DBE

6. The _____ is an IC which produces one-shot and continuous timing pulses and provides delay, interval, and cycle timer functions.
   a. MC6850
   b. MC6844
   c. MC6840
   d. MC68488

7. Which of the following ICs is used to interface the MC6800 to a floppy disk drive?
   a. MC6860
   b. MC6843
   c. MC6840
   d. MC68488

8. The _____ is used to interface the MC6800 to a telephone line.
   a. MC6843
   b. MC6860
   c. MC6844
   d. MC6821

9. When the _____ addressing mode is used, the address of the operand is obtained by adding the contents of the memory segment immediately following the opcode to the contents of the program counter.
   a. direct
   b. indexed
   c. relative
   d. immediate

10. The _____ signal controls the three-state buffers on the microprocessor end of the data bus.
    a. E
    b. BA
    c. VMA
    d. DBE

11. How many opcodes are available for the STAA instruction?
    a. one
    b. two
    c. four
    d. three

12. Which of the following instructions is used to terminate a program?
    a. INX
    b. WAI

    c. RTI

    d. STP

13. Which of the following is the maskable interrupt signal generated by a peripheral requiring service?

    a. VMA

    b. $\overline{\text{RTI}}$

    c. $\overline{\text{IRQ}}$

    d. $\overline{\text{NMI}}$

14. Main-memory ICs and I/O peripherals are addressed in the MC6800 system using

    a. polling

    b. interrupts

    c. I/O-mapped I/O

    d. memory-mapped I/O

15. Which of the following instructions will cause the contents in accumulator A to be transferred to an output peripheral whose address is 8215_{16}?

    a. INCA8215_{16}

    b. PSHA8215_{16}

    c. STAA8215_{16}

    d. LDAA8215_{16}

16. Which of the following is not included as part of the MC6800 chip architecture?

    a. ALU

    b. clock

    c. index register

    d. program counter

17. Which of the following is not a 16-bit register in the MC6800?

    a. stack pointer

    b. index register

    c. program counter

    d. instruction register

18. The condition code register monitors the contents of the:

    a. accumulator.

    b. stack pointer.

    c. instruction register.

    d. arithmetic and logic unit.

19. Which of the following is not a register used in the MC6850 ACIA?

    a. status

    b. control

    c. direction

    d. receive data

20. The _____ control signal informs the peripherals whenever the MC6800 is sending or receiving data.

    a. BA

    b. DBE

   c. TSC

   d. $\overline{\text{IRQ}}$

21. T F The 16-bit registers in the MC6800 are used to either hold or help   —
generate addresses.

22. T F Five flags are used in the condition code register.   —

23. T F The extended addressing mode is a three-byte instruction.   —

24. T F The accumulator addressing mode is a two-byte instruction.   —

25. T F The address of the operand is located in the memory byte immedi-   —
ately following the opcode when the immediate addressing mode is
used.

26. T F In the extended-addressing mode, the first instruction byte is the   —
opcode, and the second and third bytes form the address of the operand.

27. T F The LDAA instruction causes the MC6800 to transfer the contents   —
in accumulator A to memory or an output peripheral.

28. T F Memory and peripherals are synchronized with the $\phi2$ clock pulse   —
train.

29. T F The MC6800 uses memory-mapped I/O to communicate with its I/   —
O peripherals.

30. T F MC6800 instruction time is inversely proportional to the number   —
of MPU cycles necessary to perform an operation.

31. T F Instructions using the extended-addressing mode require more ex-   —
ecution time than do those using the direct-addressing mode.

32. T F Chip-address decoders must be connected to all 16 address-bus   —
lines.

33. T F The contents in the control registers of the MC6821 PIA deter-   —
mine the direction of the peripheral data-bus lines.

34. T F The MC6800 receives a nonmaskable interrupt from a peripheral   —
whenever the $\overline{\text{IRQ}}$ signal falls to a binary-zero level.

35. T F The I flag in the condition code register is set whenever the   —
MC6800 is servicing an interrupt.

CHAPTER
8

CONTROLLERS USED IN AUTOMATED MANUFACTURING

OBJECTIVES

1. Identify the two broad categories of manufacturing, discuss the characteristics of the goods produced for each, and list the types of manufacturing associated with each regarding the quantity of the product manufactured.

2. Define the terms *operation* and *process*.

3. List five ways in which controllers may be classified.

4. Draw the block diagram of an open-loop control system and discuss its principles of operation.

5. Draw the block diagram of a closed-loop control system and discuss its principles of operation.

6. Define the terms *transient response, steady-state error, stability,* and *sensitivity* as they relate to a closed-loop control system.

7. Define the terms *set-point, controlled variable, manipulated variable, disturbance,* and *system lag* as related to process controllers.

8. Define the five modes of control operation and discuss the characteristics of each.

9. Define the term *numerical control* (NC) and draw the block diagram of an NC system.

10. List six machine tools that can be numerically controlled.

11. List the major parts of a lathe and identify five operations which can be performed by this machine tool.

12. List the major parts of a milling machine and identify three types of operations it can perform.

13. Identify the *X, Y,* and *Z*-axis motions of a machine tool.

14. Describe the principles of NC programming by identifying the memory medium used to store the program and listing four methods used to prepare the program.

15. Discuss the operation of the NC controller by differentiating between point-to-point and continuous-path controllers, drawing the block diagram of a continuous-path NC controller, and discussing the function of each stage forming the controller.

16. Define the term *computer numerical control* (CNC) and describe how CNC differs from NC.

17. Define the term *direct numerical control* (DNC) and describe how DNC differs from CNC.

18. Define the term *programmable controller.*

19. Draw the block diagram of a programmable controller and discuss the function of each component forming the system.

20. List the four types of programmable controllers as classified by size.

21. Identify the two methods used to program programmable controllers and discuss the characteristics of each.

22. List four variables performed by the programmable controller and describe how each is implemented.

23. List six factors used to evaluate the quality and performance of programmable controllers.

8-1 INTRODUCTION

The two ingredients essential to mass production are the interchangeability of standardized parts and the assembly line. Eli Whitney developed the technique for manufacturing products using standardized, interchangeable parts in 1798. Whitney had opened a firearms factory in Connecticut and was bidding for a government contract to produce 10,000 muskets for the military. To minimize production costs and obtain the contract, Whitney proposed to manufacture the firearms using standardized parts, which were interchangeable. He produced the parts for three muskets to demonstrate his technique. The parts for the muskets were placed in three piles. The muskets were assembled and fired. They were then disassembled; their parts mixed up; they were reassembled and fired. This was repeated several times. In each case, the muskets operated satisfactorily. This was a major advance in manufacturing technology. Prior to this time, parts used in a particular product were custom-made for that product. Whitney won the contract to produce muskets and the technique of using standarized parts to produce a product gradually spread throughout the manufacturing industries.

Although the concept was developed earlier, the assembly line was not used until 1912 when Henry Ford implemented it in the production of the Model T Ford automobile. Prior to this, products were manufactured by bringing all of the parts together and assembling the product at a single workstation. The assembly line, sometimes called Detroit-style manufacturing, makes use of conveyors and overhead monorails to move the basic product past a series of workstations. Workers at each station attach parts or perform work on the product as it moves past them. Assembly is completed as the product moves past the final workstation.

Control is essential to production. Machines that produce interchangeable, standardized parts must be controlled to fabricate parts having high dimensional accuracy. Tools and machines on the assembly line must be controlled and synchronized with the moving conveyor or monorail to perform their operations. This chapter deals with the control systems that regulate the machinery and processes used in manufacturing.

8-2 TYPES OF MANUFACTURING

8-2.1 Introduction

Most of us take manufacturing for granted, little realizing that everything we eat, drink, wear, or use has been processed or produced in some type of manufacturing industry. Scores of different types of techniques are employed to produce these goods. For the sake of convenience they can be grouped into two broad categories—process and discrete-product manufacturing.

8-2.2 Process manufacturing

Process manufacturing is the type traditionally used in the production of liquids and some food products. Examples include petroleum products, pharmaceuticals, chemicals, soft drinks, breakfast cereals, coffee, and sugar. These products are all characterized by the continuous nature of their essence. With the exception of the volume produced in a specific length of time, there is no single finished product entity, as illustrated in Figure 8-1. There is more or less a continuous movement and activity performed on the material from the time it enters the manufacturing system to the time that the finished product leaves the other end. This type of manufacturing is highly automated, with little human intervention occuring as the product moves through its various manufacturing stages.

Process manufacturing can be divided into two categories—continuous and batch-level. In continuous manufacturing, the plant and production equipment are highly specialized and geared to the production of a single product, although by-products may be produced from derivatives of the raw material. This is high-quantity manufacturing and in most cases, it is not feasible to shut the factory down at the end of a work shift. These plants often operate continuously, 24 hours a day, 365 days a year.

Batch-level production is employed when moderate quantities of a product are manufactured periodically or as needed. Usually, enough of the product is produced to satisfy customer demands for a particular length of time. As the product sells, another batch is manufactured before the inventory is completely depleted. Most pharmaceutical, chemical, and food products are manufactured at this level.

8-2.3 Discrete-product manufacturing

This type of manufacturing is used to describe those manufacturing industries where the final product is in the form of a discrete entity, as seen in Figure 8-2. Depending upon the quantity, equipment, and quality required, discrete-product manufacturing can be grouped into three categories—job-lot, batch-level, and mass production.

Job-lot manufacturing: Job-lot manufacturing is used when relatively few parts are produced on a one-time or periodic basis. The parts are often manufactured to the specifications supplied by the customer.

Most cities and towns have small manufacturing companies or machine shops that fabricate products on a job-lot basis. They produce custom-designed devices for their customers or fabricate parts for obsolete machines and implements. For

Figure 8-1

Process manufacturing

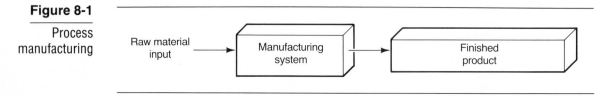

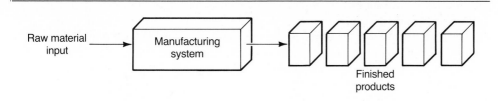

Figure 8-2

Discrete-product
manufacturing

example, a shaft breaks in a 15-year-old gravel-crushing machine owned by a paving contractor. The contractor attempts to order a replacement but is informed that the machine is obsolete and replacement parts are no longer available. Mechanics remove the shaft from the crusher and take it to a local machine shop, where a new shaft is fabricated from the dimensions of the old shaft. In this application, the machine shop performs the function of a job-lot manufacturer.

Batch-level manufacturing: Batch-level manufacturing is used when medium-size lots of a product are fabricated as needed or periodically. This is the most common type of discrete-product manufacturing. Approximately 75 percent of all products manufactured in the United States are produced in lot sizes of 50 or less. Companies utilizing batch-level manufacturing produce a variety of different types and models of products. As soon as the production run for a particular product is completed, the machines are modified to produce a different product lot.

A manufacturer of electronic test equipment, for example, produces several different models of an oscilloscope using one central assembly line. Typically, the assembly line is set up and scheduled to produce a certain number of oscilloscopes for a particular model. These instruments may be built in two days. The assembly line is altered and a particular number of a different model of oscilloscope is produced during the next three days. This process is repeated so that all of the models are eventually produced. The number of instruments produced for a particular model is determined by the current demand for that model.

Batch-level production often occurs in plants that produce products that are much more diverse than different models of oscilloscopes. A manufacturer of small implements, for instance, may produce lawn mowers, garden tractors, and snowmobiles using the same assembly line. The assembly line is set up to produce a particular number of lawnmowers. After these are built the line is modified to produce a certain quantity of garden tractors. When this task has been completed the assembly line is again changed and snowmobiles are produced.

Job-lot and batch-level manufacturing are sometimes referred to as intermittent manufacturing because the product is produced intermittently. General-purpose production machines are used with special jigs and tooling designed to facilitate the fabrication of a particular product. Often, skilled operators are required to set up and operate machines.

Mass production: Mass production is the technique employed when a very high volume of identical products are to be manufactured. Unlike job-lot and batch-level manufacturing, where generalized equipment is employed, in mass production the machines and ancillary equipment are highly specialized and dedicated to the fabrication of a single product. Since little or no machine setup or adjustment is required, the skill demands on the machine operators are lower than in the other types of manufacturing. The major activities involve placing work into the machine, removing it after the operation has been completed, and periodically checking the quality of the completed work. Production techniques utilized in the automotive and appliance manufacturing industries are illustrative of mass production. Assembly lines are set up and thousands of copies of the same model of automobile or appliances are produced.

8-3 PROCESSES

Activities performed on raw materials or a workpiece to convert it into a finished product are called operations, or processes. The two terms mean the same thing and are used interchangeably. An operation, or process, is an activity performed on the workpiece that brings it one step closer to becoming a finished part. The processes used in the fabrication of a product depend upon the type of material used, equipment available, quantity to be produced, and quality required. Some of the more commonly used processes employed in process manufacturing are heating, cooling, distilling, baking, soaking, sifting, and milling. Examples of processes utilized in discrete-product manufacturing include casting, cutting, tapping, threading, boring, drilling, annealing, forging, plating, etching, stamping, and punching.

8-4 CONTROL-SYSTEM THEORY

8-4.1 Introduction

Control-system theory was developed during the 1930s. Controllers used in automated manufacturing have been developed from this body of knowledge. Controllers may be classified in a number of different ways to identify their method of operation and application. These include: method of operation, type of intended application, mode of operation, type of technology, and method of implementation.

8-4.2 Method of operation

Whatever the application, type of technology employed, or method of implementation, controllers can be grouped into two broad categories with respect to the method of operation—open- and closed-loop.

Open-loop control systems: An open-loop control system is one whose input is set to achieve a desired output without monitoring the output or taking corrective action if the actual output deviates from the desired output. The system is composed of an input, controller, actuator, and an output, as seen in Figure 8-3. The input represents variables such as time, temperature, speed, pressure, flowrate, displacement, acceleration, voltage, and current. Controllers provide the intelligence for the system and govern the action of the actuator. The actuator is a device that implements output control—such as a motor, cylinder, solenoid, or fluid-power

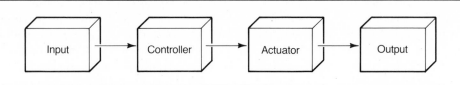

valves. The output is the process being performed. Examples of outputs in process manufacturing include the temperature of water in a hot water tank, the pH factor of a liquid, and the temperature of molten metal in a furnace. An almost endless variety of outputs exist in discrete-product manufacturing. The path of the cutting tool on a production machine as it performs cutting operations on a workpiece is a very common type of output.

Numerous types of open-loop control systems are encountered in day-to-day living. The controllers on household appliances such as washing machines, dishwashers, and toasters, all utilize open-loop control. As an illustration, consider the operation of a lawn mower. After starting the engine, the operator (input) views the height and thickness of the grass and from experience decides where the throttle (controller) should be set. This causes the engine (actuator) to turn at a particular speed and the cutting blade (output) to turn at a certain rate. As the mower travels through the grass the speed of the cutting blade varies as the grass conditions change.

Closed-loop control systems: Machines that perform operations on raw material and workpieces used in automated manufacturing must be very highly controlled to insure accurate and uniform production. Open-loop control systems often do not provide the accuracy required in either process or discrete-product manufacturing, so closed-loop control systems are usually employed wherever a high degree of accuracy is required. As illustrated in Figure 8-4, a closed-loop control system is made up of an input, error detector, controller, actuator, output, and a self-correcting negative-feedback circuit.

The key element in a closed-loop control system is the negative-feedback circuit connected between the output and input. Sensors convert the output and input variables into equivalent electrical signals, which are applied to a summing node called an error detector. The input, sometimes called the set point, represents the

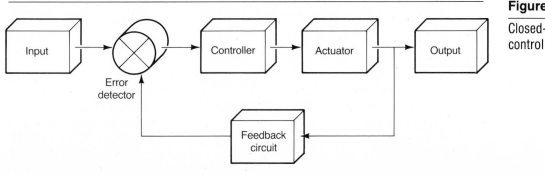

desired output, while the output signal represents the actual output. If the actual output is the same as the desired output, the feedback signal is equal in amplitude and 180 degrees out of phase with the input. The two signals cancel and the control system remains unchanged. If the actual output has deviated from the desired output, the input and feedback signals have different amplitudes and no longer cancel. The resulting difference, or error, signal is applied to the controller, causing corrective action to be taken.

Closed-loop control systems are sometimes referred to as servos, servomechanisms, and regulators. Although these terms are often used interchangeably, servos and servomechanisms are closed-loop systems used to change the position of an object. A regulator is a closed-loop system that maintains a constant output level for variables such as voltage, current, temperature, pressure, and flowrate.

Closed-loop control systems are also encountered in daily living and include heating controls for the home and automobile cruise controls for automobiles. In the case of the lawn mower previously discussed, closed-loop action occurs when the operator takes corrective action and adjusts the throttle to keep the cutting blade at a constant speed as different grass conditions are encountered. The operator senses changes in blade speed by the sound of the engine and adjusts the throttle accordingly.

Four variables associated with closed-loop control are used as criteria to evaluate the performance of the system. These include transient response, steady-state error, stability, and sensitivity.

Transient response: Transient response is a variable used to describe the output of a closed-loop system before steady-state conditions are achieved when the input changes rapidly. Depending upon the design of the system, the output will follow any one of the three paths represented by the curves appearing in Figure 8-5. The first curve represents an underdamped response. The output oscillates a few cycles before reaching the new steady-state condition. The system is overdamped when the period required for the output to reach steady-state conditions is long, as illustrated by the second curve. If the output reaches steady-state conditions in the shortest possible time without oscillations occuring, as shown by the third curve,

Figure 8-5

Transient-response
curves

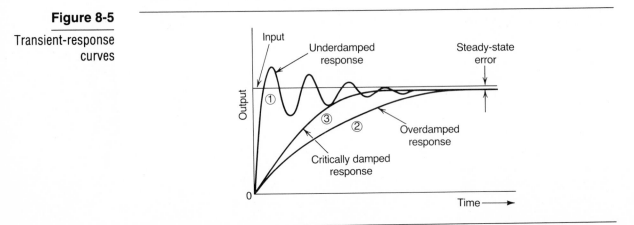

the system is critically damped. The control system is designed, selected, or adjusted for a particular type of transient response. Oscillations do not affect some manufacturing processes but cannot be tolerated in others. Likewise, some processes are not affected by the relatively long period it takes for steady-state conditions to occur when the system is overdamped, while in others, steady-state conditions must be achieved almost immediately.

Steady-state error: As seen in Figure 8-5, a slight differential between the actual and desired outputs exists after steady-state conditions have been achieved. This difference is called the steady-state error. Steady-state error is the difference between the actual output and the desired output (input) values after steady-state conditions are reached.

Stability: Stability refers to the ability of the system to reach steady-state conditions. An unstable system will cause the output to oscillate above and below the desired value, as evidenced by the underdamped response curve appearing in Figure 8-5.

Sensitivity: Sensitivity is the ratio of the percentage change in the output to the percentage change in the input. This variable identifies the effectiveness of the system in correcting for small input changes.

8-4.3 Type of intended application

Controllers are used in both process and discrete-product manufacturing. Controllers used in discrete-product manufacturing are mainly of the numerical and programmable types, which are discussed in Sections 8-5 and 8-8, respectively. Process controllers are the type primarily used in the process manufacturing industries. This is also the type of control system used in such nonmanufacturing applications as heating and air conditioning, hot-water systems, and automobile cruise controls. In addition, principles of process controllers may be incorporated into controllers employed in discrete-product manufacturing.

Either open- or closed-loop operation may be used in process controllers. Closed-loop control is by far the more common, however. Although some are custom designed, process controllers are for the most part, standardized, and may be purchased for almost any application. Terminology unique to this type of control system is identified below.

Set-point: This is the system input, or reference, selected by the operator.

Controlled variable: This is the variable that must be maintained at a constant value as established by the set point.

Manipulated variable: Sometimes called a control agent, this is a variable that causes the controlled variable to be maintained at the set-point value.

Disturbance: This is an outside influence that causes the controlled variable to deviate from the set-point value.

System lag: This refers to the time required for the manipulated variable to cause the controlled variable to return to the set-point value after a disturbance has occurred.

To facilitate an understanding of the variables just defined, a process-control system appears in Figure 8-6. Water from a hot water tank, heated by steam, is used to wash and sterilize bottles in a soft-drink bottling company.

In the morning, at the start of a new work shift, the desired temperature (set point) is selected. The set-point signal is compared with the actual water temperature (controlled variable) as indicated by the signal coming from the sensor. If the two temperatures are not equal an error is created, which is sent to the controller. The controller causes a solenoid (actuator) to open the control valve, letting a full head of steam (manipulated variable) flow through the piping jacket heating the water and raising its temperature. When the water reaches the set-point value the solenoid is deactuated, closing the control valve. The time that elapsed between the time that the desired temperature was selected and the time that the water reached that temperature is called the system lag. Assuming no heat loss from the tank, the control valve remains closed until water is taken from the tank to wash the bottles. Cold water (the disturbance) enters the tank to replace the hot water, which lowers the water temperature. This creates an error voltage that causes the valve to open, allowing steam to again raise the water temperature.

8-4.4 Mode of operation

Depending upon the type of control and output required, several different controller modes of operation are available. These include ON–OFF, proportional, proportional plus integral, proportional plus derivative, and proportional plus integral plus derivative.

ON–OFF mode: The controller causes the actuator to be either fully actuated or deactuated in the ON–OFF mode of operation. Controlling activity is achieved by the period of the ON–OFF cycling action. The thermostat used to control the temperature in buildings uses this control mode. When heating the building, the thermostat causes the furnace to come on and operate full force until the temperature in the building rises to the set-point value, at which time the thermostat causes the furnace to be shut down. The furnace remains off until the building temperature falls below the set-point value, at which time the furnace comes on again. The thermostat as a controller cannot vary the amount of heat produced by the furnace. Instead, it controls the cycling period. This is the mode of operation used by the hot water system in Figure 8-6.

Proportional mode: The manipulated variable can be maintained in any intermediate position between the fully on and off states when proportional control is utilized. The position of the control device is proportional to the controlled variable. Whenever the set-point value is changed or a disturbance forces the controlled variable to change from the set-point value, the controller causes the manipulated variable to decrease the error signal. If the proportional-control mode were employed in the control system seen in Figure 8-6, a motor would have been con-

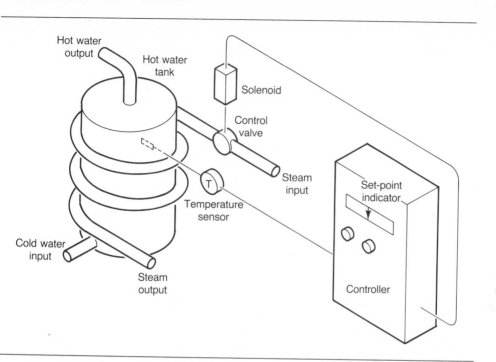

Figure 8-6

Process-control
system

nected to the control valve to regulate the flow of steam (the manipulated variable). The valve would have been fully open when the system was turned on at the beginning of the shift and then slowly closed as the water (the controlled variable), approached the set-point value.

Proportional control is a relatively simple controlling technique that works well in many applications. The disadvantage of this type of control is that if a long-term or steady-state disturbance occurs, the manipulated variable cannot make the controlled variable return to the set-point value. Proportional-mode controllers require an error signal to control the manipulated variable. The amplitude of the error signal decreases as the controlled-variable signal approaches the set-point value. Eventually, the amplitude of the error signal is insufficient to control the manipulated variable and a point of equilibrium is reached. This creates a new steady-state

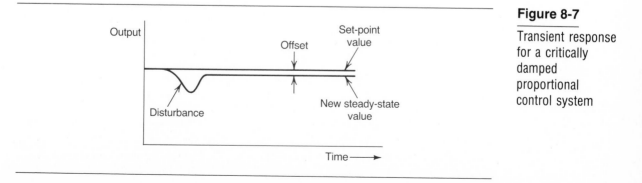

Figure 8-7

Transient response
for a critically
damped
proportional
control system

Figure 8-8

Proportional
control circuit

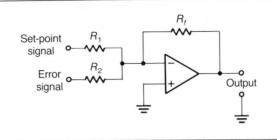

output signal which is slightly less than the set-point value, as illustrated in Figure 8-7. The difference between the set-point and new steady-state values is called an offset. The offset may or may not be acceptable depending upon the particular controller application.

An operational amplifier connected as a summing amplifier can be used to achieve proportional control, as seen in Figure 8-8.

Proportional plus integral mode: This control mode may be used in applications where an offset cannot be tolerated and the disturbance is of long duration or permanent. The circuit shown in Figure 8-9 can be employed to provide proportional plus integral control. The operational amplifier appearing in the first stage is connected as an integrator. The output of the integrator is connected to one of the inputs of a second operational amplifier, connected as a summing amplifier. The charging action of the capacitor in the feedback loop of the integrator and the summing action of the summing amplifier eliminates the offset. Because of the reactance of the capacitor and the RC time constant utilized in the integrator, the offset is eliminated for long-term disturbances only. The controller will operate as a conventional proportional controller for medium- and short-term disturbances.

Proportional plus derivative mode: For short-term disturbances, an operational amplifier can again be used to eliminate the offset. In this application, the opera-

Figure 8-9

Proportional plus
integral control
circuit

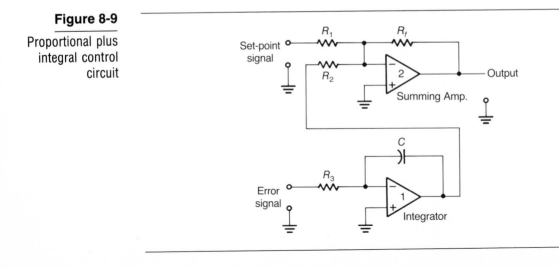

Figure 8-10

Proportional plus derivative control circuit

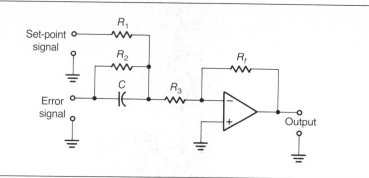

tional amplifier is connected as a differentiator, as shown in Figure 8-10. The differentiator circuit provides an output only while the controlled variable is changing or for disturbances of short duration. The controller functions as a proportional controller for medium- and long-term disturbances.

Proportional plus integral plus derivative mode: This controller incorporates the features of the three previously discussed proportional-control modes into one unit. It can be operated in the proportional mode if the disturbances are of medium duration or it can be operated in the proportional plus integral or proportional plus derivative modes if the disturbances are of long or short duration respectively.

8-4.5 Type of technology

Controllers may utilize either analog or digital technology. Digital controllers were introduced to discrete-product manufacturing with the advent of numerical control in the 1950s, and computer numerical control during the 1960s, for machine-tool control. Programmable controllers developed during the 1970s have further enhanced the application of digital technology in discrete-product manufacturing.

Controllers used in process manufacturing have traditionally employed analog technology. Many of these controllers utilize the proportional modes of operation, as it is easy to implement this type of control with analog circuits. This is beginning to change, however. High resolution A/D converters are making it possible to obtain very acurate process control with digital circuits.

8-4.6 Method of implementation

Controllers may be either hard-wired or programmable. Prior to the development of numerical-controlled machine-tool controllers and programmable controllers, controllers used in discrete-product manufacturing were hard-wired. These had to be rewired at considerable time and expense whenever changes in the product being manufactured occured. Although hard-wired controllers continue to exist, most of the controllers used in discrete-product manufacturing are programmable. In these, software determines the controlling functions. Changes are made in software rather than hardware to effect changes in controlling activities.

Controllers used in continuous-process manufacturing are, for the most part, hard-wired. Because the product never varies in this type of manufacturing, there is little need to use controllers which are programmable. Programmable controllers are used in batch-level process manufacturing, however. These controllers provide the flexibility necessary to alter the activities performed when different products are processed using the same machinery. The programmable controller is often used as the controller in this type of manufacturing.

8-5 NUMERICAL CONTROL

8-5.1 Introduction

Numerical control (NC) is a type of machine-level programmable automation in which the operation of machine tools is controlled by binary-coded instructions. First demonstrated in 1952 at the Massachusetts Institute of Technology (MIT), numerical control has provided the means for implementing digital technology in discrete-product manufacturing. Numerical control allows for the manufacture of parts at a faster rate, a higher degree of accuracy, and a reduced cost, especially in job-lot production. Manufacturing efficiency is increased substantially. The average machine tool uses only about 20 percent of available machine time to remove material. Numerical control increases this to as much as 80 percent. NC is a very flexible approach to manufacturing. A program that includes all of the information required to perform the operations on a workpiece is loaded into the controller. The machine performs the operations automatically. After the operations have been performed on the required number of workpieces, the program is removed from the controller

Figure 8-11

The NC system

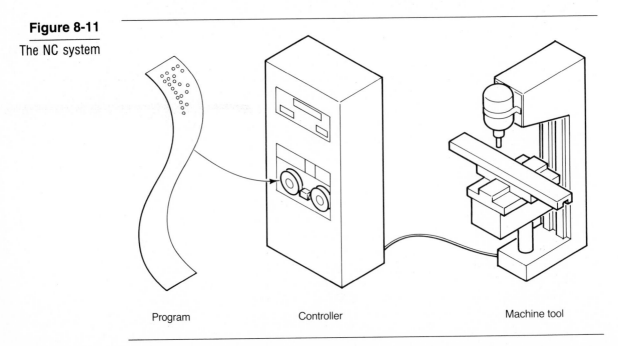

Program   Controller   Machine tool

and stored for further use. A new program is loaded into the controller and the machine is ready to perform operations on a different type of workpiece.

The NC system is composed of a machine tool, program, and controller, as illustrated in Figure 8-11. Machine tools are used to shape material, especially metal. They perform operations on a workpiece, making it take the shape of the desired part. The program contains the instructions and dimensional data that enable the controller to regulate the motions of the machine tool.

8-5.2 Machine tools

Major general-purpose machine tools that can be numerically controlled include the lathe and milling machine. Numerous more specialized machines—such as grinders, drilling machines, riveting machines, spot-welding machines, insertion machines, and plasma-jet-erosion cutters—may also be numerically controlled.

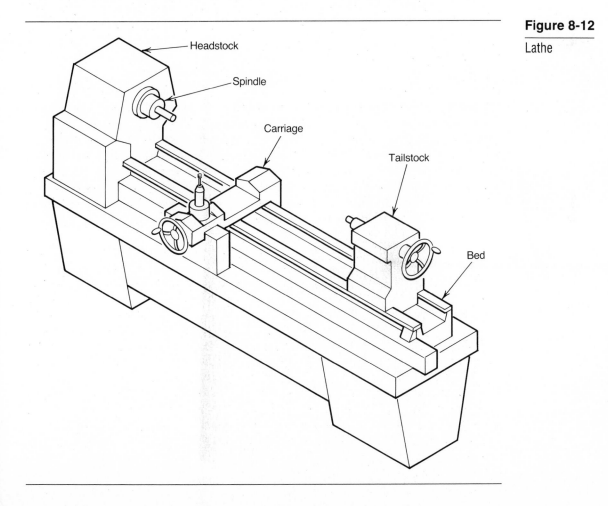

Figure 8-12

Lathe

Figure 8-13

Method of
mounting for
turning operations
on long
workpieces

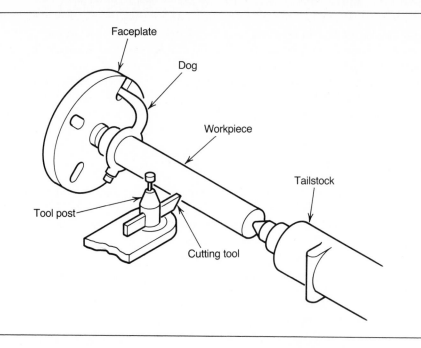

Faceplate

Dog

Workpiece

Tailstock

Tool post

Cutting tool

The lathe: This is the oldest and most commonly used machine tool. It is sometimes called an engine lathe because it was originally driven by a steam engine. A
typical lathe is depicted in Figure 8-12. Major components include the headstock
(which contains a spindle), tailstock, carriage, and bed. A cutting tool mounted in a
tool post attached to the top of the carriage cuts the rotating workpiece. The cutting
tool can be moved parallel to and across the workpiece.

The method used to mount the workpiece in the lathe depends upon its length
and the type of operation to be performed. For straight and taper turning opera-

Figure 8-14

Method of
mounting for
turning operations
on short
workpieces

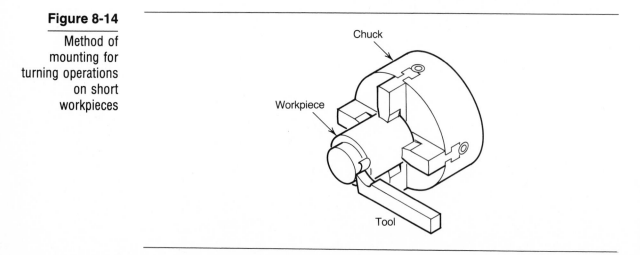

Chuck

Workpiece

Tool

tions on long workpieces, the workpiece is mounted between the head- and tail-stocks and driven by a "dog" attached to a faceplate, as seen in Figure 8-13. The workpiece is clamped in a chuck which is mounted on the headstock for straight and taper turning on short workpieces, and for drilling, boring, reaming, and thread cutting operations, as shown in Figure 8-14. The tools are held in the tail-stock assembly for drilling, boring, reaming, and thread cutting operations, as illus-trated in Figure 8-15. The operator must change the tool whenever a different size or type of tool is required. To eliminate the need for changing tools, a turret lathe can be used. On this lathe, the shape of the toolholder is square or hexagonal. Tool mounts are attached to its faces, as seen in Figure 8-16. The toolholder can be ro-tated to any one of its tool positions whenever a tool change is required.

More different operations can be performed on a workpiece by a lathe than on any other general-purpose machine tool. Common operations include straight cut-ting, taper cutting, facing, drilling, boring, reaming, tapping, knurling, and thread cutting. Two actions occur whenever a machine tool performs a metal-removing op-eration. One, either the workpiece or tool must be rotated; two, pressure must be exerted between the cutting tool and the workpiece. In lathe work the workpiece is rotated and the cutting tool is fed into and alongside of the material. Feed and speed are both very important for optimum machining time and quality. Feed is the rate at which the cutting tool moves into the workpiece for each complete rev-olution. The amount of feed required depends upon the type of workpiece material and the depth of the cut. Speed refers to the rate at which the workpiece is cut. In lathe work, speed is usually measured in feet per minute. Lathe cutting speed is

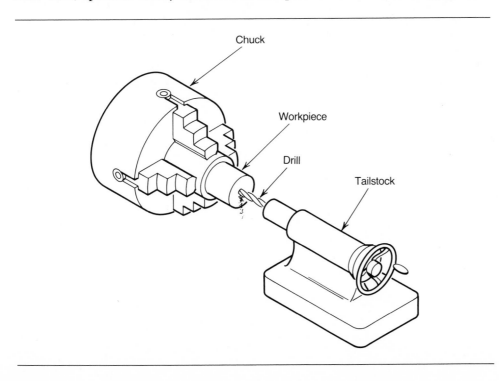

Chuck

Workpiece

Drill

Tailstock

Figure 8-15

Method of mounting tool bit for drilling, tapping, boring, and reaming operations

Figure 8-16

Turret tool holder

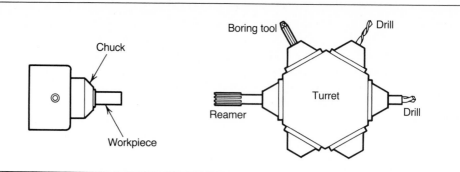

affected by the amount of feed, depth of cut, type of cutting tool, type of workpiece material, and whether a rough or finish cut is being made. Recommended feeds and speeds can be found in tables appearing in most machinery handbooks.

The milling machine: This machine performs its operations through a stationary-mounted rotating cutting tool called a cutter, which is mounted on a spindle. Cutting action occurs when the workpiece is moved against the cutter. There are two types of milling machines—horizontal and the vertical. They are named according to the position of the spindle. The spindle is mounted horizontally on a horizontal milling machine and vertically on a vertical milling machine.

A horizontal milling machine is illustrated in Figure 8-17. Major components include the overhead spindle, to which the cutter is mounted, and a movable table.

Figure 8-17

Horizontal milling machine

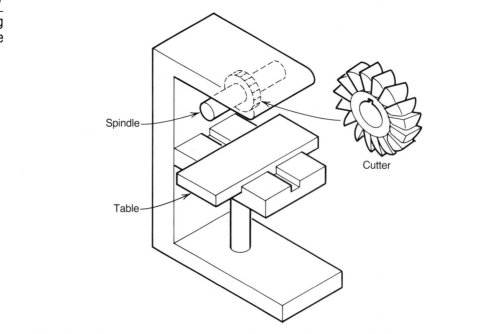

Figure 8-18

Vertical milling machine

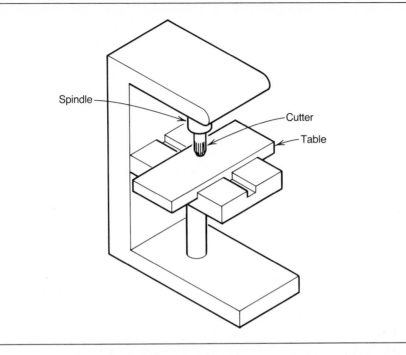

Spindle

Cutter

Table

The table has three axes of movement—longitudinal, transverse, and vertical. The workpiece is clamped to the table and the table positioned for proper cutter location and depth. Cutting action occurs when the workpiece is forced to move against the cutter.

The vertical milling machine is similar to the horizontal mill except that the spindle is mounted vertically, as shown in Figure 8-18. In addition to table movement, the spindle can be moved vertically on many of these machines.

The milling machine is a very versatile production machine. Common milling operations include the production of flat surfaces, angular surfaces, and grooves. In addition, drilling, boring, reaming, and tapping operations can be performed on vertical milling machines.

Applications of NC to machine tools: The lathe cutting tool is mounted on the carriage. Screw-feed mechanisms, such as that shown in Figure 8-19, allow the cutting tool to be positioned for the desired cutting path. Screw-feed mechanisms attached to the table of a milling machine cause the table to be positioned for the desired cutting action.

Initially, NC machines were made from conventional machine tools whose screw-feed mechanisms were coupled to small motors, as seen in Figure 8-20. To some extent this situation continues to exist. Retrofitting kits may be purchased to convert these machines to numerical control. Most NC machines in use today are specifically built for that type of operation, however. These machines have heavier duty spindles, carriages, and screw-feed mechanisms because of their increased production use. Depending upon the weight to be moved, motors coupled to the

Figure 8-19

Screw-feed
mechanism

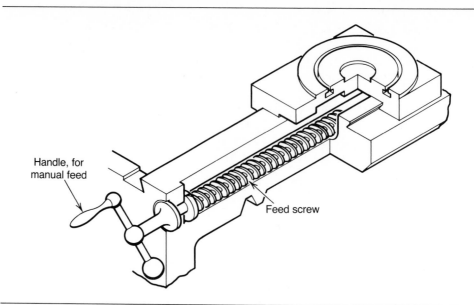

Figure 8-20

Motor coupled to
screw mechanism
for numerical
control

screw mechanisms may be either electric or hydraulic. Electric drive is usually employed for mechanisms requiring 10 horsepower or less while hydraulic motors are used for applications requiring more than 10 horsepower.

Sensors and limit switches are used to provide the NC controller with cutting-tool position data and to start and stop cutting action. Angular digital encoders,

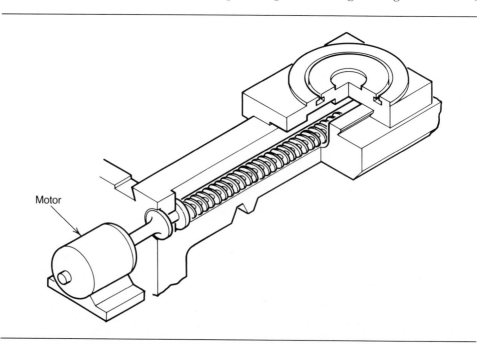

LVDTs, and tachometers may be attached to screw mechanisms, carriages, and tables to provide angular and linear displacement data. Limit switches are commonly used to stop machine motion at the end of the cutting path. In the case of surface grinders, where a back-and-forth motion is required, limit switches can be used to change the direction of table motion.

The concept of numerical control is based on the ability of the machine tool to locate positions on a workpiece identified by rectangular-coordinate dimensions. Numerically controlled machine tools may have two or more axes of movement. The Electronics Industry Association (EIA) has defined 14 different axes of movement in its NC RS-267 numerical-control standards. The more common of these are listed below.

Z-axis motion: Z-axis motion is parallel to the machine spindle, as seen in Figure 8-21. The spindle holds the cutting tool on milling, drilling, boring, and tapping machines. On lathes, the spindle holds the workpiece.

X-axis motion: X-axis motion is horizontal and parallel to the surface holding the workpiece.

Y-axis motion: Y-axis motion is perpendicular to both the X and Z axes.

A, B, and C motions: A, B, and C are angles that identify rotary movement about the axes parallel to the X, Y, and Z axes, respectively.

U, V, and W motions: These are slide motions parallel to the X, Y, and Z axes, respectively.

No single machine is able or needs to produce all of these motions. Most machines produce motion in the X, Y, and Z planes. These are called three-axis NC

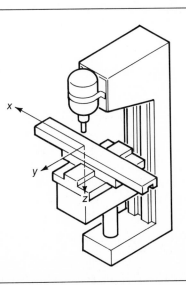

Figure 8-21

Machine-tool axis motion

machines. Only the most complex machines produce more than four axes of motion.

8-5.3 The program

Although diskette or magnetic tape may be used, punched tape is the most common type of memory medium used in numerical control. Data stored on the tape are either in the form of ASCII or EIA code. Two methods are used to program NC machines—manual and computer-assisted, sometimes called computer-aided, programming. Manual programming is commonly used for programming two-axis machines and for simple operations performed on three- and four-axis machines. The programmer develops the program by studying the engineering drawings and preparing a manuscript written in NC machine language. The manuscript includes the operations that must be performed on the workpiece, their sequence, coordinates identifying where the operations are to occur, feed and cutting speeds, and ancillary functions such as coolant application and tool clearances for workpiece holders. The manuscript is reproduced in tape form using a tape punch. The resulting program is verified by placing it into the controller and making a trial run on soft metal, wood, or styrofoam. This provides a means of identifying any programming errors and checking for collision avoidance between the cutting tool and workpiece holders without the risk of damaging the machine or ruining a workpiece.

Preparing the program manually and punching the tape becomes tedious if a large number of operations are required. Computer-assisted programming is used to simplify programming in these applications. The computer utilized to accomplish this is not part of the NC system. A separate computer is required.

Several different computer-assisted programming languages are available. These include APT, ADAPT, COMPACT II, ACTION, SPLIT, NUFORM, and UNIAPT. APT (*Automatic Programmed Tool*) is the language most commonly used. This language was developed at M.I.T. during the developmental stages of numerical control. The programmer uses English-language statements to describe the part appearing on the engineering drawing.

Depending upon the availability of programming personnel and computers, three different techniques are available for implementing computer-assisted programming—time-sharing, in-house, and an external service agency. Time-sharing is used when an on-site computer is not available. The program is written in an NC programming language and typed in ASCII code on a time-sharing terminal. The same basic programming technique is employed for in-house programming except that an on-site computer is employed. Private programming businesses that write and develop programs for customers are available in many cities. These businesses may be used by a manufacturer as a source for all NC program development or to reduce programming overloads.

8-5.4 The NC controller

NC machine controllers are commonly referred to as machine control units (MCUs). Two types of MCUs are utilized in NC systems. These are the point-to-point, which is used for two-dimensional positioning, and the continuous-path, sometimes called contour, which is used in some two- and all three-dimensional applications. Point-to-point controllers are used in NC machines that perform drilling, boring, tapping, reaming, countersinking, spot welding, punching, and riveting operations. In these applications, the MCU causes the machine tool to locate a point on the workpiece, as identified by predetermined X and Y coordinates, and perform the operation. In an application such as drilling holes in a flat workpiece, the spindle is programmed to move from a reference point, often called the home position, to the first hole location. After locating the point, the spindle is lowered and the hole is drilled. The spindle is retracted and moves to the next location and the hole is drilled. This procedure is repeated until all of the holes have been drilled. The primary prerequisites in drilling and other point-to-point operations are accuracy and speed. The spindle must be positioned accurately at the proper X and Y coordinates and, after the drilling operation has been completed, must be moved to the next set of coordinates as rapidly as possible. The path that the spindle follows in moving between coordinates is determined by the controller; it is not part of the NC program. Usually, the most direct route is taken. This is not necessarily a straight-line path, however.

In point-to-point NC machines, the cutting tool is in contact with the workpiece only at the specified coordinates. This is not the case for continuous-path NC machines such as lathes and milling machines. The cutting tool is in continuous contact with the workpiece as it moves between coordinates. The MCU on such machines must accurately position the cutting tool and provide a precise path for the tool to follow as it moves from one coordinate to another.

The technique used in moving the cutting tool between coordinates is called interpolation. Three interpolation techniques are utilized—linear, circular, and parabolic. Linear interpolation involves moving the tool from one set of coordinates to another in a straight line. This is the technique used in straight-line and taper cutting. Circular interpolation is used when arcs, curves, and circle shapes must be cut. Parabolic interpolation is employed for making free-form cuts.

MCUs may use either open- or closed-loop control. Open-loop control is the type traditionally employed in point-to-point controllers. In these devices, three variables are of paramount importance. Two, positional accuracy and speed, have already been identified. The third variable is repeatability. Not only must the tool be positioned on the workpiece accurately and rapidly, it must be placed at the same point on each successive workpiece. Positioning accuracies as great as 0.0001 inch are possible with excellent repeatability on well-maintained machines using open-loop control. DC stepping motors can be used to drive the screw mechanisms on light-duty NC machines, as illustrated in Figure 8-22. These motors can be accurately controlled without the use of displacement sensors and feedback circuits. Although only a single motor is shown in Figure 8-22, a separate motor is required for each different screw mechanism.

Closed-loop controllers are used in continuous-path and in some point-to-point machines. The diagram of a closed-loop NC system appears in Figure 8-23. Two feedback paths are employed. One provides positioning data to the controller while the other supplies velocity data. Together, they provide the controller with the data

Figure 8-22

Open-loop NC
system

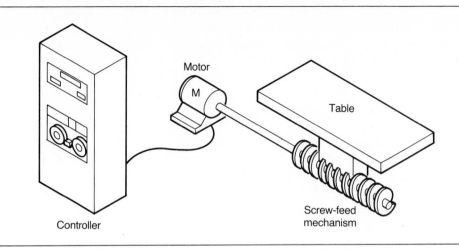

required to position the milling-machine table holding the workpiece and guide it during the machining operations. A separate drive motor is required for each axis of table movement.

The devices and circuits forming a continuous-path MCU vary, depending upon the manufacturer and application. In general, the controller consists of a control panel, tape reader, data-decoding-and-control unit (DDCU), feedrate-control unit

Figure 8-23

Closed-loop NC
system

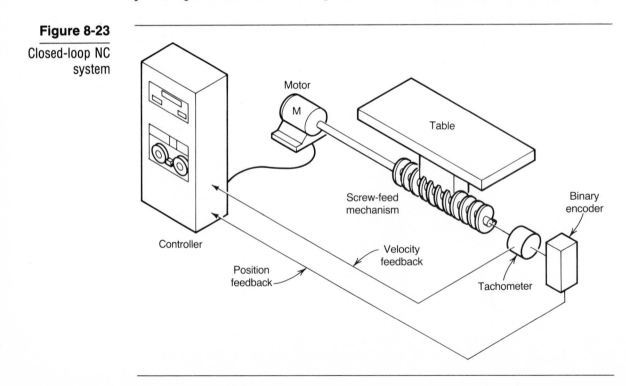

(FCU), interpolator unit (IU), servo unit (SU), and ancillary unit (AU), as seen in Figure 8-24. The function of each of these stages is discussed below.

The control panel: The control panel contains the starting and stopping switches, system status lights, and manual override switches that let the operator take control of the machine if required.

The tape reader: This mechanism is the vehicle by which the program stored on the tape is placed into the controller. Two types of tape readers are available— mechanical and optical. The mechanical tape reader uses switch-lever contacts to detect the presence of holes, while the optical reader employs a light source and photodetectors. The optical tape reader is more commonly used as it has fewer mechanical parts to wear and is more reliable. The rate at which the tape reader is capable of reading alphanumeric characters is extremely important in continuous-path controllers. The controller must not be slowed by waiting for data to be loaded into it. Most tape readers used in continuous-path controllers employ a reading speed of 200 to 300 characters per second (c/s).

Data-decoding-and-control unit: The ASCII- or EIA-coded program data are sent in serial form from the tape reader to this unit and are initially stored in a buffer register for decoding purposes. The decoded information forms control signals representing position, feedrate, and ancillary data, which are sent to other units in the MCU.

Feedrate-control unit: Machine feedrate speed, included as part of the program, is decoded by the DDCU and is sent to the feedrate-control unit. The FCU converts the decoded feedrate-speed information into a control signal that makes the machine move at the programmed rate. This signal is sent to the interpolator unit.

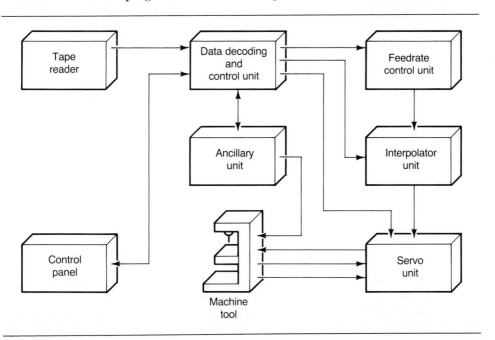

Figure 8-24

Continuous-path machine control unit

Interpolator unit: This is the most important unit in the MCU. Here feedrate-speed and positioning data are combined to provide continuous-path or contour machining. Inputs to the IU include position data from the DDCU and feedrate-speed data from the FCU. The interpolator unit provides the servo unit with a rate/position signal that causes multiaxis machine motion to occur.

Servo unit: The servo unit receives rate/position data from the IU and converts them into control signals, which make the machine move. Velocity feedback obtained from tachometers, and position feedback developed from angular binary encoders, potentiometers, and LVDTs are also applied to the SU for reference purposes. The servo unit's output is a driving signal for the motors connected to the machine-screw mechanisms.

Ancillary unit: This unit develops control signals from the decoded information coming from the DDCU to implement ancillary functions such as turning coolant on and off, rotating a turret, or changing a tool, if an automatic tool changer is employed. Control signals generated within the AU are sent to the appropriate machine-tool areas.

8-6 COMPUTER NUMERICAL CONTROL

Computer numerical control (CNC) is an NC system using a dedicated computer, usually a microprocessor, as its controller. The weakest links in the NC system are the tape and tape reader. The tape is the memory medium and has to be completely recycled each time a part is produced. This results in excessive tape wear when operations are performed on a large number of workpieces. Substituting mylar tape for paper tape eliminates the problem of tape wear but the problem of the tape reader remains. Its electromechanical composition has always been a source of problems.

The physical appearance of NC and CNC machines are similar and the program is punched onto a tape for CNC just as it is for NC. The tape is cycled through the reader only once in a CNC system, however. As the tape is cycled through the tape reader, the program is loaded into the controller memory. As many parts as desired may be produced while the tape remains idle, eliminating, to a very large extent, the wear on the tape reader.

CNC provides other advantages in addition to reducing tape reader wear. In an NC system the controller is hard-wired. Any changes made in the machine tool require modification of the controller. Machine-tool changes can be incorporated into a CNC controller by simply reprogramming it. Many CNC systems perform self-diagnosis and monitoring functions. On some models, messages appearing on the video monitor alert the operator whenever a tool change is required.

Comptuer numerical control was developed in 1970. Minicomputers were frequently used as the controller. As microprocessors were developed, engineers began incorporating them into the controller as a replacement for the minicomputer.

The functions performed by a five-axis CNC machine can be controlled by an MC6800 or 8085 microprocessor, for example. This has substantially decreased the cost of CNC.

8-7 DIRECT NUMERICAL CONTROL

Direct numerical control (DNC), sometimes called down numerical control, is a type of numerical control in which a central computer controls several separate CNC machines. The diagram of a DNC system appears in Figure 8-25. Major components include a central mini- or mainframe computer, a machine-interface unit (MIU) to interface the central computer and machine tools, the machine tools, and a bidirectional serial bus, which connects the MIU to the machine tools.

DNC allows a higher level of manufacturing automation. CNC provides for the automation of a single machine, whereas several machines can be controlled using DNC. Because of the use of a larger computer, DNC systems have greater computational capabilities than do CNC systems. There is no need for tapes and tape readers in DNC. The central computer causes the program to be downloaded into the controller memory of the individual machines as the need arises. The complete program is not downloaded at one time. Because it needs a larger computer, DNC is more expensive than CNC and is best suited for large manufacturing companies. A major disadvantage of DNC is that if a fault occurs in the computer, the complete system is shut down. Two computers are sometimes employed to provide redundancy. One computer is active while the other remains on standby. If a fault occurs in the active computer, the standby machine is automatically switched into active status.

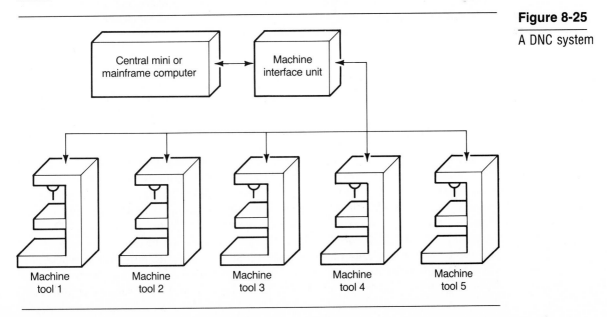

Figure 8-25

A DNC system

8-8 PROGRAMMABLE CONTROLLERS

8-8.1 Introduction

A programmable controller (PC), sometimes called a programmable logic controller (PLC) or programmable process controller (PPC), is a general-purpose programmable controller whose controlling function is effected through software. It is one of the more important links in the chain of technologies required for total plant-level automation.

The programmable controller was developed by the manufacturers of industrial controllers at the request of engineers from General Motors Corporation to replace hard-wired relay controllers used to control sequential machine-tool operations and materials handling equipment. Before the programmable controller was introduced in 1969, automobile model changes required extensive assembly-line down time while electricians rewired, or reprogrammed, scores of hard-wired relay control panels to effect the changes needed in the manufacturing processes. With programmable controllers, all that is required when an automobile model change dictates a modification in assembly line operations is a change in the controllers' program.

The growth of applications for programmable controllers has been slow but steady since they were introduced into the automobile manufacturing industry—originally, to replace electromagnetic relays. Upgraded versions that emulated timing and counter functions were soon available. By the mid-1970s their use was common in other mass-production industries. At the same time they were being implemented in batch-level process-manufacturing applications. Early model programmable controllers operated solely in the ON–OFF mode. More recent PCs are capable of proportional, proportional plus integral, proportional plus derivative, and proportional plus integral plus derivative control. This flexibility has made them useful as CNC controllers in many applications.

In addition to controlling production machinery and processes, many of the most current models perform a number of different record-keeping activities—such as monitoring the number of parts produced, the number of parts rejected, and the downtime for a particular machine for each shift.

8-8.2 PC architecture

Programmable controllers are made up of three major sections. As shown in Figure 8-26, these include the central processor unit (CPU), input/output section, and programming device. The CPU performs the controller logic functions and contains a microprocessor, RAM, ROM, and rectifier power supply. The input/output section contains the interfacing circuits that connect the CPU to the I/O devices. The programming device is the vehicle used by the programmer to place instructions and data into the CPU. Programmable controllers operate in an industrial environment in which the ambient temperature may vary considerably and in which excessive vibration, dust, and strong electromagnetic fields may be present. The CPU and I/O section are housed in well-shielded steel cabinets, which are grounded to pro-

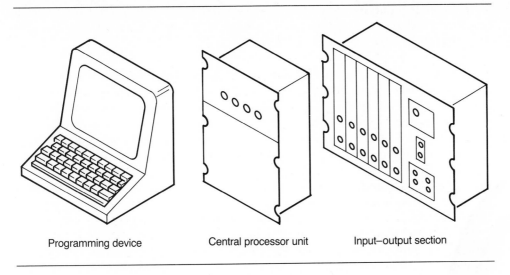

Figure 8-26

Programmable
controller

Programming device   Central processor unit   Input–output section

tect and isolate the circuits from the environment as much as possible. Optical isolation is usually employed in the I/O interface circuits.

The block diagram of a programmable controller appears in Figure 8-27. Most
PCs utilize 8-bit microprocessors although 16-bit devices are employed in large systems. RAM and ROM size typically range from a low of 256 bytes for small PCs to
65 k-bytes for the large types. Microcomputers require large amounts of RAM and
small quantities of ROM, but the opposite is usually the case with programmable
controllers. A large portion of the program is permanently stored in ROM while
RAM is used to hold the variables that pertain to a particular controlling application.

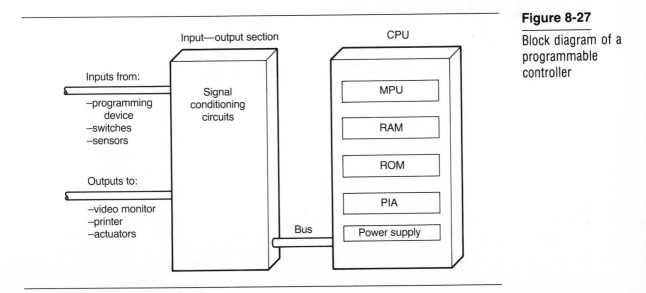

Figure 8-27

Block diagram of a
programmable
controller

The power supply usually consists of a bridge rectifier connected to a voltage regulator. A battery is often used as a backup power source for RAM to retain program data if the main power is disrupted.

One programmable controller can control the operation of more than one machine or process—the only limiting factors being the number of I/O ports and memory capacity. PCs with many different I/O-port quantities are available and range from approximately 6 to 10,000. For purposes of size identification, programmable controllers have been grouped into four categories, as listed in Table 8-1.

Transmission lines connected to the input ports of the I/O section carry signals that come from a variety of sources. Push-button switches, limit switches, and sensors are commonly used as input devices. Input variables are often nonelectrical in many control applications. Sensors are needed to convert variables such as temperature, humidity, displacement, strain, acceleration, pressure, flowrate, and liquid level into equivalent electrical signals. Signal-conditioning circuits such as amplifiers, filters, and A/D converters are needed to change these signals into usable digital data.

Output devices normally include actuators such as solenoids and motors. These may be grouped into two categories— those whose power ratings do not exceed the output rating of the programmable controller, and larger actuators whose power requirements exceed the output rating of the controller. Small devices may be connected directly to the output port, as seen in Figure 8-28(a), while a relay or contactor is connected between the output port and actuator for higher power devices, as shown in Figure 8-28(b). In this application, the output port is connected to the control coil of the relay or contactor. Indicator lights, a video monitor, and printer are often connected to output ports to monitor and record the status of the system.

8-8.3 Programming PCs

Relay-ladder logic has been the method most frequently used to program programmable controllers. This is the language used to program hard-wired relay logic controllers. This task traditionally has been performed by electricians or plant maintenance personnel. Utilizing the same language in programmable controllers eliminates the need for these individuals to learn a new programming language. In addition, relay-ladder logic is an efficient language. Only a single byte of memory storage capacity is needed to set the state of a normally closed relay, whereas it

Table 8-1

PROGRAMMABLE-
CONTROLLER
CLASSIFICATION
BY SIZE

SIZE	NUMBER OF I/O PORTS
Micro PCs	6–63
Small PCs	64–255
Medium PCs	256–999
Large PCs	1000 and above

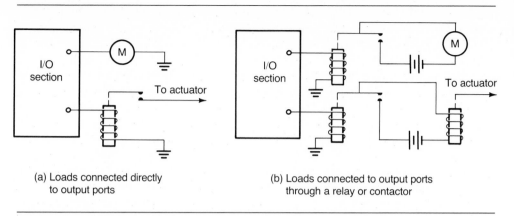

Figure 8-28

Method of
connecting
actuators to a
programmable
controller

(a) Loads connected directly
to output ports

(b) Loads connected to output ports
through a relay or contactor

takes 6 to 12 bytes of memory to achieve the same result using BASIC or FOR-TRAN.

To develop the program, the programmer makes a relay-ladder logic diagram of the circuit needed to facilitate the control function. The program is then loaded into the controller through a keypad or keyboard which serves as the programming device. Relay-ladder symbols are superimposed over some of the keys, as illustrated in Figure 8-29. The programmer selects the proper symbols by depressing the appropriate keys. Usually, a video monitor is utilized to monitor the program as it is being loaded into the controller. The symbols appear on the monitor and can be positioned where desired. As more and more of the program is entered, the logic diagram appearing on the monitor takes the shape of the diagram originally developed by the programmer. To increase the efficiency of the program, special instructions are often included. These include jump instructions, which allow portions of the program to be bypassed; jump-to-subroutine instructions, which permit subroutines to be stored in any convenient memory segment; and sequencer instructions, which allow the PC to emulate the functions of a drum switch.

Although relay-ladder logic has been the major language used in programming programmable controllers, higher level languages are used in many of the more current systems. Approximately half of the controllers presently being produced use relay-ladder logic while the other half utilize a higher level language such as

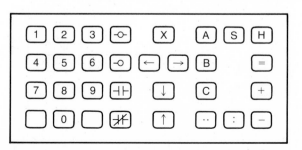

Figure 8-29

Programmable-controller device
symbols

BASIC or FORTRAN. The programmable controller can operate only in the ON–OFF mode when relay-ladder logic is employed. The use of higher level languages allows the PC to operate in the proportional modes extending its applications. In addition, some programmable controllers may be used to perform production record-keeping activities, and may even have self-diagnostic troubleshooting routines. None of these features are possible if relay-ladder logic is utilized.

8-8.4 Facilitating the controller function

Programmable controllers are used as replacements for hard-wired controllers employing relays, timers, counters, and drum switches. For the most part, the techniques used to emulate these devices are accomplished through software, although some microprocessor vendors produce support-integrated circuits to perform some of these functions. The techniques commonly used to perform the control activities are listed below.

Relays: The relay is replaced with the programmable controller program. Relay logic is similar to that of a logic gate. The contacts are either closed or open representing on and off conditions respectively.

Timers: Programmable controllers can be used to provide ON-delay, OFF-delay, interval, and cycle-timer functions. These can be achieved through either software or hardware techniques. The microprocessor in the CPU is programmed to provide these functions when the software technique is employed. The disadvantages of this technique include need for additional memory to hold the timer program and the additional time it takes the microprocessor to perform the timer activities. An integrated circuit called a programmable timer is used in the hardware technique. This device performs the timer functions while allowing the microprocessor to effect other activities. The manufacturers of many microprocessors build a programmable-timer IC for this purpose. Motorola produces the MC6840 Programmable Timer Module (PTM), which can be used with several of the MC6800 MPU family members. The 8253 Programmable Interval Timer (PID) is manufactured by Intel to be used with their 8080/8085 devices for the same purpose. This timing feature is built into some microprocessors such as the MC6801, MC6805, 8048, and 8049.

Counters: The CPU clock is the source of pulses for timing applications. The technique used to achieve counter functions is similar to that employed for timing except that an external source of pulses from a PTM or PID are used rather than the CPU clock.

Drum switches: The sequential nature of drum switches is simulated through software.

8-8.5 Programmable-controller specifications

Several factors are used for evaluating the quality and performance of programmable controllers when selecting a unit for a particular application. These are listed below.

Number of I/O ports: This specifies the number of I/O devices that can be connected to the controller. There should be sufficient I/O ports to meet present requirements with enough spares to provide for moderate future expansion.

Output-port power ratings: Each output port should be capable of suppling sufficient voltage and current to drive the output peripheral connected to it.

Scan time: This is the speed at which the controller executes the relay-ladder logic program. This variable is usually specified as the scan time per 1000 logic nodes and typically ranges from 1 to 200 milliseconds.

Memory capacity: The amount of memory required for a particular application is related to the length of the program and the complexity of the control system. Simple applications having just a few relays do not require significant amounts of memory. Program lengths tend to expand after the system has been in use for a while. It is advantageous to acquire a controller that has more memory than is presently needed.

Vendor support: Items such as maturity of the product, warranty service, and training are extremely important. Answers to questions pertaining to the reliability of the controller in similar applications; type and length of warranty provided; and training for those who will be installing, maintaining, and programming the controller affect the selection of a particular product.

Ability to be integrated into higher levels of automation: Although it may not be a requirement at the present time, the controller should be able to be connected into a system communications network for higher levels of automation.

Summary

1. Manufacturing may be grouped into two broad categories—process and discrete-product.

2. Process manufacturing is the type used in the production of liquids and some food products.

3. Batch-level production is used when moderate quantities of a product are to be manufactured.

4. Discrete-product manufacturing includes job-lot, batch-level, and mass production.

5. An operation, or process, is an activity that brings raw material or a workpiece one step closer to becoming a finished product.

6. There are two types of controllers as identified by their method of operation—open- and closed-loop.

7. An open-loop control system is one whose input is set to obtain a desired output without monitoring the output or taking corrective action if the actual output deviates from the desired output.

8. Closed-loop control systems are sometimes called servos, servomechanisms, or regulators.

9. Transient response describes the output of a closed-loop control system before steady-state conditions are reached.

10. Steady-state error is the difference between the actual and desired output values after steady-state conditions have occurred.

11. The set-point in a process controller is the system reference selected by the operator.

12. The time required for the manipulated variable to cause the controlled variable to return to the set-point value after a disturbance has occured is called system lag.

13. The ON–OFF mode causes the load connected to the output of a controller to be fully actuated or deactuated.

14. The position of the actuator is proportional to the controlled variable when the proportional mode is used.

15. Numerical control is a type of machine-level automation in which machines are controlled by binary-coded instructions.

16. The major components forming a lathe are the head-stock, tailstock, and carriage.

17. The workpiece is held by a chuck for drilling, boring, and reaming operations performed on a lathe.

18. A turret lathe may be used when multiple tools are required to perform lathe operations.

19. The milling machine performs its operations when the workpiece, mounted on a table, is moved against a rotating cutting tool.

20. Operations creating flat surfaces, angular surfaces, and grooves may be produced on a milling machine.

21. The NC punched-tape program may be prepared manually or by computer-assisted programming.

22. Point-to-point controller applications include drilling, boring, tapping, reaming, inserting, and spot-welding.

23. The technique used in moving the cutting tool between points in continuous-path controllers is called interpolation.

24. Computer numerical control is an NC system that uses a microprocessor as the CPU in the controller.

25. Direct numerical control is a type of NC in which a central computer controls several CNC machines.

26. A programmable controller is a general-purpose controller whose controlling function is implemented through software.

27. The programmable controller is made up of a CPU, I/O section, and programming device.

28. Relay-ladder logic has been the method most often used to program programmable controllers, although higher level programming languages may be utilized.

29. Scan time is the speed at which the programmable controller executes relay-ladder logic diagrams.

30. Switches and sensors are often used as devices connected to the input ports while motors and solenoids are commonly connected to the output ports of a programmable controller.

Chapter Examination

1. Draw the block diagram of an open-loop control system.

2. Draw the block diagram of a closed-loop control system.

3. Draw the block diagram of a continuous-path NC machine control unit (MCU).

4. Draw the block diagram of a programmable controller.

5. Discuss the major differences between an NC controller and a programmable controller.

6. A programmable controller is used to control the heating, ventilation, and air conditioning in a building. Using the information presented in this and previous chapters, draw a block diagram depicting the major units in the control system (including the interface circuits) and identify typical inputs and outputs.

7. List the four types of programmable controllers as classified by size.

8. An electronics test equipment vendor manufacturers three models of oscilloscopes, two models of pulse generators, and three models of function generators at a particular facility. The production line is usually set up to produce 30 units of each product at a time. This is an example of:
   a. mass production
   b. job-lot manufacturing
   c. continuous manufacturing
   d. batch-level discrete-product manufacturing

9. Which of the following is *not* a lathe operation?
   a. boring
   b. cutting tapers
   c. cutting grooves
   d. cutting flat surfaces

10. Which is *not* a factor used to evaluate the performance of a programmable controller?
   a. scan time
   b. vendor support
   c. memory capacity
   d. programming ease

11. Which of the following is *not* a method of classifying controllers?
   a. mode of operation
   b. method of operation

    c. number of I/O ports

    d. method of implementation

12. Which of the following is *not* a type of discrete-product manufacturing?

    a. mass

    b. job-lot

    c. continuous

    d. batch-level

13. _____ manufacturing is utilized when a relatively few number of parts are to be produced on a one-time or periodic basis.

    a. Mass

    b. Batch

    c. Job-lot

    d. Continuous

14. _____ is the ratio of the percentage change in the output to percentage change in the input of a closed-loop controller.

    a. Stability

    b. Slew rate

    c. Sensitivity

    d. Steady-state error

15. Which of the following is *not* a variable associated with closed-loop control systems?

    a. stability

    b. set-point

    c. sensitivity

    d. transient response

16. The time required for the manipulated variable to cause the controlled variable to return to the set-point value after a disturbance has occurred defines

    a. stability.

    b. scan time.

    c. system lag.

    d. sensitivity.

17. The semiconductor manufacturing industries are a type of

    a. mass production.

    b. continuous process manufacturing.

    c. batch-level process manufacturing.

    d. batch-level discrete-product manufacturing.

18. A central computer is used to control the operation of five lathes, three milling machines, a drilling machine, and four robots. This is an example of a(n) _____ system.

    a. NC

    b. PC

    c. CNC

    d. DNC

19. _____ is an outside influence which causes the controlled variable to deviate from the set-point value in a process controller.

    a. Noise

    b. Disturbance

    c. Transient voltage

    d. The manipulated variable

20. T F Batch-level production is used when moderate quantities of a product are to be produced.   ——

21. T F The principal element in a closed-loop control system is the feedback loop—connected between the output and input.   ——

22. T F Stability is the ability of a control system to reach steady-state conditions.   ——

23. T F Controller sensitivity is the difference between the actual and desired outputs after steady-state conditions have been achieved.   ——

24. T F The controlled variable causes the manipulated variable to be maintained at the set-point value in a process controller.   ——

25. T F The proportional plus integral mode is used in those control applications where an offset cannot be tolerated and the disturbance has a short time duration.   ——

26. T F A tape reader is not required in computer numerical control (CNC).   ——

27. T F CNC is a type of numerical control in which a central computer controls several separate NC machines.   ——

28. T F Programmable controllers were developed to replace hard-wired controllers used to govern sequential machine-tool operations and materials-handling equipment in mass-production manufacturing applications.   ——

29. T F Transient responses in closed-loop control systems include underdamped, overdamped, and critically damped.   ——

30. T F The controller output oscillates a few cycles before reaching a new steady-state condition whenever the system is overdamped.   ——

31. T F Process controllers are the type used to control production machine tools.   ——

32. T F The tape reader is the least reliable component in a direct numerical control (DNC) system.   ——

33. T F The feedrate-control unit (FCU) in the MCU converts the decoded feedrate-speed information into a control signal that makes the machine move at the programmed rate.   ——

34. T F Point-to-point controllers are usually employed in NC lathes.   ——

35. T F A controller is used to automatically lower and raise the gates in a dam to control the quantity of water stored. A liquid-level sensor is used to monitor the depth of the water. The controller is utilizing open-loop control.   ——

36. T F A closed-loop control system used in the guidance circuitry of a rocket used to launch communications satellites into orbit would probably use proportional plus integral plus derivative control.   ——

CHAPTER
9

MANUFACTURING AUTOMATION

OBJECTIVES

1. List the three components forming an industrial robot.

2. List six applications for industrial robots.

3. List four methods of classifying industrial robots.

4. Identify the four types of industrial robots as classified by the type of motion they produce.

5. Identify the four types of industrial robots as classified by lifting capacity.

6. Identify the two types of industrial robots as classified by the type of control utilized.

7. Identify the three types of industrial robots as classified by the type of power supply utilized.

8. List seven different robot axes of movement and describe the type of motion produced by each.

9. Identify five devices used as robot end effectors.

10. Discuss the characteristics of and describe the relative advantages and disadvantages of hydraulic, pneumatic, and electric power supplies used in robots.

11. List three ways in which a robot can be programmed and discuss the advantages and disadvantages of each.

12. Identify four types of automated materials-handling (AMH) equipment and describe the general applications for each.

13. Discuss the purpose of machine vision systems; list their general applications; and describe the principles of operation of a typical system.

14. Define the terms *resolution*, *image processing speed*, *discrimination*, and *accuracy* as they relate to machine vision systems.

15. Identify four techniques used to identify objects automatically.

16. List the major components forming a bar-code scanning system.

17. List six types of bar-code symbols.

18. Discuss the purpose of computer-aided design (CAD); list the components forming a CAD system; and identify the types of drawings produced.

19. Describe the purpose of computer-aided engineering (CAE) and discuss the significance of CAE and CAD in automated manufacturing.

20. Describe the purpose of computer-aided manufacturing (CAM) and identify the components forming a typical CAM system.

21. Define the term *computer-integrated manufacturing* (CIM); identify the components forming a CIM system; and discuss the significance of CIM in automated manufacturing.

22. Define the term *flexible manufacturing*

system (FMS) and identify the components forming a typical FMS.

23. Define the term *local area network (LAN)*.

24. Define the term *protocol*.

25. List three LAN topologies and discuss the characteristics of each.

26. List the seven communication layers forming the ISO/OSI model.

27. Identify two techniques used to access a LAN.

28. Discuss the applications for and identify the IEEE standards for MAP, TOP, and MAN.

29. List three factors that must be considered when selecting a transmission-line medium for a LAN application.

30. Identify three transmission-line mediums used in LANs and discuss the characteristics of each.

31. Describe how light is propagated through an optical fiber.

32. Define the term *electrical noise*; identify the two major sources of noise; and discuss the procedures used to minimize the effects of noise in manufacturing data communications.

33. Differentiate between baseband and broadband communications and identify LAN protocols using each.

American manufacturers are under a great deal of pressure to increase the efficiency of their operations. These industries compete on the world market with manufacturers from other countries. In many of these countries labor costs are much lower than they are in the United States and in some countries the production costs of some goods are subsidized by the government. The only alternatives available to many manufacturers operating in the U.S. are to automate, move their operations overseas, or go out of business. This chapter deals with the techniques utilized by the manufacturing industries to automate their production operations. Automation may occur at the machine, factory floor, or plant level. Various techniques are used to facilitate automation. Some, such as CNC and programmable controllers, were discussed in the last chapter. These primarily permit machine-level automation. Industrial robots, automated materials-handling equipment, computer-aided drafting and engineering, and communications between the various computer-operated machines are all devices and techniques that bring about factory-floor and plant-level automation.

9-2 INDUSTRIAL ROBOTS

9-2.1 Introduction

An industrial robot is a CNC machine primarily used to perform operations where the work may be hazardous, monotonous, or difficult for humans. Typical operations include spray painting, welding, materials handling, machining, and assembly—with spray painting, welding, and materials handling being the most widely used. Materials handling covers a variety of activities and includes such things as removing parts from a conveyor and placing them on a pallet; removing parts from a pallet and placing them on a conveyor; and loading workpieces onto and unloading workpieces from machine tools such as lathes, milling machines, and grinders. These are called pick and place applications. Machining activities are primarily limited to drilling and cutting activities at the present time. Assembly operations are the fastest growing applications for robots, especially in electronics manufacturing, where the robot can be used to load printed-circuit boards. In addition, a few robots are capable of performing inspection operations.

9-2.2 Types of robots

A robot is composed of a manipulator, controller, and power supply. The manipulator, sometimes called an arm, is the mechanical device which performs the work. The controller provides the signals that govern the movement of the manipulator, while the power supply provides the energy to move it.

Dozens of different types of robots are available to perform the operations previously mentioned. They can be classified in several different ways, including their lifting capacity, type of control employed, kind of power supply used, and the type of motion they produce. These are discussed below.

Lifting capacity: This classification groups robots according to their manipulator weight-lifting capability and includes small, intermediate, large, and heavy duty. Small robots are those which can lift up to 25 pounds; intermediate robots can lift a maximum of 300 pounds; large robots can lift loads up to 500 pounds; and the heavy-duty types can lift in excess of 500 pounds.

Type of control employed: Two types of robots are available based on type of control—point-to-point and continuous path.

Type of power supply: This classification refers to the type of power used to drive the manipulator. Driving power may be electric, hydraulic, or pneumatic.

Type of motion produced: There are four types of robots classified by the type of motion they produce. These include the polar, cylindrical, Cartesian, and revolute. The manipulator moves in three different axis directions for all four types, as seen in Figure 9-1. The manipulator of the polar robot swivels around horizontally, pivots up and down vertically, and transverses in and out, as shown in Figure 9-1(a). The manipulator of the cylindrical robot swivels around horizontally and transverses up and down and in and out, as illustrated in Figure 9-1(b). The Cartesian robot appears in Figure 9-1(c). The manipulator transverses back and forth in the horizontal plane, up and down in the vertical plane, and in and out. The movement of the revolute robot is similar to that of the human arm. Three axes of movement are possible, as seen in Figure 9-1(d).

9-2.3 The manipulator

The kind of motion produced by the manipulator is dictated by the type of robot. The movement produced by the revolute robot is the most complex. This is the most frequently used robot because of the versatility of its manipulator. The manipulator is divided into several sections to provide dexterity, as shown in Figure 9-2. These include a torso that joins the base, a shoulder joint, a lower arm, an elbow joint, an upper arm, a wrist, and an end effector. Actuators in the form of electric motors, fluid-power motors, or cylinders, are connected to the joints to provide movement. These may be coupled directly to the joints or indirectly through gears or chains. Six to seven different axes of movement are possible, as indicated in Figure 9-3. These include manipulator sweep (horizontal movement about the base), shoulder swivel (vertical movement of the upper arm), elbow extension (vertical movement of the upper arm), pitch (vertical movement of the end effector at the wrist), yaw (horizontal movement of the end effector at the wrist), and roll (a circular end-effector motion.) A seventh axis movement is made possible by mounting the robot on a platform and allowing it to move back and forth in the horizontal plane.

A number of different types of end effectors are available for most robots. These include mechanical grippers, hooks, scoops, vacuum cups, and electromagnets. Mechanical grippers are the most versatile. Several different techniques are employed to facilitate the gripper function. The gripper must be capable of grasping a

Figure 9-1

Robot types
classified by
motion

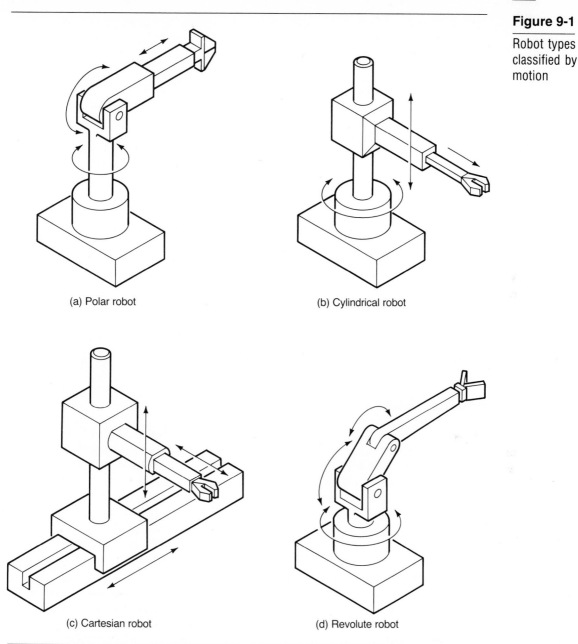

(a) Polar robot

(b) Cylindrical robot

(c) Cartesian robot

(d) Revolute robot

part and holding it in position without applying undue pressure and damaging it. It should also be capable of handling parts of various sizes and shapes. Vacuum cups may be used when the robot must pick up thin sheets of metal, glass, plastic, or similar materials. Electromagnets mounted on the end of the manipulator can be used to pick up sheets of ferrous materials.

Figure 9-2

The revolute robot
manipulator

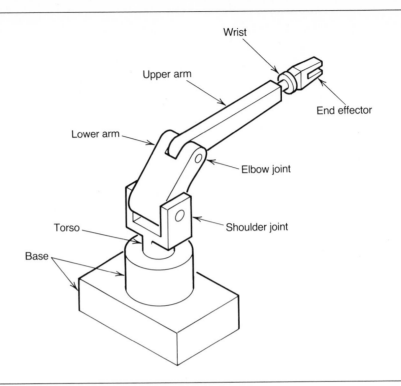

9-2.4 The controller

The controller generates the control signals that operate the actuators to produce manipulator motion. Like other CNC machines, the motion control of robots may be either point-to-point or continuous. Point-to-point robots are the more common and are used in pick-and-place, machine-loading, and spot-welding applications. In these robots the programmer writes each position into the controller memory and the sequence is read back during the work cycle. Open-loop controllers are normally used in point-to-point robots. Memory requirements are minimal as only the position data are required for each operation. The path the manipulator follows when going from one point to another is inconsequential, although the most direct path is the one usually followed.

Robots used in spray painting, continuous welding, machining and assembly operations, and in some materials-handling applications usually employ continuous controllers. In these robots, the manipulator follows a series of closely spaced points simulating smooth continuous movement. The closed-loop mode is usually used. These are sometimes called servo robots. Servo robots require considerably more memory than do the point-to-point type, since all the points in the manipulator must be loaded into memory during the programming phase.

The architecture used in robotic controllers varies. In the simpler types the controller includes an MPU, main memory, and I/O interfacing for the sensors and ac-

Figure 9-3

Manipulator axes movement

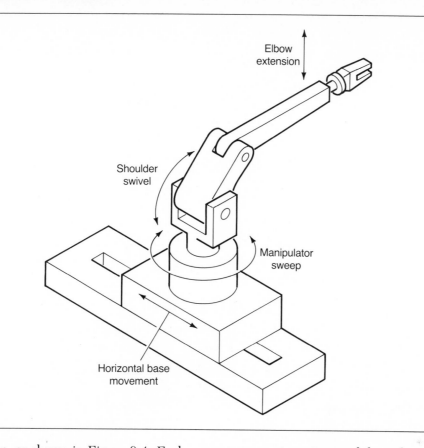

Elbow extension

Shoulder swivel

Manipulator sweep

Horizontal base movement

tuators, as shown in Figure 9-4. Each sensor represents an input while each actuator represents an output. A robot having seven axes of movement requires seven separate outputs. The interfacing techniques discussed in Chapter 6 are used to connect the sensors and actuators to the microprocessor. Although almost any microprocessor can be used, processors that are self-contained microcomputers, such as the Motorola MC6801, are popular. The MC6801 has the same architecture as the MC6800, and in addition, has an on-board clock, RAM, ROM, and I/O interfacing. Manipulator control may be distributed in more complex robots. In these a mainframe, mini-, or microcomputer may be used as the CPU with a separate microprocessor used with each actuator.

9-2.5 Actuator power supplies

Hydraulic, pneumatic, or electric power may be used to energize the actuators to move the manipulator. The circuit employed in hydraulic robots includes a power unit, plumbing, valves, and actuators. Depending upon the application, either motors or cylinders may be employed as actuators.

The main factory air supply is usually used to drive pneumatic robots. Airflow to

Figure 9-4

Robot controller

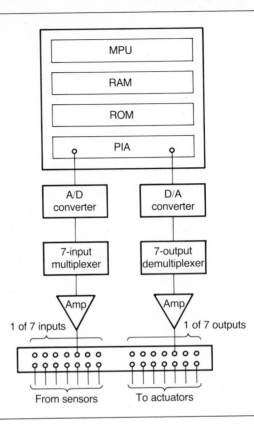

the pneumatic actuators is controlled by solenoid valves mounted on a manifold connected to the main air-supply line. Like hydraulic robots, either motors or cylinders can be utilized as actuators. Both conventional and stepping motors may be used as actuators in electric robots. Conventional motors are smaller and develop more torque than do stepping motors with equivalent horsepower ratings. Except for the simplest types of pick-and-place robots, positional encoders are necessary to allow the controller to track the position of the manipulator when conventional motors are used. Angular positional encoders are usually not required when DC stepping motors are utilized.

Factors affecting the choice of a power supply include lifting capacity, positioning speed and accuracy, and the operating environment. Hydraulic power can be employed in any size robot but is primarily used in the large and heavy-duty types. The characteristics of hydraulic robots include large lifting capacity, fast positioning speed, and good positioning accuracy. They sometimes cannot be used in environments where cleanliness is a prerequisite because of the possibility of a valve leaking or transmission line breaking. Pneumatic powered robots are the slowest, least accurate, and least expensive. Some small and intermediate, and a few large robots utilize this type of drive. Electric-motor drive can be used in almost any size of robot but is mainly used in small- and intermediate-size machines. These robots are cleaner, smoother, quieter, and more accurate than are the fluid-power types. Fluid-power robots are often preferred, and may have to be used, in environments such as spray painting booths or in the handling of chemicals, where an electric arc could create a fire or explosion.

9-2.6 Programming the robot

A number of different programming techniques and languages are available for robot applications. Robots may be programmed using a teach pendent, by leading the manipulator through its required movements (sometimes called lead-through programming), and by off-line programming.

A teach pendent is a hand-held keypad that is connected to the controller. The operator positions the manipulator by depressing the proper keys until it is positioned in the desired location. The operator then loads positional data into the controller's memory. This procedure is continued until all the different manipulator positions required for a particular operation have been stored in memory. The controller uses these points to create the path the manipulator follows when performing its operations.

When the lead-through technique is used, the operator places the controller into a teach or learn mode and physically guides it through its required movements. During this time positional data are continually loaded into memory. When the programming phase has been completed, the controller is placed in the operate mode and the robot is able to repeat the movements. On large robots, where it is difficult to move the manipulator, a teaching handle is attached to the manipulator and the robot is led throught the desired program sequence.

The teach-pendent and lead-through programming techniques are both popular and have the advantage of simplicity. The operator need not be a skilled programmer. Both are excellent if motion and position are the only functions required of the robot. This is the type of programming most frequently used for pick-and-place, welding, and spray-painting applications. In higher level automation applications, motion and position data form only a fraction of the total information required for the robot to perform its operations. Touch and vision sensors, communication, and interfacing with other CNC machines are as important as motion and position in these types of applications. As much as 90 percent of the programming for some robots is concerned with factors other than motion and position. In these, off-line programming is utilized.

Because of the diversity of the operations performed, differences between controllers, and differences in driving power-requirements, it has been difficult to develop a few standardized robot languages like those used for computers and other CNC machines. As a result, a wide variety of off-line programming languages exist. In many cases a language has been created for use with a particular robot. Most have some elements of commonality, however. A few languages allow the robot to be programmed from the data base developed during the design phase of the product. Some of the more commonly used languages appear in Table 9-1.

Table 9-1

OFF-LINE ROBOT
PROGRAMMING
LANGUAGES

AL (*Arm Language*)
AML (*A Manufacturing Language*)
BAM-8
FORTH
Karel
MCL
PLAW (*Programmed Language for Arc Welding*)
RAIL
Robocam
Robotalk
ROL (*RObot Language*)
ROPS (*Robot Off-line Programming System*)
VAL

9-2.7 Sensors used in robots

Almost any of the sensors discussed in Chapter 4 can be used on robots. Position sensors, such as binary angular encoders, are required to measure manipulator-joint position. Strain gages may be mounted on the end effector to control gripper pressure. Machine vision systems, discussed in Section 9-4, can be connected to provide the robot with sight to give it the ability to avoid obstacles and extend its capabilities to include picking up parts that are randomly positioned, identifying parts, inspecting parts, and assemblying parts.

9-3 AUTOMATED MATERIALS HANDLING

Automated materials handling (AMH) is essential to automated manufacturing. Raw material must be delivered to the first workstation and the resulting workpiece, called work-in-progress, transferred between workstations as the ensuing processes turn it into a finished product. If assembly is required, parts must be obtained from the warehouse and delivered to the assembly point. After the product has been assembled and inspected, it is placed in the warehouse as inventory or sent to the shipping department. Depending upon the type and size of plant, a number of techniques are utilized to accomplish the automatic movement of materials. These include automated storage and retrieval systems (AS/RS), conveyors, overhead monorails, and automatic vehicles. AS/RS are used to store parts in the warehouse and take them out when needed, as illustrated in Figure 9-5. Conveyors and monorails are excellent for moving material in mass-production manufacturing but lack the flexibility required in batch-level discrete-product manufacturing. Automatic vehicles are used instead. A variety of automatic vehicles are available for almost any type of moving activity and include automatic forklifts, called pallet trucks; wagons pulled by automatic tractors, called automatic guided vehicles (AGVs), as seen in Figure 9-6; and small single-unit carriers, referred to as agile autonomous

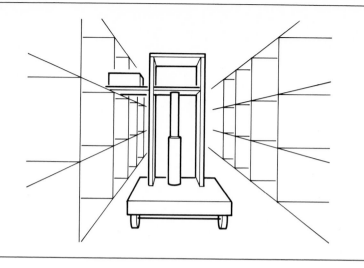

Figure 9-5

Automated storage
and retrieval
system

vehicles (AAVs). AGVs are mainly used to carry large and heavy parts. They are not self-loading and pallet trucks are used to load and unload the wagons. A number of different methods are used to steer AGVs and include rail-guide, wire-guide, and light-guide. Rail-guided AGVs run on a central rail attached to the factory floor. Wire-guided AGVs are guided by a wire embedded in a narrow channel in the floor through which RF current is fed. This creates a magnetic field which is detected by a sensing coil mounted on the AGV. The detected signal is processed and sent to the guidance mechanism of the tractor causing the tractor to follow the wire. Either a fluorescent dye is painted on the floor or an luminescent tape is laid on the floor

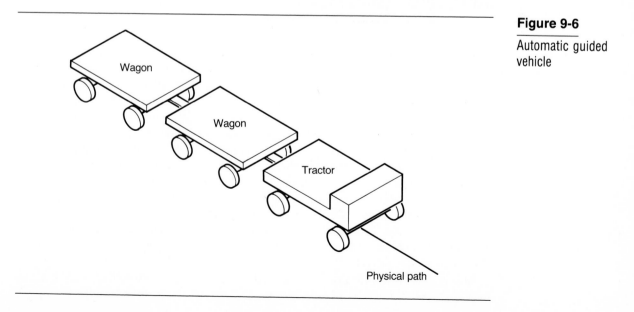

Figure 9-6

Automatic guided
vehicle

Figure 9-7

Agile autonomous
vehicle

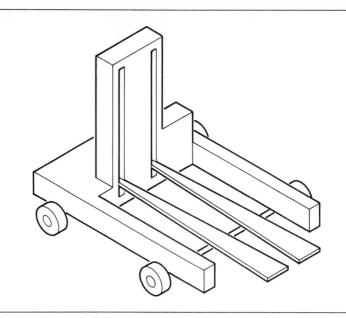

for light-guided AGVs. A photodetector mounted on the tractor allows it to follow the dye or tape.

Agile autonomous vehicles, such as the one in Figure 9-7, do not follow a physical path as do AGVs, but are guided by an RF wave which is transmitted to the vehicle. Most of these vehicles can load and unload themselves.

9-4 MACHINE-VISION SYSTEMS

Sometimes called computer vision, machine vision enhances the capability of robots and other CNC machines such as automatic assemblers. Robots equipped with machine vision can remove parts from a conveyor, check parts for missing or broken pieces, sort parts, and position parts in relationship to other parts. Although machine-vision systems differ, thay all contain a camera and CPU. The camera scans the scene, and solid-state detectors connected in a matrix configuration in the camera convert light reflected from the object being viewed into electrical impulses, called pixels. The camera scans the scene from left to right and top to bottom. Each scene that has been created by the scanning action is called a frame. Scanning speeds vary but most systems utilize a frequency of 30 frames per second. The number of photodetectors connected in the matrix determines the detail of the final image. Typically, these are connected in a 256×256 grid, or a 484×320 grid, which allow 65,536 or 154,880 pixels respectively to be generated for each frame. The analog pixel signals are converted into digital signals that represent an image of the scene being viewed.

Depending upon the application, two types of vision systems are available: binary and gray scale. The binary system is strictly black and white with no gray in between. The voltage level for each pixel is either a binary zero or one. On the other hand, pixels in the gray-scale system may have as many as 256 different level values. Because of this, additional memory and a more powerful microprocessor is required in the CPU of a gray-scale system. The additional data handling requirements can be reduced by using a technique called windowing. A window is an electronic mask that is placed around a small portion of the image. Only those pixels enclosed within the window, or mask, are processed.

Several different viewing techniques are available. The technique selected primarily depends upon the application. If, for example, a robot is inspecting parts for missing or broken pieces, an image of a known good part is loaded into CPU memory for a standard. The image of each part that is picked up and inspected by the robot is compared with the standard. If the part is complete and unbroken, its image coincides with the standard and the robot accepts the part. If the part is broken or pieces are missing, the two images are not the same and the part is rejected. A robot can be used to check printed circuit boards in a similar fashion. The image of an acceptable board is loaded into CPU memory. Solder-pad location, conductor width, and spacing between adjacent conductors can all be checked as the robot picks up the boards, one at a time, and compares the image of each board with the image of the standard board stored in memory.

Criteria used to evaluate machine-vision systems include resolution, image processing speed, discrimination, and accuracy. These are discussed in the following paragraphs.

Resolution:　This variable identifies how well the system reproduces a recognizable image. Resolution is proportional to the number of pixels forming the image.

Image processing speed:　This is a measure of the number of binary bits that can be processed in one second. Image processing speed is affected by the type of illumination, image complexity, accuracy required, and whether windowing is utilized. Typical vision systems are capable of recognizing or inspecting 2-12 items per second.

Discrimination:　This is the ability of a machine-vision system to recognize variations in the light intensity reflected from a part being scanned. Gray-scale vision-system discrimination is superior to binary vision-system discrimination.

Accuracy:　Machine-vision accuracy represents the percentage of correct matching or identification decisions made by the machine vision system regarding a lot size of parts being examined.

9-5 BAR-CODE SCANNERS

Automated manufacturing requires that some form of automation be utilized to automatically identify and keep track of parts in inventory and work-in-progress as it moves through its various manufacturing stages. This information forms part of the computer database and is used for such activities as balancing material between workstations, controlling production output, keeping inventory to an absolute minimum, and measuring production efficiency. Numerous automatic identification techniques are available, including bar-code scanning, radio frequency identification (RFID), surface acoustical wave (SAW), and optical character recognition (OCR). Bar-code scanning is the type most commonly used because of its simplicity and accuracy. Studies show that one error occurs in three million identification scans in bar-code scanning. OCR, where the chances of an error occuring are one in every 10,000 identification operations, has the next highest accuracy.

Bar-code scanning was first used in supermarkets. Its application rapidly spread to other types of retail stores, wholesale distributors, and manufacturing facilities. The bar-code scanning system consists of a bar code, a scanner containing a light source, and a decoder. The bar code consists of a series of vertical black bars of varying widths separated by white spaces of varying distances as seen in Figure 9-8. It can be printed on almost any surface and represents, in symbolic form, part characteristics such as model or serial number, manufacturer, purchase order number, price, and routing information. Several different types of scanners are available. They include the moving beam, fixed beam, hand-held wand, and combinations of these.

Numerous bar-code symbols are in use. The more common of these include the universal, or Uniform, Product Code (UPC); European article numbering (EAN); code three-of-nine (C3 of 9); Interleaved and industrial two-of five (I2 of 5); Code 128; and Codabar. The Universal Product Code is the type used in supermarkets and similar retail stores. The latter four are used in manufacturing applications.

Depending upon the application, moving-beam and fixed-beam scanners can be mounted on robot manipulators, on the arms of automatic machine loaders, or on insertion machines to identify parts. Hand-held wands are used to identify parts in receiving, inventory, shipping, and at other key locations on the factory floor.

9-6 AUTOMATION TECHNIQUES

9-6.1 Introduction

Automated manufacturing is much more than computer controlled machines. It is a philosophy that must permeate the thinking and work activities of all the company's employees. In practice, manufacturing automation is the employment of computer technology from the initial design of a product to its final checkout. All of the com-

Figure 9-8

Bar code

puter-based equipment used in the design, processing, assembly, and testing phases operates from the same database. The techniques used to automate the factory are discussed in the following sections.

9-6.2 Computer-aided drafting

Computer-aided drafting (CAD) is automation applied to engineering drawing. CAD systems vary in the types of operations they are capable of performing. Some systems are little more than automatic drafting tools while others are capable of performing a large portion of the product design activities, developing performance specifications, and determining product cost.

The CAD system consists of a CPU, which is often a high-quality microcomputer, and I/O peripherals. Systems include all or most of the components depicted in Figure 9-9. Input devices usually include a keyboard, a digitizing table, a tracing table, and a light pen. Output devices include a color video monitor, a printer, and an $X-Y$ plotter. Drawings are created by connecting together points, lines, planes, curves, and symbols, which are stored in CPU memory to form a graphics file.

Both two- and three-dimensional drawings can be created with CAD. Two-dimensional views are utilized in drawing some mechanical devices and making architectural, pipeline, and floor layouts; making electronic and fluid-power schematic diagrams; and creating flow charts. Wire-frame, surface, and solid models can be created using three-dimensional CAD systems. Wire-frame models are produced by connecting lines to identify surfaces. They require little memory and computer processing time but provide only a limited amount of information about the part. The surface model is developed from planes, curves, and cylinders which are stored in memory. Although this model more completely defines the object, it still lacks the ability to represent the solid characteristics of the part. The solid model, is by far, superior to the other models when a complete description of a solid part is required. In addition to creating two- and three-dimensional drawings, CAD systems produce drawings of parts that are cross-sectioned, rotated, and dimensioned. Some systems will even simulate motion.

9-6.3 Computer-aided engineering

Computer-aided engineering (CAE) is closely related to CAD. CAE allows product design, analysis, testing, and production planning to occur almost simultaneously. Data developed during the CAD and CAE phases are loaded into memory of a central computer and form part of a common data bank. Data stored in the data bank are used during the manufacturing phase of the product.

Computer-aided engineering workstations are similar in appearance to CAD workstations. The major difference is the CPU. CPUs used in CAE workstations are usually much more powerful. Most utilize 16- or 32-bit microprocessor CPUs. Hewlett-Packard's HP 9000 Series 320 engineering workstations, for example, utilize the MC68020 32-bit microprocessor. CAD and CAE do not replace the draftsman and engineer. Instead, they free these individuals to perform the creative aspects of design.

Figure 9-9

Components forming a CAD workstation

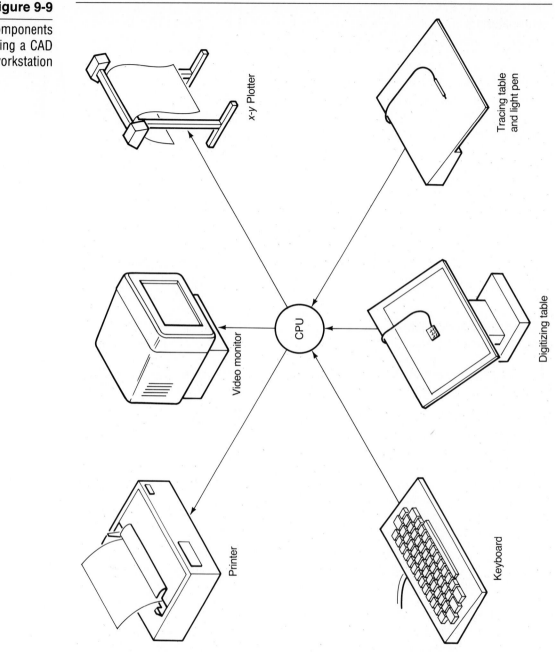

9-6.4 Computer-aided manufacturing

Computer-aided manufacturing (CAM) is factory-floor-level automation. What CAD and CAE do for the drafting and design stages of a product, CAM does for its fabrication. Computer-aided manufacturing includes all the computer numerical controlled machines, programmable controllers, robots, automated materials-handling equipment, assembly equipment, and inspection equipment—all linked together with hardware and software to manufacture a product automatically.

9-6.5 CAD/CAM

Although CAD and CAM may be used separately, the full automation power of both is realized when they are used together. Production increases of 3 to 10 times are common. The part is designed using CAD and CAE while CAM translates the design into a finished product. Data generated during the CAD and CAE phases become the starting point and forms the initial database for production, manufacturing, and industrial engineers in developing the process plans. The activities included in CAD/CAM are illustrated in Figure 9-10.

9-6.6 Manufacturing planning and control software

After a product has been designed, decisions must be made regarding the types of materials, processes, and machines required to fabricate it. The flow of materials, or work-in-progress, must also be established. Manufacturing planning and control software (MPCS) packages have been developed to aid in making these decisions. MPCS provides the coordination required between workstations, materials-handling equipment, and the warehouse to produce a product in the most time efficient and cost effective method possible.

A dozen or more different MPCS packages are available. One of the most popular of these is Manufacturing Resource Planning (MRP II). This software package includes such factors as the types of materials required to fabricate the product, the materials presently in inventory, materials that must be obtained, and assembly line scheduling.

9-6.7 Computer-integrated manufacturing

Computer-integrated manufacturing (CIM) is an extension of CAD/CAM which integrates CAD, CAE, computer numerical controlled machine tools, programmable controllers, robots, automated materials-handling equipment, automatic assembly, automated inspection, automated warehouse, and automated business-office activities with software, as illustrated in Figure 9-11. CIM is plant-level automation. A common computer database is used for all manufacturing and ancillary activities including design, drafting, manufacturing planning, parts production, assembly,

Figure 9-10

CAD/CAM activities

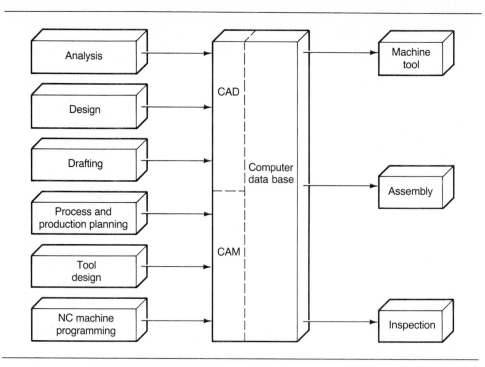

inspection, and business operations. All operate from the same specifications, draw-ings, and schedules. The data requirements for CIM are enormous. The data bank in a typical medium-size facility contains approximately 25 million pieces of infor-mation, most of which are generated by CAD and CAE. Items forming the data-base include design criteria, drawings, dimensions, product specifications, bills of material, part-forming instructions, assembly instructions, quality control, inven-tory, purchase orders, and shipping records.

9-6.8 Flexible manufacturing systems

Mass production dates back to the days of Henry Ford, who initiated Detroit-style high-volume mass production. Mass production has greatly increased the standard of living in the United States and has made possible the owning of automobiles, home appliances, and electronic entertainment equipment for the majority of the residents of this country. Most of what is produced is not manufactured using mass production techniques, however. Approximately 75 percent of all products manu-factured in the U.S. are in batches of 50 or less. This level of production is much more expensive than that of high-quantity mass production. Machine tools are per-forming operations only about seven percent of available machine time due to time lost in changing holders, jigs, tools, and loading and unloading workpieces onto and off of the machine. This is contrasted with mass production with its highly special-

Figure 9-11

Components
forming CIM

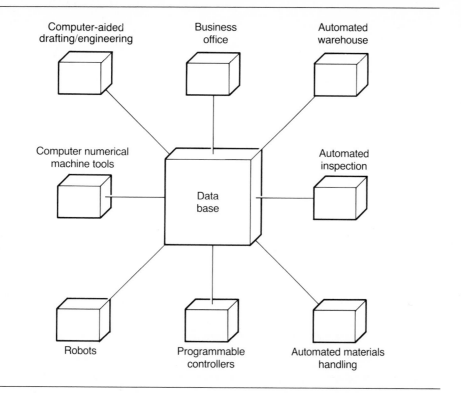

ized machines and assembly lines which allow machines to perform processes approximately 50 percent of available machine time.

Flexible manufacturing systems (FMS) were developed to automate, increase the efficiency, and lower the cost of batch-level discrete-product manufacturing. The concept of FMS originated in England during the late 1960s, was eventually introduced in this country, and has slowly evolved to its present state. FMS is CAM used in batch-level manufacturing. FMS integrates CNC machines, programmable controllers, robots, automatic vehicles, automated assembly stations, and automated inspection stations.

The architecture of an FMS is determined by the types of products to be manufactured. In its usual format, the system is composed of 6 to 12 assorted CNC machines and an automatic guided vehicle that will deliver workpieces to any machine in the system in any sequence required. Robots or other automatic loading and unloading mechanisms load and unload the machines. A magazine capable of holding the tools required to perform the operations is mounted on each machine. A robot or automatic arm changes the tools in the sequence desired.

The FMS appearing in Figure 9-12 is made up of three milling machines, two lathes, robots to load and unload the machines, an automatic guided vehicle, an inspection machine, an assembly station, and a central computer. The automatic guided vehicle carries the raw material and workpieces from the loading area to the

Figure 9-12

An FMS

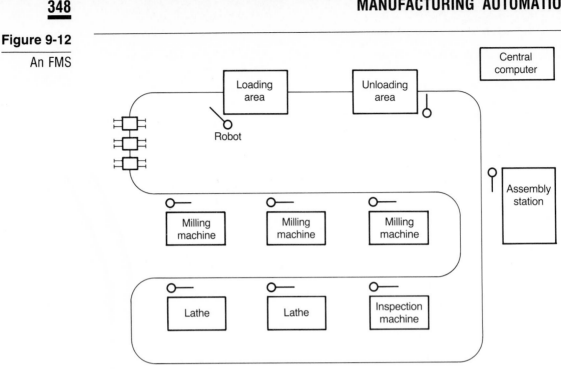

milling machines and conveys the work-in-progress between the machine tools, the inspection machine, the assembly station, and to the unloading area.

FMS has the potential to revolutionize batch-level manufacturing and raise its production level to that approaching mass production. Some experts think that the development of FMS signifies the beginning of a second industrial revolution.

9-6.9 CIM and FMS applications

Machine-tool manufacturers in the United States, Japan, Great Britain, and several Western European countries produce machine tools designed to be integrated into CIM and FMS installations. The United States leads the rest of the world in the development and production of automated factory-floor machine tools. Japan, however, leads the other countries in the implementation of these machines. Major automated-machine-tool manufacturers in the U. S. include Cincinnati Milacron, Kerney and Trecter, White Consolidated, and Bridgeport.

There are few fully implemented CIM systems in the United States at the present time. There are several reasons for this. Manufacturing plants in this country are often older than their counterparts overseas and retrofitting is expensive. Stockholders and management are often reluctant to invest the large sums necessary to automate a facility even though the payback period is usually relatively short. Labor often resists change. Automation implies a reduction in the number of production people. In a typical CIM or FMS installation a few technicians are available to maintain the machines and repair breakdowns. Machines are loaded and unloaded and parts move through the assembly process automatically. These machines may operate one, two, or three shifts per day. Manpower is often required to bring materials out to the loading area and to take the finished product away from the unloading area, however. This activity is usually performed at night.

CIM is often appropriate for both large and moderate-size manufacturing facilities. General Motors is an example of CIM utilized in large plants. This company has made a major commitment to total factory-level automation by upgrading many existing facilities and building new ones.

Allen-Bradley, a manufacturer of electric-motor starters and controllers, is an example of a company that has successfully implemented CIM in a moderate-size facility. This company installed a CIM system in their electric-motor-starter plant in Milwaukee to meet foreign competition. Twenty six machines are used in assembly, testing, and packaging. The installation can produce two sizes of motor starters in 125 variations and can manufacture 600 starters per day[21].

Cone Drive Operations of Ex-Cell-O Corporation in Michigan is another example of a company that has utilized CIM in a moderate size plant. This company manufactures worm gears. Overseas competition and escalating labor and materials costs mandated making changes for the company to continue to compete in a world market. This company was able to reduce costs, decrease manufacturing time, and improve quality with CIM[5].

As with CIM, there are relatively few FMS installations in this country. As previously mentioned, FMS has the potential to raise batch-level production levels to that approaching mass production. General Electric, for instance, utilizes FMS to produce 2000 different models of its basic meter at its Somersworth, New Hampshire facility. Total factory output is approximately one million meters per year[13].

Deere and Company, an agricultural manufacturer, increased its production efficiency considerably by installing FMS in its Waterloo, Iowa facility. Tractor cabs and bodies produced at the Waterloo plant are jointed with chassis and engines shipped from other Deere and Company facilities. Each part, such as the engine, transmission, and wheels is automatically sent to a specific tractor ordered by a dealer. Parts are automatically retrieved from inventory and delivered to the assembly line at the exact time they are needed. Using FMS, Deere and Company is capable of producing tractors in half the time while cutting inventory costs by approximately 50 percent [13].

9-7 LOCAL AREA NETWORKS

9-7.1 Introduction

All of the CNC machines, programmable controllers, robots, and AMH equipment on the factory floor in an automated factory must be connected together to allow the machines to communicate with one another. This communications link is called a local area network (LAN). A LAN is a private, on-site communications system composed of interfacing hardware, software, and a transmission line that connects the computer-operated equipment together. Because the LAN links equipment that is located on the premises, telephone lines cannot be used as the transmission-line medium. The physical length on a LAN usually does not exceed three miles.

Computer-based equipment connected to a LAN is called a workstation, or node. The technique used to exchange data over a LAN is called protocol. Protocol establishes the rules and standards necessary for communications to occur between workstations and includes such parameters as data-transmission speed, addressing technique, data priority, and binary logic levels.

9-7.2 Topology

The term topology is used to describe the method used to connect workstations to the LAN. Three LAN topologies have been developed for factory-floor, engineering, and office applications and include the star, bus, and ring.

A central computer is employed in the star topology, as shown in Figure 9-13. Communication can occur between the computer and any workstation or between workstations with the data routed through the computer. Although simple in concept, this topology has the disadvantage that if the computer develops a fault, the entire system is disabled. It also requires more transmission line than do the other topologies. A separate transmission line must be run from the computer to each workstation. For these reasons, the star topology is primarily limited to small-office, CAD, and CAE applications.

The bus toplogy is formed from a single bus to which individual workstations are connected via branch buses, as seen in Figure 9-14. This LAN utilizes distributed control. Workstations can communicate with one another without going through a central computer. Any workstation connected to the network can send data to any other station. Each station examines the data address to determine if it is the intended receiver. This is the simplest LAN to implement, as additional workstations can be added by connecting them to the main bus. It is also the most reliable. A fault occuring at an individual workstation does not cause the entire system to fail.

Figure 9-13

Star topology

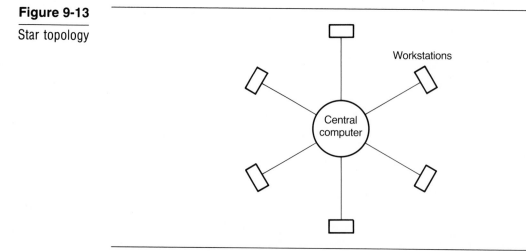

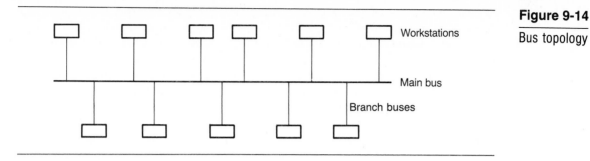

Figure 9-14

Bus topology

The diagram of a ring LAN appears in Figure 9-15. In this system each workstation is connected in series with another. Like bus topology, this LAN utilizes distributed control. A central computer is not required. Unlike the bus LAN, where each workstation can generate and receive data independently, messages originating from workstations connected in a ring LAN circulate in either a clockwise or counterclockwise direction and may have to pass through other stations before arriving at their destination. Whenever data arrive at a station, the station examines the address and, if that station is the intended receiver, it decodes the data for its use. If it isn't, the data are passed on to the next station.

It is more difficult to add workstations to the ring LAN than to the bus type, as the transmission line has to be opened and the equipment connected to both ends of the opened line. A bypass network is often connected across each workstation. If a fault develops at a particular station, that station is bypassed, allowing the system to continue to operate without the faulty station.

9-7.3 Types of LANs

One of the major obstacles to implementing manufacturing automation is the inability of computer-based equipment to communicate with one another. Approximately 200 vendors produce equipment that can be connected to a LAN. Companies wishing to implement CAM or CIM often find that after they have purchased the automated equipment and connected the various pieces to the network only part of the equipment will communicate with each other. Too often, in an effort to promote

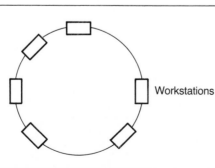

Workstations

Figure 9-15

Ring topology

the sales of their own products, vendors have paid little attention to making their equipment compatible with the products of other vendors.

Dozens of different LANs have been developed to link microcomputers together, minicomputers to mainframes, and microcomputers to mainframes. Popular examples include Ethernet and StarLAN. Ethernet was developed as a joint effort between Digital Equipment Corporation, Intel, and Xerox. This LAN employs a bus toplogy in which up to 1024 workstations can be connected. A number of different vendors produce interface circuits that allow their products to be connected to the Ethernet system. The Ethernet IBM PC Card, for example, allows the IBM personal computers to be incorporated into an Ethernet LAN. A typical Ethernet installation appears in Figure 9-16.

StarLAN was developed by the American Telephone and Telegraph Company (AT&T) and allows the AT&T 3B series PC6300, and IBM PC microcomputers to communicate with one another. Ethernet, StarLAN, and the other LANs are important because they provide the means for linking computer-based equipment.

Figure 9-16

An Ethernet installation

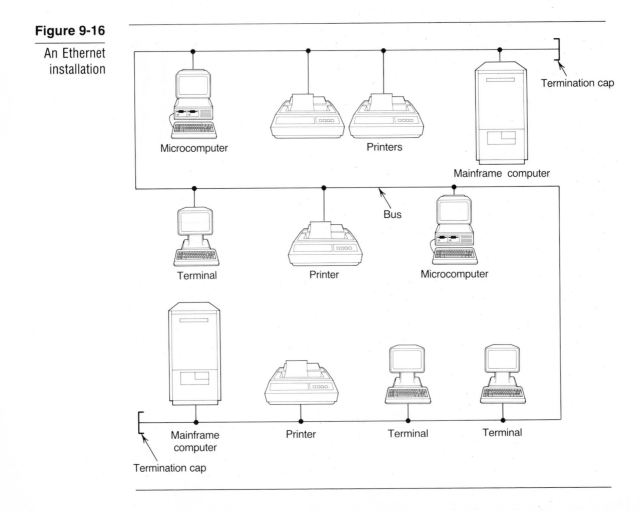

None has developed into an industry-wide standard, however. Most are more or less proprietary in nature and have met with only limited success. For the most part, they allow communications between equipment produced by the same vendor or between compatable equipment. They have primarily been used to link office machines or CAD/CAE workstations together.

9-7.4 LAN interface standards

To assist in the development of an industry-wide standard LAN, the International Standards Organization (ISO) has developed a model for open systems interconnection (ISO/OSI). This model separates the activities of a LAN into a seven-layer hierarchy as seen in Figure 9-17. Each layer represents a module that is responsible for implementing particular networking services to the next higher layer or module.

The physical layer is the lowest level module. It deals with the mechanical and electrical interconnection parameters and includes the transmission-line connectors, binary voltage levels, data-transmission rates, and impedances. The data-link layer is concerned with message acknowledgment, possession, detection, and sequencing. The techniques employed to route data and interface the workstations to the LAN are developed at the network layer. The transport layer deals with network addressing and the procedures for entering and exiting the LAN. The methods used in initiating and terminating communications are established at the session layer while activities at the presentation layer are concerned with formatting, editing, and display. At this layer the data are demodulated and restructured. The application layer is the highest level. Events occurring here manage the lower layer modules. In general, the lower layers deal with techniques which govern the transfer of data between workstations while the activities which occur at the higher layers are concerned with information processing.

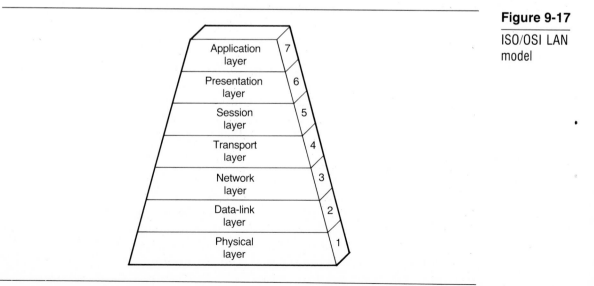

Figure 9-17

ISO/OSI LAN model

The Institute of Electrical and Electronics Engineers (IEEE) has developed a set of standards at the physical and data-link layers of the ISO/OSI model which are known as IEEE 802. At the physical layer, the standard defines the mechanical and basic electrical parameters required for a vendor's workstation to be connected to the LAN. Protocol is defined at the data-link level.

9-7.5 Accessing techniques used in distributed LANs

Some method must be employed in distributed LANs to allow the LAN to be accessed in a systematic fashion by any workstation and to prevent multiple stations from sending data simultaneously. Two accessing methods—contention-based and deterministic—have been developed for this purpose. Several techniques are available to implement each method. One of the more popular contention-based accessing techniques is carrier sense multiple access with collision detection (CSMA/CD). This is also known as the IEEE 802.3 accessing standard. When CSMA/CD is employed, a workstation desiring to transmit data first listens to the network. If the network is free, the station begins its transmission. Whenever two or more stations transmit data simulatneously, a "collision" of the data occurs, which results in additional activity appearing in the LAN. This is detected by the transmitting stations causing them to terminate transmission until the network becomes free. Again, the stations listen to the network and attempt to transmit. If they transmit at the same time they are forced to get off the line. This process continues until one of the stations gets on the LAN first. CSMA/CD is primarily used in applications where relatively few stations are connected to the LAN, such as a small office. In this type of application data are transmitted more or less intermittently; the timing when data are transmitted is not critical; and access delays may be tolerated.

Deterministic accessing is usually employed in large CAD and CAE systems and in factory-floor CAM applications. In these cases, the data transfer is more lengthy, and more constant, and excessive delays cannot be tolerated. Several different techniques have been developed to implement deterministic accessing. The type most frequently used is the control "token," sometimes called the token-passing technique. The control token is a code that continually moves through the bus or ring LAN from station to station, as illustrated in Figure 9-18. A station may transmit or receive data only while the token is at that station. If the station has nothing to transmit, it passes the token on to the next station. The IEEE standard for the token-passing bus LAN is IEEE 802.4 while that for a ring LAN is IEEE 802.5.

9-7.6 Manufacturing Automation Protocol

Concerned about the lack of communications standards for factory-floor computer-based equipment, General Motors appointed a corporate-level task force in 1980 with the goal of developing a protocol that would become a standard for CAM applications in their many manufacturing facilities. Known as the Local Communications Networks Users Group, this task force developed a protocol patterned after

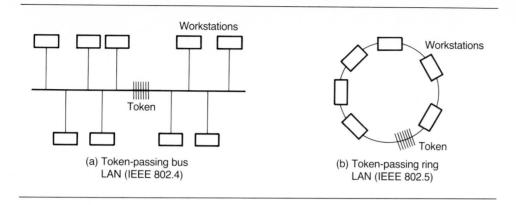

Figure 9-18

Deterministic
accessing by
control token

the seven-layer ISO/OSI model. Called Manufacturing Automation Protocol (MAP), the system is a token-passing bus LAN based on the IEEE 802.4 standard. In 1982 General Motors officials mandated that all future CAM equipment purchased by the company must have the capability of being connected into a LAN which utilizes MAP. Because General Motors was such a large user of CAM equipment, the major CAM equipment vendors soon began developing the interface hardware and software necessary to make their equipment MAP compatable. Other large manufacturers soon followed General Motors' lead in utilizing MAP in their automated-manufacturing facilities.

MAP is based on the ISO/OSI model and is divided into a seven-layer hierarchy as shown in Figure 9-19. At the present time interfacing techniques have been developed for only the first five layers. This is sufficient for many CAM applications, however.

MAP is an evolving technology and will take several years to reach maturity. Although not yet an industry-wide manufacturing protocol, it is being accepted by more and more vendors and CAM users. Recently, a World Federation of MAP Users Group composed of over 1600 United States, Canadian, European, and Japanese companies has been formed to help establish a single world-wide standard factory-floor communications protocol. A typical MAP network appears in Figure 9-20.

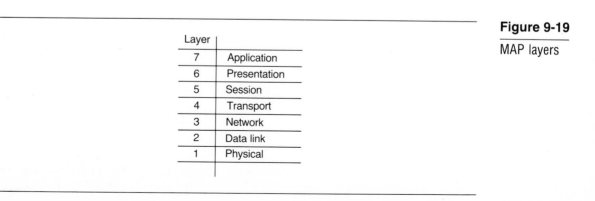

Figure 9-19

MAP layers

Figure 9-20

A MAP installation

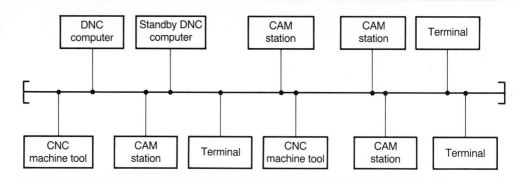

9-7.7 Technical Office Protocol

Most large manufacturing companies began automating their office and engineering operations as the need arose. Boeing Company, an aircraft manufacturer is a typical example. Over a period of several years, Boeing acquired computers for accounting, word processing, inventory control, and CAD/CAE applications. Some of the computers operated independently while others were linked with LANs to form cells of office automation. By 1980, Boeing's data-communication needs had increased by so much that the company needed to link the various computers and automated cells together. Faced with the lack of an industry-wide communcations standard that allowed multivendor computers to be linked together, Boeing developed a set of protocol standards for office, CAD, and CAE communications. This protocol standard is called Technical Office Protocol (TOP). Like MAP, TOP is based on the seven-layer ISO/OSI model but uses the IEEE 802.3 standard. Developmental work continued through the mid-1980s with other companies and agencies such as the National Bureau of Standards (NBS) becoming involved. Because TOP and MAP are both based on the seven-layer ISO/OSI model, many aspects of both protocol systems are similar even though TOP is designed for office and engineering communications and MAP is used for CAM equipment. Developmental activities for both protocols were merged in 1986 to form a MAP/TOP user group. In addition to Boeing, General Motors and several other large manufacturing companies have begun using TOP. TOP has the potential to become the universal industry-wide standard for linking computer-based office equipment together.

In 1987, work was initiated in developing a protocol to link MAP and TOP together. When this protocol is defined and the hardware and software developed, total plant-level automation, or CIM, will be possible using multivendor computer-based equipment. Manufacturing operations will be completely controlled from a single integrated and standardized database.

9-7.8 Metropolitan-Area Net

Just at a time when local area networks are beginning to mature, investigation has begun into developing a super LAN which can be used to link several LANs to-

Figure 9-21

MAN

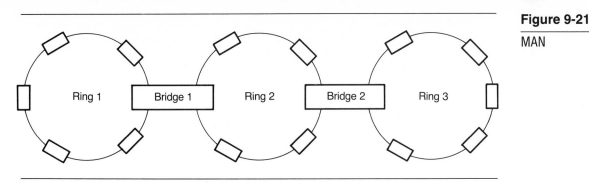

gether. Called Metropolitan-Area Net (MAN), this communcations link will allow for the integration of several LANs belonging to a corporation which might be scattered about a geographical location of up to 25 miles. A task force formed by the IEEE has initiated the development of a protocol standard known as IEEE 802.6. The proposed MAN system consists of a number of rings connected together by bridges that transfer data from one ring to another, as shown in Figure 9-21.

9-7.9 Transmission-line mediums

Three factors must be considered in planning for a LAN—topology, method of access, and the transmission-line medium that links the workstations and through which data are sent. Three types of transmission-line mediums have been borrowed from the fields of radio and television communications for this purpose—twisted-pair, coaxial cable, and optical fiber. Several factors have to be taken into consideration when selecting a transmission-line medium for a particular LAN application. These include transmission-line losses, bandwidth of the data to be transmitted, and noise. Transmission-line losses include copper and radiation losses. Copper losses are caused by the resistance of the line and can be kept to a minimum by keeping the LAN as short as possible and using wire of sufficiently large cross-sectional area. Radiation losses occur when the data sent through the LAN are sufficiently high in frequency to cause the circular electromagnetic field about the line to be radiated into space. Radiation losses can be minimized by using a twisted-pair transmission line, coaxial cable, or optical fiber.

Twisted-pair transmission line: This transmission line is based on the principle that the electromagnetic fields about a pair of conductors which connect a source to a load are 180 degrees out of phase, as shown in Figure 9-22. Twisting the conductors together causes the fields to oppose one another and cancel. A twisted-pair transmission line appears in Figure 9-23. This has been a popular transmission medium because of its simplicity and low cost. Its major disadvantages have been its limited bandwidth and susceptibility to receiving external electromagnetic fields.

Coaxial-cable transmission line: An improved transmission line is shown in Figure 9-24. This medium is composed of a center conductor, a braided outer conductor, a polyethylene dielectric to separate the two conductors, and a plastic outer covering. The braided outer conductor is usually grounded at both the source and

Figure 9-22

Figure 9-22

The magnetic fields about a conductor-pair

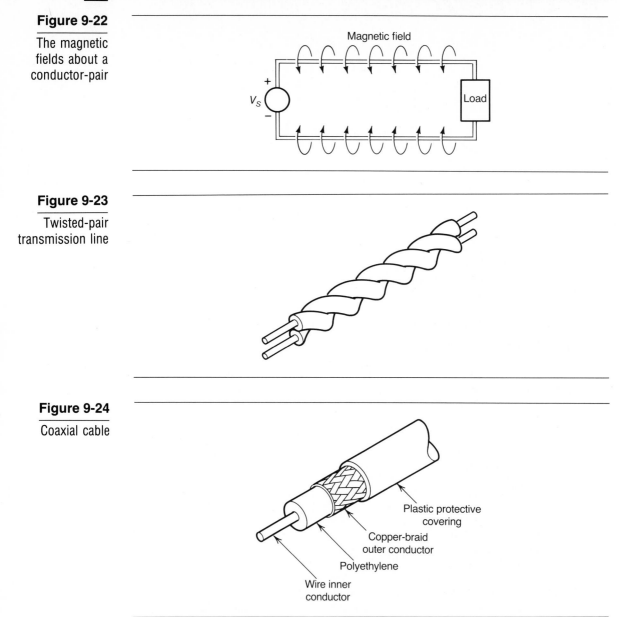

Figure 9-23

Twisted-pair transmission line

Figure 9-24

Coaxial cable

receiving ends of the system. This eliminates radiation losses by confining the electromagnetic field to the inner area of the cable. Coaxial cable is considerably more expensive than twisted-pair but is capable of carrying a much broader bandwidth of data and is not affected by external electromagnetic fields.

Optical-fiber transmission line: The transmission-line medium in optical-fiber lines is a glass cable that carries light energy. As illustrated in Figure 9-25, the medium is composed of a solid fiber core, cladding, and a protective cover. The core is

Figure 9-25

Optical-fiber
transmission line

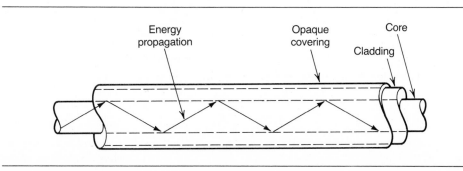

made of highly refined glass, or in some cases plastic, whose diameter ranges from approximately 3 to 200 μm, depending upon the wavelength of the light to be transmitted. The cladding surrounding the fiber is made of a less dense glass material. The refraction coefficient at the surface where the fiber meets the cladding is sufficiently high to allow light to be reflected off the sides of the fiber. An opaque covering encases the cladding and provides mechanical strength and protection for the cable.

Light is introduced into the core at an angle and travels at that angle to the outside surface area of the core. When it encounters the less dense cladding material, it is refracted back in the opposite direction until it is refracted off of the opposite side. This process continues, with the energy being propagated through the fiber in a zig-zag fashion, as seen in Figure 9-26.

The unit that converts the electrical energy into an equivalent optical signal is called the transmitter. This stage is composed of a solid-state light source that generates an optical carrier, a circuit to modulate the carrier, and an interfacing circuit, as shown in Figure 9-27. The light source is usually an LED or LD. LEDs are especially popular because of their simplicity but have a low output power, typically 10 μW to 1 mW, which limits their range to approximately 3000 feet. For longer distances the laser diode is usually employed.

An optical detector is connected to the other end of the optical fiber to change

Figure 9-26

Radiant-energy propagation in an optical fiber

Figure 9-27

Optical transmitter

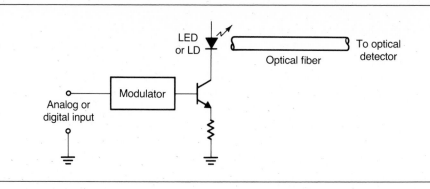

the light signal into an equivalent electrical signal. The detector usually consists of a photodiode, a photo PIN diode, or phototransistor. The output of the detector is connected to an amplifier to increase the signal amplitude, as shown in Figure 9-28. Although considerably more expensive than coaxial cable, optical-fiber transmission lines have the advantages of having lower transmission-line losses, a very wide bandwidth, and extremely small size, and are not affected by external electromagnetic fields. Bandwidth is a measure of how much data a transmission line can carry. Twisted-pair transmission line typically has a maximum bandwidth of 1 MHz; depending upon the type, coaxial cable may have a bandwidth as great as 450 MHz; while optical-fiber transmission line bandwidth is in the vicinity of 30 GHz.

9-7.10 Electrical noise

Noise is any form of electrical energy that interferes with the reception of information. Noise may be generated within the system or may come from an external source. Internal noise is generated by the circuits in the workstations connected to the LAN and includes thermal and transistor noise. Thermal noise is small voltages produced by thermal interaction between free electrons and moving ions in a conductive medium. Although present in all devices through which current flows, it is most noticeable in resistors. This type of noise is proportional to temperature, and the noise voltage produced has a very wide bandwidth—spreading across most of the electromagnetic spectrum. This was first studied by J. B. Johnson in 1928 and is often referred to as Johnson noise. Since the frequencies of the noise signals appear

Figure 9-28

Optical detector

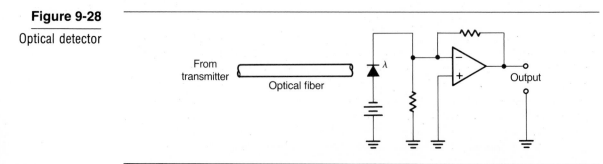

across most of the spectrum, it is also called white noise as it is a composition of all signal frequencies in the optic section of the electromagnetic spectrum.

Transistor noise is that which is introduced by semiconductors. Sometimes called shot noise, it is caused by the discrete-particle characteristics of the current carriers present in semiconductor devices. These drift and diffuse through the semiconductor material in a random fashion. Semiconductor devices also generate thermal noise. The total noise created by these devices is the sum of the shot and thermal noise. Internal noise cannot be entirely eliminated but its effects can be minimized by using well-designed circuits.

External noise is created by electromagnetic fields that move across equipment and transmission lines, causing noise voltages to be induced. This noise comes from both natural and man-made sources. Natural sources include the stars, sun, and lightning discharges. Although they affect radio communications, they have little effect on data communications within a factory. Man-made noise is primarily created by arcing and can be especially bothersome in some manufacturing facilities. Motor and generator brushes, arc welders, relays, and other switching devices are all sources of man-made noise. The only protection against this type of noise is shielding. Controllers and other electronic equipment located on the factory floor must be mounted in metallic cases which are grounded. Coaxial cables whose outer shield is grounded or optical fibers are used as transmission lines.

9-7.11 Types of transmission utilized in LANs

Either analog or digital data may be sent through a LAN using baseband or broadband transmission. Baseband transmission is more limited than broadband and supports only a single channel. It may be used with any of the three LAN topologies. Because of its single-channel capability and resulting limited bandwidth, twisted-pair transmission line is usually used as the LAN transmission-line medium. The data may be sent directly through the line or it may modulate an RF carrier. TOP employs baseband transmission.

Several channels of information may be carried simultaneously when broadband transmission is utilized. Information in the form of digital data, sensor signals, video, and audio may all be carried at the same time. This is made possible by modulation and frequency division multiplexing (FDM) and is the same technique as is used in cable television to carry multiple channels. Coaxial cable and optical fiber may both be used as the transmission-line mediums. Coaxial cable is by far the more common because of its lower cost. In addition, the technology of connecting workstations to a coaxial LAN is more fully developed than that of a fiber-optic LAN. Developmental work is progressing on an optical fiber MAP LAN called MAP 3.0.

Any one of the several modulation methods used in radio communications may be used to modulate the RF carrier if coaxial cable is used as the transmission line. This includes amplitude modulation (AM), frequency modulation (FM), phase modulation (PM), frequency shift keying (FSK), pulse amplitude modulation (PAM), pulse width modulation (PWM), pulse position modulation (PPM), and pulse code modulation (PCM). With the exception of frequency and phase modulation, these same modulation methods may be utilized when an optical-fiber transmission-line medium is employed. Frequency and phase modulation are difficult to achieve since the frequency and phase of an optical source are constant. Broadband transmission may be used with either the bus or ring topologies. MAP utilizes this type of transmission.

Summary

1. An industrial robot is a CNC machine designed to perform operations where the work may be repetitious or hazardous for humans.

2. An industrial robot is composed of a manipulator, a controller, and a power supply.

3. Industrial robots are used in spray-painting, welding, machining, assembly, inspection, and materials-handling applications.

4. Industrial robots may be categorized as to the type of motion they produce, lifting capacity, type of control, and kind of power supply utilized.

5. Robots grouped by the type of motion produced include polar, cylindrical, Cartesian, and revolute.

6. Robots grouped by lifting capacity include small, intermediate, large, and heavy duty.

7. Hydraulic, pneumatic, and electric power may be used to drive the devices which actuate the manipulator.

8. Robotic axes movement includes manipulator-sweep, shoulder-swivel, elbow-extension, pitch, yaw, roll, and horizontal base movement.

9. Three different methods are used to program robots—the teach pendent, the lead-through technique, and off-line programming.

10. Automated materials-handling equipment such as AS/RS, conveyors, monorails, and automatic vehicles form an integral part of automated manufacturing. They store, retrieve, load, unload, and move materials automatically.

11. Machine-vision systems give robots and some other CNC machines "sight" and allow these machines to sort, check, and position objects.

12. Bar-code scanning is used to identify and keep track of inventory and work-in-progress.

13. Numerous automated manufacturing planning and control software (MPCS) packages have been developed. One of the more common is manufacturing resource planning (MRP II).

14. Computer-aided drafting (CAD) and computer-aided engineering (CAE) are computer technology applied to the drawing and design phases of manufacturing. Data developed during these phases form a large part of the computer database used in the manufacture of the product.

15. Computer-aided manufacturing (CAM) is factory-floor automation and includes the application of the computer in all phases of product fabrication. CAM includes the CNC machines; programmable controllers; automated materials-handling, assembly, and inspection equipment; processes; and software required to produce a product.

16. CAD (and CAE) merged with CAM forms CAD/CAM. The product is designed using CAD and CAE and is fabricated using CAM. Data generated during the CAD and CAE phases is used to program the machines employed to fabricate the part.

17. Computer-integrated manufacturing (CIM) is plant-level automation that integrates CAD, CAE, CAM, and business office activities. A common computer database is used for all manufacturing and ancillary activities.

18. Flexible manufacturing systems (FMS) are factory-floor automation systems designed for discrete-product batch-level production.

19. The local area network (LAN) is an on-site communications link made up of hardware, software, and a transmission line that allows computer-based workstations to communicate with one another.

20. A protocol provides a standard for such parameters as addressing, transmission speeds, prioritizing, and sequencing of data in a LAN.

21. A topology is a configuration used in connecting workstations to a LAN. The three topologies used in a LAN are the star, bus, and ring.

22. Distributed control is used in bus and ring LANs. A central computer is not required.

23. The two accessing techniques employed in a distributed-control LAN are contention-based and deterministic.

24. CSMA/CD is a contention-based accessing technique.

25. The control token is a deterministic accessing technique.

26. IEEE-802 is a standard that defines the ISO/OSI model.

27. MAP is a control-token protocol based on IEEE 802.4 which is being used in some LANs to link multivendor CAM equipment.

28. TOP is a protocol based on IEEE 802.3 used in some LANs to link multivendor computer-based office equipment and CAD/CAE workstations.

29. MAN is a super LAN that links several smaller LANs.

30. Factors that must be considered in planning for a LAN include topology, access, and transmission-line medium.

31. The three transmission-line mediums used in LANs are the twisted-pair, coaxial cable, and optical fiber.

32. The transmission medium in an optical fiber is a glass cable that carries radiant energy from the transmitter to the detector, or receiver.

33. The light source in an optical transmitter is usually an LED or LD.

34. The detector in an optical receiver may be a photodiode, photo PIN diode, or a phototransistor.

35. Electrical noise is any type of electrical energy that interferes with the reception of data.

36. Noise may be generated by electronic circuits or come from an outside source. Internal noise includes thermal and transistor noise. External noise sources include stars, the sun, lightning discharges, and man-made arcs.

37. Data in a LAN may be transmitted in baseband or broadband form.

38. Only a single signal, block of data, or channel may appear on the LAN at any instant when baseband transmission is employed.

39. Multiple signals, blocks of data, or channels may be carried by a LAN when broadband transmission is employed.

40. Modulation and FDM is required when broadband transmission is used.

Chapter Examination

1. List the components forming a CIM system.
2. What is an FMS?
3. List the components forming a typical FMS.
4. What is a LAN?
5. Define the term *protocol*.
6. List three methods used to program robots and discuss the advantages and disadvantages of each.
7. List two advantages of using hydraulic power rather than electric power to drive a robot.
8. List two advantages of using pneumatic power rather than electric power to drive a robot.
9. List two advantages of using electric power rather than fluid power to drive a robot.
10. Identify the two types of robot controllers and list the general applications for each.
11. Define the term *computer-aided manufacturing* (CAM).
12. _____ is a contention-based protocol based on IEEE 802.3.
    a. MAN
    b. MAP
    c. CAP
    d. TOP
13. Which of the following is not a type of robot as classified by lifting capacity?
    a. tiny
    b. large
    c. heavy duty
    d. intermediate

14. _____ is a super LAN based on the IEEE 802.6 standard.
    a. MAN
    b. MAP
    c. CAP
    d. TOP

15. _____ is a token-bus protocol based on IEEE 802.4.
    a. MAN
    b. MAP
    c. CAP
    d. TOP

16. Which of the following is not a factor when selecting a transmission-line medium for a LAN application.
    a. noise
    b. topology
    c. bandwidth
    d. number of nodes

17. Which one of the following automatic identification techniques is the most accurate?
    a. SAW
    b. OCR
    c. RFID
    d. bar code

18. Which of the following is (are) used to automatically store and retrieve parts in a warehouse?
    a. AGVs
    b. AAVs
    c. AS/RS
    d. pallet trucks

19. Which of the following is not a type of bar code symbol?
    a. UPC
    b. EAN
    c. AMH
    d. C3 of 9

20. _____ in a machine-vision system identifies how well the system reproduces a recognizable image.
    a. Accuracy
    b. Resolution
    c. Discrimination
    d. Processing detail

21. The _____ layer in the ISO/OSI LAN model is called the session layer. It is here that methods used in initiating and terminating communications are established.
    a. third
    b. fifth

    c. sixth

    d. seventh

22. Which of the following is not a method used to program a robot?

    a. on-line

    b. off-line

    c. lead-through

    d. teach pendent

23. T F TOP is a protocol used to combine several LANs into a super LAN.   —

24. T F MAN is based on the IEEE 802.3 standard.   —

25. T F MAP utilizes a bus LAN and is accessed by a control token.   —

26. T F LEDs or LDs are used as light detectors when an optical fiber is employed as a LAN transmission-line medium.   —

27. T F Baseband transmission is utilized with TOP.   —

28. T F Shielding is the only effective method of reducing the effect of external noise on communication links in manufacturing plants.   —

29. T F Positioning sensors must be used whenever DC stepping motors are utilized as actuators in electric robots.   —

30. T F AAVs are primarily used in a warehouse to store and retrieve parts automatically.   —

31. T F Yaw is a circular end-effector motion on a robot.   —

32. T F A typical machine-vision system is made up of a light source coupled to a photodetector via an optical-fiber transmission line.   —

33. T F AGVs are usually used to carry large and heavy parts on the factory floor.   —

34. T F Manipulator sweep is a vertical movement of the robot arm about the shoulder.   —

35. T F Most machine-vision systems scan the scene being viewed 30 times per second.   —

36. T F MAP utilizes broadband transmission.   —

37. T F AAVs require a physical path on the factory floor to guide them.   —

38. T F Light energy is propagated through an optical-fiber transmission line by the movement of photons in the cladding layer.   —

39. T F modulation and frequency-division multiplexing (FDM) are both required in baseband transmission.   —

40. T F An FMS is composed of highly specialized CNC machines, conveyors, and monorails.   —

ANSWERS

CHAPTER 1

1. See Figure 1-3(c)
2. Knife, toggle, slide, push-button, rotary, limit, mercury contact
3. Fuses, circuit breakers, GFIs
4. SCS - see Figure 1-64
   diac - see Figure 1-66
   SCR - see Figure 1-52
   triac - see Figure 1-65
5. Motor driven, thermal, electronic, dashpot
6. 11,994 Ω
7. To decrease the switching time of the transistor
8. 100 A
9. 150 A
10. 100 A
11. NEMA 5
12. b
13. d
14. c
15. a
16. d
17. c
18. c
19. a
20. c
21. T

22. F
23. F
24. F
25. T
26. F
27. F
28. F
29. T
30. T
31. F
32. F
33. T
34. F
35. T
36. T
37. F
38. F
39. F
40. F
41. F
42. T
43. T
44. F
45. F

CHAPTER 2

1. 1200 rpm
2. 9.583%
3. 100 Hz
4. 17.857 A
5. 1.857 A
6. Torque is created by the interaction of the electromagnetic fields produced by the armature and field windings, as illustrated in Figure 2-16.
7. Many motors will overheat if operated with a low voltage. Undervoltage protection devices disrupt the line voltage if the voltage drops below a predetermined level.
8. (a) See Figure 2-13.
   (b) See Figure 2-10.
   (c) See Figure 2-14.
   (d) See Figure 2-18.

     (e) See Figure 2-19.
     (f) See Figure 2-20.
  9. b
10. b
11. c
12. c
13. d
14. c
15. a
16. b
17. d
18. d
19. c
20. a
21. b
22. d
23. b
24. a
25. F
26. T
27. T
28. F
29. T
30. T
31. T
32. F
33. T
34. F
35. F
36. F
37. F
38. F
39. F
40. T

CHAPTER 3

1. 12,010.5 lbs
2. 113.235 psi
3. gear, vane, piston

4. gear: 1000–1500 psi; vane: 500–1500 psi; piston: 2500–10,000 psi

5. housing, piston, piston rod, piston rings, packings

6. manual, mechanical, pilot pressure, electrically

7. This pump has a set of mesh gears which rotate whenever the prime mover is turned on. A partial vacuum is created by the disengaging teeth. This causes oil to be pulled from the reservoir into the pump housing. The oil fills the evacuated space and is carried along with the rotating teeth until forced out of the pump by the action of the engaging teeth.

8. This type of valve has three ports. Depending upon the required application, two ports can be used as inputs while the other serves as an output or one port can be used as an input and the other two used as outputs.

9. The baffle prevents the oil returning from the system from being immediately recirculated again. This provides time for the oil to cool, contaminants to settle, and air to be purged.

10. Hydraulic motors produce more horsepower and torque than do the pneumatic type. Both can be operated submerged in a liquid, are explosion proof, and can be operated in a stalled condition indefinitely without damage. The pneumatic motors produces less horsepower and torque but is cleaner than the hydraulic types. The remaining characteristics are similar to hydraulic motors.

11. The lubricator causes a light oil to be atomized and injected into the air line to prevent packings and seals from drying out and to prevent metal-to-metal contact from occuring in actuators.

12. Four-way direction control

13. Check

14. See Figure 3-37

15. T
16. F
17. T
18. F
19. F
20. T
21. F
22. F
23. T
24. F
25. F
26. F
27. F
28. T
29. T
30. F

CHAPTER 4

1. A transducer is a device that converts one form of energy into another.
2. A sensor is a transducer which converts a physical phenomenon into a variable which can be measured.
3. Strain is the change in the shape of an object caused by stress.
4. Float-operated rheostat, differential pressure transducer, capacitive liquid-level sensor, and sonic liquid-level measurement.
5. Thermal excitation, electron collision, photon collision
6. Thermocouple, RTD, thermistor, solid-state temperature sensor
7. Potentiometer, angular digital encoder, synchro
8. A synchro is a rotary transducer that converts angular displacement into an AC voltage or an AC voltage into angular displacement.
9. The synchro is composed of a transmitter and receiver. Both have a rotor and three stator windings. Each rotor is connected to a single-phase voltage source while the transmitter and receiver stator windings are connected together. A torque is created by the interaction of the rotor and stator magnetic fields in the receiver, which causes the receiver rotor to turn whenever the transmitter rotor is turned.
10. Bellows, Bourdon tube, diaphragm
11. a
12. b
13. b
14. d
15. c
16. c
17. a
18. b
19. c
20. d
21. b
22. c
23. c
24. b
25. a
26. d
27. a
28. c
29. c
30. c
31. b

32. a
33. c
34. c
35. c
36. a
37. c
38. c
39. a
40. T
41. T
42. T
43. F
44. T
45. F
46. F
47. F
48. T
49. T
50. F
51. F
52. T
53. F
54. T
55. F

CHAPTER 5

1. A microprocessor is a programmable integrated circuit that is a computer CPU fabricated on a single semiconductor substrate.
2. The key to the versatility of the microprocessor is its programmability. Through programming, the MPU can be made to perform arithmetic and logic operations and to imitate numerous circuit functions.
3. See Figure 5-2.
4. The address, control, and data buses make up the bus structure in a microprocessor. The address bus carries the address of the memory segment or I/O peripheral device to be accessed. The control bus handles the control and timing signals required to fetch and execute the program instructions. Program instructions and data travel over the data bus.
5. The system clock provides the timing pulses that synchronize the operation of the circuits within the microprocessor, memory, and I/O peripherals.

6. a
7. a
8. d
9. c
10. a
11. a
12. a
13. d
14. a
15. b
16. c
17. F
18. T
19. F
20. T
21. F
22. T
23. T
24. F
25. F
26. F
27. F
28. F
29. F
30. T
31. T
32. F

CHAPTER 6

1. Start and stop bits are used to identify the beginning and ending of a binary word in asynchronous serial transmission. Clock pulses are not used to synchronize the data with the system. Synchronizing pulses are transmitted at approximately 100-byte intervals in synchronous serial transmission. An extra-transmission-line pair is required as clock pulses are utilized.

2. Asynchronous serial transmission requires only a single transmission line-pair since clock pulses are not used. Asynchronous serial transmission is slower, since start and stop bits have to be transmitted with each binary word.

3. A parity generator examines the number of binary ones appearing in the serial binary word. If there is an even number, the generator inserts a binary one

into the most significant bit position to make the total number of binary ones odd. If there is an odd number of binary ones, a binary zero is inserted into the most significant bit position.

4. See Figure 6-14

5. An address decoder for each memory or I/O peripheral IC is connected to the address bus. The output of the decoder is connected to the enabling input terminal of the three-state buffer connected to the particular IC. As the program is executed, the particular device which is to send or receive data is addressed. The address decoder detects the address and provides a Chip Select (CS) or Chip Enable (CE) signal that is sent to the enabling input terminal of the three-state buffers. The buffers are turned on causing the data bus lines of the addressed IC to be connected to the system data bus.

6. c
7. a
8. d
9. b
10. b
11. b
12. c
13. b
14. c
15. a
16. b
17. a
18. c
19. a
20. T
21. T
22. F
23. F
24. F
25. F
26. T
27. T
28. F
29. F
30. T
31. F
32. F
33. T
34. F

35. T
36. F
37. T
38. F
39. F
40. F

CHAPTER 7

1. b
2. c
3. a
4. c
5. c
6. c
7. b
8. b
9. c
10. d
11. d
12. b
13. c
14. d
15. c
16. b
17. d
18. a
19. c
20. a
21. T
22. F
23. T
24. F
25. F
26. T
27. F
28. T
29. T
30. F
31. T

32. F
33. F
34. F
35. T

CHAPTER 8

1. See Figure 8-3
2. See Figure 8-4
3. See Figure 8-24
4. See Figure 8-27
5. Programming techniques and memory mediums differ. The NC controller is designed to be used with a particular machine tool whereas a programmable controller is designed for general-purpose applications.
6. Figure 8-30 goes here
7. Micro, small, medium, large
8. d
9. c
10. d
11. c
12. c
13. c
14. c
15. b
16. c
17. c
18. d
19. b
20. T
21. T
22. T
23. F
24. F
25. F
26. F
27. F
28. T
29. T
30. F
31. F

32. F
33. T
34. F
35. F
36. T

CHAPTER 9

1. Components forming CIM include CAD, CAE, CAM, programmable controllers, automated materials handling equipment, automated assembly equipment, automated inspection equipment, automated office equipment, software, and LANs.

2. FMS is factory floor automation used in batch-level discrete product manufacturing.

3. Components forming a typical FMS include CNC machines, automatic tool changers, robots, automatic vehicles, automated assembly machines, and automated inspection machines—all linked together with a LAN.

4. A LAN is composed of interfacing hardware, software, and a transmission line that links together computer-based workstations and allows them to communicate with one another.

5. Protocol is a set of rules and standards that allows equipment connected to a LAN to communicate with one another.

6. The teach pendent, lead-through method, and off-line programming are techniques used to program industrial robots. The teach pendent and lead-through method both have the advantage of programming ease. Programming is limited to motion and position activities only, however. The program may be developed from CAD and CAE and may include activities other than motion and position when off-line programming is utilized. The major disadvantage of off-line programming is that a higher skill level of programming is required.

7. (a) greater lifting power
   (b) it is safer in certain environments

8. (a) less expensive
   (b) it is safer in certain environments

9. (a) greater accuracy
   (b) cleaner, quieter, smoother

10. (a) point-to-point and continuous path
    (b) point-to-point robots are used in pick and place, machine loading, insertion, and spot welding applications. Continuous path robots are used for spray painting, continuous welding, machining, and assembly operations.

11. CAM is automation at the factory floor level. It includes the processes, CNC machines, robots, AMH equipment, automated assembly equipment, automated inspection equipment, and software required to fabricate a product.

12. d
13. a
14. a
15. b
16. d
17. d
18. c
19. c
20. b
21. b
22. a
23. F
24. F
25. T
26. F
27. T
28. T
29. F
30. F
31. F
32. F
33. T
34. F
35. T
36. T
37. F
38. F
39. F
40. F

REFERENCES

CHAPTER 1

1. Allen-Bradley Company, *Industrial Catalog #107*, Milwaukee, WI.
2. Busman Manufacturing, *Buss® Fuses Catalog*, Earth City, MO.
3. Computing and Software, Inc., *Diode and SCR D.A.T.A. Book*, 27th Edition, Orange, NJ.
4. Driscoll, E. F., 1976 *Industrial Electronics: Devices, Circuits, and Applications*, Chicago, IL. American Technical Society.
5. Globe-Union Inc., *Centralab Catalog #201*, Milwaukee, WI.
6. Humphries, J. T. and Sheets, L.P., 1983 *Industrial Electronics*, North Scituate, MA., Breton.
7. J-B-T Switches, *J-B-T Catalog #201*, New Haven, CT.
8. Maloney, T. J., 1986 *Industrial Solid-State Electronics, Devices and Systems*, Englewood Cliffs, NJ., Prentice-Hall.
9. Miller, R. M., 1978 *Industrial Electricity*, Peoria, IL., Bennett.
10. Nadon, J. M., et al., 1984 *Industrial Electricity*, Albany, NY., Delmar.
11. Omron Corporation of America, *Relays Catalog*, Schaumburg, IL.
12. Omron Corporation of America, *Switches Catalog*, Schaumburg, IL.
13. Omron Corporation of America, *Timers Catalog*, Schaumburg, IL.
14. Patrick, D. and Fardo, S., 1986 *Industrial Electronics Devices and Systems*, Englewood Cliffs, NJ., Prentice-Hall/Reston.
15. Sigma Instruments Inc., *Stock Relay Catalog*, Braintree, MA.
16. Struters-Dunn Company, *Distributor Stocked Relays Catalog*, #c/1010, Pitman, NJ.
17. Watt, J. H., 1970 *American Electricians Handbook*, 9th ed., New York, NY., McGraw-Hill.

CHAPTER 2

1. Allen-Bradley Company, *Bulletin 1334*, Milwaukee, WI.
2. Allen-Bradley Company, *Bulletin 1350*, Milwaukee, WI.
3. Critchlow, A. J., 1985 *Introduction to Robots*, New York, NY., Macmillan.
4. Herman and Alerich, 1985 *Industrial Motor Control*, Albany, NY., Delmar.
5. I&CS, October 1986 *Selecting Electrical Adjustable Speed Motor Drives*, by Jerry J. Pollack, pp 43–46.
6. Machine Design, February 1986, *More Angles with Five-Phase Steppers*, by Jay Batten, pp. 60–65.
7. Nadon, J. M. et al., 1984 *Industrial Electricity*, Albany, NY., Delmar.
8. PCIM, February 1987 *Power Interface Circuits for Motor Drives*, by J. Myers and H. Abramowitz, pp. 24–32.
9. Production Engineering, February 1984, *Efficient Speed Control of AC Motors*, by James Hayes, pp. 62–65.
10. Superior Electric Company, *Design Engineer's Guide to DC Stepping Motors*.

CHAPTER 3

1. Hedges, C. S., 1965 *Industrial Fluid Power Text*, vol. 1, Texas, Womack Machine Supply Company.
2. Oster, J., 1969 *Basic Fluid Power: Hydraulics*, New York, NY., McGraw-Hill.
3. Stewart, H. L., 1977 *Hydraulic and Pneumatic Power for Production*, 4th ed., New York, NY., Industrial Press.
4. Womack Education Publications, 1970 *Fluid Power in Plant and Field*, Texas.

CHAPTER 4

1. Allocca, J. A. and Stuart, A., 1984 *Transducers Theory and Application*, Englewood Cliffs, NJ., Prentice-Hall/Reston.
2. Cockrell, W. D., 1958 *Industrial Electronics Handbook*, New York, NY., McGraw-Hill.
3. Driscoll, E. F., 1976 *Industrial Electronics: Devices, Circuits, and Applications*, Chicago, Il., American Technical Society.
4. Humphries, J. T. and Sheets, L.P., 1983 *Industrial Electronics*, North Scituate, MA., Breton.
5. Murphy, J. T. and Smott, R.C., 1982 *Physics Principles and Problems*, Columbus, OH., Merrill.
6. Omega Engineering Incorporated, 1986 *Complete Flow and Level Measurement Handbook and Encyclopedia*, Stamford, CT.
7. Omega Engineering Incorporated, 1986 *Complete Pressure and Strain Measurement Handbook and Encyclopedia*, Stamford, CT.

8. Omega Engineering Incorporated, 1986 *Complete Temperature Measurement Handbook and Encyclopedia*, Stamford, CT.

9. Prensky, S. D., 1971 *Electronic Instrumentation*, 2nd ed., Englewood Cliffs, NJ., Prentice-Hall.

10. Sensors, July 1985 *Use of Variable Reluctance Transducers*, by Jeff Gartner, pp. 38–43.

11. Sensors, December 1985 *Coriolis-Based Mass Flow Measurement*, by Alan Young, pp. 6–10.

12. Sensors, January 1986 *Aerospace Temperature Sensor Application*, by Randal A. Gauthier, pp. 31–36.

13. Sensors, March 1986 *New Pressure Sensor Technology*, by Jim Fitzsimmons, pp. 18–20.

14. Sensors, May 1986 *Humidity Sensors*, by Bronislan Lainer, pp. 28–36.

15. Sensors, July 1986 *Solid-State Pressure Sensors*, by Frank Perrino, pp. 44–46.

16. Woolvet, G. A., 1977 *Transducers in Digital Systems*, England, Peter Peregrinus Ltd.

CHAPTER 5

1. Computers and Electronics, October 1984 *Motorola's Muscular 68020*, by Les Solomon, pp. 74–76.

2. Digital Design, September 1986 *MC68030 Manages Memory*, by Hunter Scales, pp. 129–131.

3. Digital Design, October 1986 *A Closer Look at MC68030 Memory Management*, by C. Parrott and B. Moyer, pp. 72–74.

4. Electronics, September 18, 1986 *First Look at Motorola's Latest 32-bit Processor*, pp. 71–75.

5. Electronics, October 21, 1985 *Intel's 80386 Runs Multiple Operating Systems*, pp. 50–52.

6. Electronics Week, April 15, 1985 *Intel Takes Wraps Off 386*, pp. 18–19.

7. Mini-Micro Systems, July 1984 *Rair Extends Microcomputing with 80286-based Supermicro*, by Keith Jones, p. 83.

8. Riesgo, Ray E., 1985 *The 68000 Microprocessor*, San Diego, CA., Computer Systems Associates, Inc.

9. *16-bit Microprocessor User's Manual*, 3rd ed., Motorola, Englewood Cliffs, NJ., Prentice-Hall.

CHAPTER 6

1. Artwick, B. A., 1980 *Microcomputer Interfacing*, Englewood Cliffs, NJ., Prentice-Hall.

2. Brey, B. B., 1984 *Microprocessor Hardware Interfacing and Applications*, Columbus, OH., Charles E. Merrill.

3. Eggebrecht, L. C., 1983 *Interfacing to the IBM Personal Computer*, Indianapolis, IN., Howard W. Sams.

4. Hordeski, M. F., 1985 *Microprocessor Sensor and Control Systems*, Englewood Cliffs, NJ., Prentice-Hall.

5. IC&S, December 1986 *What's Happening with A/D and D/A Converters?*, by Peter Cleveland, pp. 21–24.

6. Lesea and Zaks, 1978 *Microprocessor Interfacing Techniques*, Berkeley, CA., Sybex.

7. Staugaard, A. C. Jr., 1982 *Individual Learning Program in Microprocessor Interfacing*, Benton Harbor, MI., Heath Company.

CHAPTER 8

1. Allen-Bradley Company, *Fundamentals of Numerical Control*, Highland Heights, OH.

2. Bukstein, E., 1961 *Industrial Electronics Measurement and Control*, Indianapolis, IN., Howard W. Sams.

3. Cox, R. A., 1983 *Technician's Guide to Programmable Controllers*, New York, NY., Delmar.

4. Electronics, May 5, 1983 *Programmable Controller Stays On Line by Relying on Backup*, by Radhey Khanna.

5. Hordeski, M. F., 1985 *The Design of Microprocessor, Sensor, and Control Systems*, Englewood Cliffs, NJ., Prentice-Hall/Reston.

6. IC&S, May 1983 *Programmable Controllers: A Technology Update*, by Peter Cleveland.

7. IEEE Transactions on Industrial Applications, November/December 1982 *Monitoring and Control of Textile Processing Lines: Utilizing the Power of Programmable Controllers*, by Darrel Glanker.

8. Production Engineering, July 1983 *PC Versatility Goes On and On, and On*, by Leslie Jasany.

9. Production Engineering, April 1985 *Trends in Control—The PC Moves Up*, by Thomas Bullock.

10. Schmidtt, N. M., and Farwell, R.F., 1983 *Understanding Electronic Control and Automation Systems*, Dallas, TX., Texas Instruments Learning Center.

11. Tooling and Production, July 1984 *Selecting Programmable Logic Controllers*, by David Johnson.

12. Water/Engineering & Management, January 1984 *Understanding and Applying the Programmable Controller*, by Anthony Arnone.

CHAPTER 9

1. Automation News, December 2, 1985 *Bar Coding Techniques Essential for Factory Inventory Control*, by C. H. Knowles and M. L. Sanyor, pp. 19, 20, 26.

2. Automation News, December 2, 1985 *Symbol Talks Bar Coding Scanners and Standards*, by Robert Malone, p. 20.

3. Baker, Donald G., 1985 *Fiber Optic Design and Applications*, Englewood Cliffs, NJ., Prentice-Hall/Reston.

4. Cheo, Peter K., 1985 *Fiber Optics Devices and Systems*, Englewood Cliffs, NJ., Prentice-Hall.

5. CIM Technology, Winter 1985 *The 1985 LEAD Awards Recognizing Excellence in CIM*, by Rita R. Schreiber, pp. 21–23.

6. CIM Technology, Spring 1986 *MAP Gives Deere and Company A Competitive Edge*, p. 15.

7. CIM Technology, Spring 1986 *MAP Pilots: Promises and Pitfalls*, by Paul W. Accampo, pp. 19–23.

8. Critchlow, Arthur J., 1985 *Introduction to Robots*, New York, NY., Macmillan.

9. Digital Design, August 1986 *Battle of the Buses—Can We Talk?*, by Dave Wilson, pp. 34–38.

10. Electronics, October 16, 1986 *Data Communications*, by Robert Rosenburg, pp. 88–89.

11. ESD, January 1987 *Dense Chips Ease Networking*, by Denny Cormier, pp. 44–47.

12. ESD, March 1987 *Wanted Communications Software for Multiprocessing*, by Richard W. Boberg, pp. 72–76.

13. Fortune, February 21, 1983 *The Race to the Automatic Factory*, by Gene Bylinski, pp. 52–64, © 1983 Time Inc., All rights reserved.

14. Hartley, John 1984 *FMS At Work*, United Kingdom, IFS Ltd.

15. Hewlett-Packard Journal, September 1986 *High-Performance SPU for a Modular Workstation Family*, by Jonathon J. Rubinstein, pp. 12–16.

16. Holland, John R. *Flexible Manufacturing Systems*, Dearborn, MI., Society of Manufacturing Engineers (SME).

17. I&CS, November 1986 *MAP: A Standard for Survival*, by Lewis I. Solomon, pp. 57–59.

18. I&CS, November 1986 *Tools to Implement MAP Reach "Critical Mass"*, by Tom Balch, pp. 61–64.

19. Industry Week, November 10, 1986 *TOP, Born of Boeing's Dilemma*, by Mark L. Goldstein, pp. 94–97.

20. Material Handling Engineering, May 1985 *Machine Vision: Eyes of the Automated Factory*, by Gene Schwind, p. 117.

21. Material Handling Engineering, July 1985 *Allen-Bradley Puts Its Automation Where Its Market Is*, by Bernie Knill, pp. 62–66.

22. Material Handling Engineering, October 1986 *Integrated Manufacturing*, p. 10.

23. Robotics Age, June 1985 *Recognizing Parts That Touch*, by Scott O. Roth.

24. Robotics Age, September 1985 *Karel: A Programming Language for the Factory Floor*, by M. Ward and K. Stoddard, p. 10.

25. Robotics Age, September 1985 *Forth for Robot Control*, by George Dooley, p. 97.

26. Sensors, April 1986 *Not Just Vision*, by Roberto Toth, p. 2.

27. Sensors, June 1986 *The Fundamentals of Machine Systems*, by Stepanie Henkel, pp. 2–8.

28. Sensors, December 1986 *Vision Sensing in Electronic Hardware*, by Larry Werth, pp. 16–27.

INDEX